GENERALISATION OF GEOGRAPHIC INFORMATION: CARTOGRAPHIC MODELLING AND APPLICATIONS

GENERALISATION OF GEOGRAPHIC INFORMATION: CARTOGRAPHIC MODELLING AND APPLICATIONS

Edited by

William A. Mackaness
Institute of Geography
The University of Edinburgh, Scotland, UK

Anne Ruas
The COGIT Laboratory, IGN-France
Paris, France

L. Tiina Sarjakoski
Department of Geoinformatics and Cartography
Finnish Geodetic Institute, Finland

ICA ACI

ELSEVIER

Published on behalf of the International Cartographic Association by Elsevier

Amsterdam – Boston – Heidelberg – London – New York – Oxford
Paris – San Diego – San Francisco – Singapore – Sydney – Tokyo

Elsevier
The Boulevard, Langford Lane, Kidlington, Oxford OX5 1GB, UK
Radarweg 29, PO Box 211, 1000 AE Amsterdam, The Netherlands

First edition 2007

British Library Cataloguing in Publication Data
A catalogue record for this book is available from the British Library

Library of Congress Cataloging-in-Publication Data
A catalog record for this book is available from the Library of Congress

ISBN-13: 978-0-08-045374-3
ISBN-10: 0-08-045374-0

For information on all Elsevier publications
visit our website at books.elsevier.com

Contents

Preface vii

Acknowledgements xi

List of Contributors xiii

1. Understanding Geographic Space 1
   by W. A. Mackaness

2. Conceptual Models of Generalisation and Multiple Representation 11
   by L. T. Sarjakoski

3. A Synoptic View of Generalisation Operators 37
   by N. Regnauld & R. B. McMaster

4. Modelling the Overall Process of Generalisation 67
   by L. Harrie & R. Weibel

5. Evaluation in the Map Generalisation Process 89
   by W. A. Mackaness & A. Ruas

6. Database Requirements for Generalisation and Multiple Representations . . . 113
   by S. Mustière & J. van Smaalen

7. A Real-Time Generalisation and Map Adaptation Approach for Location-Based Services 137
   by T. Sarjakoski & L. T. Sarjakoski

8. Experiments in Building an Open Generalisation System 161
   by A. Edwardes, D. Burghardt & M. Neun

9. A Data Warehouse Strategy for on-Demand Multiscale Mapping 177
   by E. Bernier & Y. Bédard

10. Relevance of Generalisation to the Extraction and Communication of Wayfinding Information 199
    by M. Sester & B. Elias

11. 3D Building Generalisation 211
    by L. Meng & A. Forberg

12. Characterising Space via Pattern Recognition Techniques: Identifying Patterns in Road Networks 233
by F. Heinzle & K.-H. Anders

13. Generalisation of Geographical Networks 255
by R. Thomson & R. Brooks

14. A Prototype Generalisation System Based on the Multi-Agent System Paradigm 269
by A. Ruas & C. Duchêne

15. Managing Generalisation Updates in IGN Map Production 285
by F. Lecordix & C. Lemarié

16. Automated Generalisation in a Map Production Environment – the KMS Experience 301
by P. West-Nielsen & M. Meyer

17. Observations and Research Challenges in Map Generalisation and Multiple Representation 315
by W. A. Mackaness, A. Ruas & L. T. Sarjakoski

Consolidated Bibliography 325

Author Index 357

Subject Index 367

Preface

There is an interesting tension between the coupled technologies of geographical information systems (GIS), geovisualisation and work in map generalisation. GIS technology has created a fundamental paradigm shift in how we manage, interact with, and visualise geographic information. Somewhere in this paradigm shift, the human cartographer appears to have been absented from the process and somehow replaced by colour ramps, choropleth mapping and symbol charts. Furthermore, some have argued that the need for high quality cartographic portrayal has been obviated by developments in dynamic interaction techniques and exploratory data analysis methods (typified by so-called "Geographical Exploration Systems"). But we would argue that irrespective of the media or degree of interaction, it remains the case that the power of maps lie in their ability to abstract geographic space, and that different levels of abstraction reveal different patterns and properties inherent among the geographic phenomena being represented. The ability to abstract data being ever more important in today's information society – in which the volume of data exceeds our insatiable appetite for more! In response to the changing paradigm of map use, this field of research has drawn on a large number of complimentary disciplines and the expertise of various research groups. Research in this field now draws on expertise in exploratory data analysis, interface design, cognitive ergonomics, and agent based methodologies. It also draws extensively on expertise in databases, surface and topological modelling and semantic reference systems in order to reason about space; and these are just a few! At its most fundamental, generalisation is about modelling geographic space – applying a set of techniques in order to derive different abstractions at varying levels of detail. Within highly automated environments, there exists an opportunity to model the process of map design in a way that goes well beyond selection from a set of drop down menus. This book presents systems that begin to embody notions of cartographic design, reflecting a shift of balance in the cartographic decision making process from the user to the machine. In such contexts, we begin to ask: what is the optimal balance in that decision making? How do we design interfaces that support intuitive specification of map needs, and how is the generalisation process seamlessly embedded within interactive, mobile or exploratory systems? How are solutions deemed to be correct, how are they evaluated, how is their quality assessed?

It is clear that any development in modelling the process of map generalisation cannot be separated from data modelling and data storage issues. From a database perspective, we might argue that any given map is nothing more than an instantaneous "slice" through a (set of) databases. How we store the data, its dimensionality (3D, or spatiotemporal data), the type and detail of attribution: all these factors have a huge impact on the level of generalisation achievable, and the speed with which solutions can be generated. Both the title and the contents of this book reflect the importance of data analysis techniques and of database issues more generally in deriving generalised output. The panacea for this research is a multiple representation database, in which data are stored as different objects of the same geographic entity or phenomenon and linked across different resolution levels. One of the major advantages of multiple representation databases is that the data are updated only at the finest scale, but from which we can propagate those updates, derive and then store multiple representations of phenomenon at increasing levels of granularity. Navigating among these different levels of detail enables composition of any number of maps (varying in theme and scale) either as a basis for exploration or production. Such maps can be tailored to support any number of tasks, across a range of media, in support of a variety of tasks (both visual and non visual).

Structure

Our motivation for writing this book stems from a desire to make researchers and practitioners aware of advances in the field of map generalisation and multiple representation and to provide a detailed description of the state of the art. In writing this book we seek to promote the fields of map generalisation and multiple representation, to demonstrate its broadening application, and to emphasise its relevance to the exploration and representation of geographic information. The structure of the book gives emphasis to both the theoretical as well as the practical, attempting to demonstrate the critical role of the former in achieving the latter. Collectively we attempt to offer an exposition of the critical components required to design and implement autonomous solutions in automated cartography in support of various activities in the Geosciences. But this book does not cover every aspect of this research field. For example we do not include a comprehensive description of all aspects of cartographic symbolisation and design (this would make the book too voluminous), neither do we present cell based (raster) solutions to map generalisation. We intentionally focus on vector based models given their complete dominance in the field of automated mapping. In some cases we refer to other literature rather than give detailed explanation of well-established techniques. In this sense the book should be viewed as a compliment (not replacement) to existing texts in the field, whilst giving additional emphasis to ideas of multiple representation. Textbooks worthy of note that relate specifically to work on map generalisation and multiple representations include Buttenfield and McMaster (1991); João (1998); Müller et al. (1995a); McMaster and Shea (1992); and Ruas (2002a). In combination with these and other texts on this topic, it is hoped that the reader will gain a clear picture of the breadth of cognate disciplines contributing to this field, as well as a thorough understanding of how these various methodologies are being applied in the real world.

The purpose of this text is threefold, to present the theoretical advancements in the field, to show how developments have advanced solutions to the point that they are being incorporated into existing commercial solutions, and to provide an overview of future research challenges. The opening chapter discusses the importance of scale in the context of geographical understanding. The following chapters (Chapters 2 to 6) offer an exposition of the critical components required for the design and implementation of autonomous systems of design. They look at ways of modelling the complex decision making process of cartographic design within automated environments, demonstrating the need for techniques equivalent to the hands and eyes of the human cartographer – from the initial selection of data, to synthesis through to the evaluation of solutions. The sophistication of that solution is increasingly dependent upon database technologies capable of supporting a range of spatial analysis techniques, and the storage of intermediate results. The middle set of chapters (Chapters 7 to 11) explore the application of map generalisation techniques in a variety of contexts: in support of delivery of information from map services to mobile devices, in extracting and communicating wayfinding information, and in their application to 3D generalisation. The later set of chapters (Chapters 12 to 16) examines the most current thinking behind generalisation, and examines the application of those techniques in commercial and industrial environments. The concluding chapter (Chapter 17) attempts to highlight key findings distilled from the various contributions. Overall the book seeks to provide detailed explanation and illustration of methodologies at the forefront of research in this field. Because of the strong interlinkages between the chapters, we felt a consolidated reference list was a more efficient presentation of supporting literature. The index (ordered alphabetically and by subject matter) provides an alternate method by which the reader can explore the contents of this book.

Collectively therefore, the book should provide a strong footing for those contemplating research in this area for the first time, but also pointing to literature and exciting ideas at the forefront of this field, and from across the spectrum of cognate disciplines. This text will be very relevant to researchers working in Geographical Information Science, in Informatics and Computer Science, and more specifically those interested in the modelling and portrayal of geographic information at a range of scales (or levels of detail). This includes researchers in the fields of geovisualisation and database design. The specific inclusion of chapters detailing the application of the theory in map production environments makes this text highly relevant to practitioners, to research and production engineers working in publishing and national mapping organisations. To that end we hope the book encourages further engagement between industry and research communities. This is the fifth book published on behalf of the International Cartographic Association. Other books in this series include Guptill and Morrison (1995), Moellering and Hogan (1997), Moellering et al. (2005), Peterson (2003), and Dykes et al. (2005).

Front Cover

The maps on the front cover are from the IGN-France BD Carto® database. This vector database was digitised from mapping at 1 : 50 000 scale and consequently cannot be used to produce 1 : 100 000 mapping scale without considerable cartographic editing. The image in the background (from a region in the south of France, near Roquebillière) illustrates what happens when you display 1 : 50 000 data at 1 : 100 000. The map struggles to precisely convey the road network, the interconnectivity of road junctions, and the sinuosity of roads. The solution (superimposed within a hand held device), was derived by a system currently being used in production at IGN, France. The solution was created using automatic generalisation and label placement techniques. The incorporation of an automated solution into their existing map production environment was just one deliverable from the Carto2001 Project – a project that sought to exploit the research being undertaken within the organisation. The Carto2001 project reflects a vision of a single detailed database that is kept current through frequent small updates, and from which multiple products, varying in theme and detail, are automatically derived and integrated with other information. The specific aim of the Carto2001 project was to derive the 1 : 100 000 map series (called TOP100) from the BD Carto® database using automatic generalisation techniques. Using this approach, it took 50 hrs to process a single TOP100 map sheet (typically covering an area of 11 616 square km.) with a further 100 hrs of manual interactive work to complete the derivation process. This compares with 1200 hrs required to accomplish the same task using an interactive system that did not incorporate generalisation techniques capable of autonomous application. A similar story can be told in the case of text placement; automatic label placement took 12 hrs of processing and 160 hrs of manual interactive work to complete the work, compared with 800 hrs required for interactive placement of text.

Our placement of the map within a hand held device reflects some artistic licence! But it is easy to imagine a number of technical routes by which this image could be transmitted to such a device, and a number of projects have demonstrated methods by which users can interact with such data – in particular, in combination with other sources of data (thematic and/or non spatial). In arriving at this overall design, we wanted to particularly stress two points of view. The first point is that the quality of solutions generated by map generalisation technology is now of sufficient quality and robustness that they can be incorporated into map production systems. There are many yardsticks by which we might measure the quality, advancement, or value of research in this field, but it is certainly significant that these algorithms are able to generate high quality solutions, with a high degree of self-evaluation sufficient to enable major cost savings in production – in particular, to alter the cognitive ergonomics between the user and the machine such that it is the machine that is doing more of the decision making (rather than simply responding to the actions of the interactive user).

The second idea we wanted to convey in this image is wrapped up in ideas of ubiquitous computing, and the way in which technology is fundamentally changing how we structure, manipulate and access (geographical) information. We wanted to stress the value of generalisation research beyond the static map; its value in supporting the modelling of geographical space (for example in populating multiple representation databases), and underpinning interaction methodologies (such as intelligent zoom); the idea of being able to interactively view geographic information at multiple levels of detail. In this context there are very important links with cognate disciplines and a variety of research themes (particularly visualisation and exploratory data analysis). Many disciplines are technology driven – the technology creating a paradigm shift in how we interact with, and make decisions about, the world around us. No surprise then, that this is a truly interdisciplinary field, drawing on expertise in databases and data modelling, spatial analyses techniques, and given the complexity of the art and science of cartography, the field of Artificial Intelligence.

Acknowledgements

This book reflects the efforts of many individuals and organisations. It is a collaborative effort that stretches back to 2000 when ideas of a book on map generalisation were first discussed by the then chairs of the ICA Commission on Map Generalisation, Robert Weibel and Dianne Richardson. This publication was prepared in co-operation with, and partially funded by, the International Cartographic Association (ICA). We thank Robert McMaster, Chair of the ICA Publications Committee, for his strong and continuing support throughout this project. Many of the contributing authors are active participants in the ICA Commission on Map Generalisation and Multiple Representation. A huge debt of thanks is owed to all those who have participated in the various activities of the Commission, and to all those who have helped to foster interest in this field, develop and share new ideas, and to drive so dynamically this inter disciplinary field of research. Specific mention goes to the reviewers of the chapters contained in this book, for their rigorous review and in the overall shaping of the text; To Barbara Buttenfield, Jean-Paul Donnay, Elsa Maria João, Menno-Jan Kraak, Christelle Vangenot, Lassi Lehto, Liqiu Meng, Martien Molenaar, Byron Nakos, Peter van Oosterom, Nicolas Regnauld, Dianne Richardson, Monika Sester, Sabine Timpf, John Stell, and Jantien Stoter. We also appreciate the help of Jaakko Kähkönen and Grzegorz Koroluk in working on some of the figures, Omair Chaudhry for his work on the references, and Francois Lecordix for preparation of the maps on the front cover of this book. We also thank Mrs. Kirsti Filén for her help in creating the consolidated reference list.

As a backdrop to those specifically named, the editors wish to give special thanks to the various research groups, agencies and software companies supporting research of this type. The ICA and many other organisations support and encourage the development and sustainability of networked communities of researchers characterised by a willingness to share ideas. ICA Commissions worthy of note would include the ICA Commission for Maps and the Internet (ICA, 2005b), and Visualisation and Virtual Environments (ICA, 2005c). Special thanks is also due to many researchers who have presented and participated in the workshops and tutorials organised by the ICA Commission on Generalisation and Multiple Representation (ICA, 2005a).

List of Contributors

Karl-Heinrich Anders
Institute of Cartography and Geoinformatics
University of Hannover
Appelstraße 9a, D-30167 Hannover
Germany
karl-heinrich.anders@ikg.uni-hannover.de

Yvan Bédard
Centre de recherche en géomatique
Pavillon Louis-Jacques Casault
Université Laval
Québec G1K 7P4, Canada
yvan.bedard@scg.ulaval.ca

Eveline Bernier
Centre de recherche en géomatique
Pavillon Louis-Jacques Casault
Université Laval
Québec G1K 7P4, Canada
eveline.bernier@scg.ulaval.ca

Rupert Brooks
Atlas of Canada, Geomatics Canada
Ottawa, Ontario K1A 0E9, Canada
now at Centre for Intelligent Machines
McGill University
Montreal, Quebec H3A 2A7, Canada
rupert.brooks@mcgill.ca

Dirk Burghardt
Department of Geography
University of Zurich
Winterthurerstr. 190, CH-8057 Zurich
Switzerland
burg@geo.unizh.ch

Cécile Duchêne
Institut Géographique National (IGN)
Laboratoire COGIT
2-4 av. Pasteur, F-94160 Saint-Mandé
France
cecile.duchene@ign.fr

Alistair Edwardes
Department of Geography
University of Zurich
Winterthurerstr. 190, CH-8057 Zurich
Switzerland
aje@geo.unizh.ch

Birgit Elias
Institute of Cartography and Geoinformatics
University of Hannover, Germany
Birgit.Elias@ikg.uni-hannover.de

Andrea Forberg
Technische Universität München
Weiglstr. 9, 80636 München, Germany
Forberg.Andrea@GMX.de

Lars Harrie
Lund University, GIS Centre
Sölvegatan 12, SE-223 62 Lund, Sweden
lars.harrie@nateko.lu.se

Frauke Heinzle
Institute of Cartography and Geoinformatics
University of Hannover
Appelstraße 9a, D-30167 Hannover
Germany
frauke.heinzle@ikg.uni-hannover.de

François Lecordix
Institut Géographique National (IGN)
2-4 av. Pasteur, F-94160 Saint-Mandé
France
Francois.lecordix@ign.fr

Cécile Lemarié
Institut Géographique National (IGN)
2-4 av. Pasteur, F-94160 Saint-Mandé
France
Cecile.lemarie@ign.fr

William A. Mackaness
Institute of Geography
University of Edinburgh
Drummond St
Edinburgh, Scotland, EH8 9XP, UK
William.mackaness@ed.ac.uk

Robert B. McMaster
Department of Geography
414 Social Sciences Building
267 19th Avenue South
University of Minnesota
Minneapolis, MN 55455, USA
mcmaster@umn.edu

Liqiu Meng
Chair of Cartography
Technische Universität München
Arcisstr. 21, 80333 München, Germany
meng@bv.tum.de

Marlene Meyer
Cartographic Department
Kort & Matrikelstyrelsen
Rentemestervej 8, DK-2400 Copenhagen NV
Denmark
mlm@kms.dk

Sébastien Mustière
Institut Géographique National
COGIT Laboratory
2-4 av. Pasteur, 94165 Saint-Mandé, France
sebastien.mustiere@ign.fr

Moritz Neun
Department of Geography
University of Zurich
Winterthurerstr. 190, CH-8057 Zurich
Switzerland
neun@geo.unizh.ch

Nicolas Regnauld
Ordnance Survey Research Labs
Ordnance Survey, Romsey Road
Southampton SO16 4GU, UK
Nicolas.Regnauld@ordnancesurvey.co.uk

Anne Ruas
Institut Géographique National (IGN)
Laboratoire COGIT
2-4 av. Pasteur, F-94160 Saint-Mandé
France
anne.ruas@ign.fr

Tapani Sarjakoski
Department of Geoinformatics
and Cartography
Finnish Geodetic Institute
P.O. 15, FIN-02431 Masala, Finland
Tapani.Sarjakoski@fgi.fi

L. Tiina Sarjakoski
Department of Geoinformatics
and Cartography
Finnish Geodetic Institute
P.O. Box 15, FIN-02431 Masala, Finland
Tiina.Sarjakoski@fgi.fi

Monika Sester
Institute of Cartography and Geoinformatics
University of Hannover, Germany
Monika.Sester@ikg.uni-hannover.de

John van Smaalen
Universiteit van Amsterdam
Institute for Biodiversity and
Ecosystems Dynamics (IBED)
Computational Bio- and Physical Geography
Kruislaan 318, 1098SM Amsterdam
The Netherlands
smaalen@science.uva.nl

Robert Thomson
School of Computing
The Robert Gordon University
Aberdeen AB25 1HG, UK
rcthomson@yahoo.com

Robert Weibel
University of Zurich
Department of Geography
Winterthurerstrasse 190, 8057 Zurich
Switzerland
weibel@geo.unizh.ch

Peter West-Nielsen
Cartographic Department
Kort & Matrikelstyrelsen
Rentemestervej 8, DK-2400 Copenhagen NV
Denmark
pw@kms.dk

W.A. Mackaness, A. Ruas and L.T. Sarjakoski (Editors)

Chapter 1

Understanding Geographic Space

William A. Mackaness

Institute of Geography, The University of Edinburgh, Drummond St, Edinburgh, Scotland, EH8 9XP, UK
e-mail: William.mackaness@ed.ac.uk

Abstract

Geographers have long understood the relevance of scale to the discernment of pattern and the derivation of meaning from geographic data. The power of the map lies in its ability to abstract space. It plays a critical role in identifying and interpreting process. Their variety reflects the breadth of application, now made broader by advances in database technology and media. Traditionally the human cartographer was tasked with selecting and symbolising some abstracted subset of reality. But technological developments in the capture, management and visualisation of geographic information have created a paradigm shift in how we create, utilise and interact with maps. This technology shift has required us to precisely examine the art and science of cartography and explore innovative ways of modelling cartographic design in the context of geographical information science and interactive techniques.

Keywords: pattern, structure, spatial cognition, geography, scale, generalisation, map interpretation

1.1. Scale, Pattern and Geographic Meaning

Ideas of scale and pattern are central to the process of interpretation in the geosciences. "All geographical processes are imbued with scale" (Taylor, 2004, p. 214), making issues of scale "intrinsic to nearly all geographical query" (McMaster and Shea, 1992, p. 1). Choosing scales of analysis, comparing output at different scales, describing constructions of scale (Leitner, 2004) are all common practices in the geosciences. We do this because we wish to know the operational scales of geographic phenomena, how relationships between variables change as the scale of measurement increases or decreases; we want to know the degree to which information on spatial relationships at one scale can be used to make inferences about relationships at other scales (Sheppard and McMaster, 2004). The scale of observation very much "scopes" the problem – providing a starting point from which we "interface" with phenomena that may occur over widely varying spatial, temporal and organisational scales. There is no single natural scale at which a geographical problem should be studied. So the selection of a particular scale (or level of detail) is often just a starting point – one that imposes a necessary and perceptual bias – "a filter through which the system is viewed" (Levin, 1992, p. 1943).

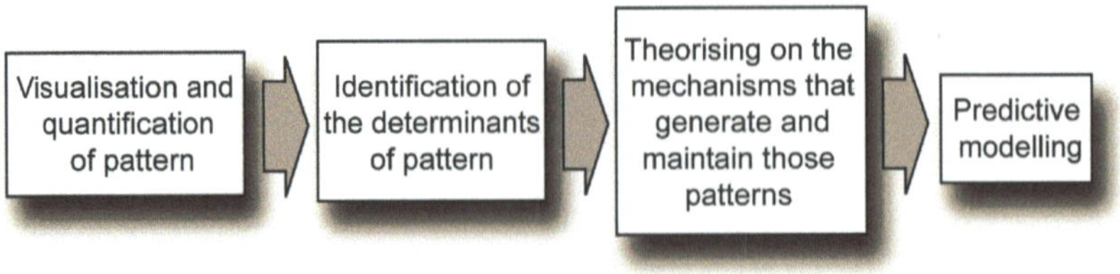

Fig. 1.1. A pathway from pattern to prediction.

The patterns that are evident at any given scale will have specific causes and consequences – those patterns are a manifestation of a collective set of processes, behaviours and responses operating at a range of scales. The hierarchical relationships among these various patterns constitutes a partonomy (Tversky, 1990) – a partonomy, like a taxonomy, is a hierarchy but based on a "part-of" relation rather than a "kind-of" relation. A set of partonomic relationships gives rise to distinctive spatial configurations that enables the categorisation of pattern and the identification of phenomena, as a first step in their interpretation.

Examination of geographical phenomena requires us to study how pattern and variability change with the scale of description and to address problems of "cross linkage" – supporting cross scale analysis in situations where relationships vary across scale (Wilbanks, 2002). Understanding of pattern in terms of the processes that produce them is the essence of science – their observation leads us to theorise about the causal processes, and to develop predictive models. "The essence of modelling is, in fact, to facilitate the acquisition of this understanding, by abstracting and incorporating just enough detail to produce observed patterns" (Levin, 1992, p. 1946). Levin goes on to argue that by viewing phenomena at coarser scales, we move from the unpredictable to behaviours that are regular enough to allow generalisations to be made – generalisations in the epistemological sense of the word! Thus there has always been an inextricable link between scale, pattern and meaning (Hutchinson, 1953) leading some to argue for a theory of scaling (Schneider, 2001). Because we are naturally spatial thinkers (Kaiser, 1993), our efforts to develop theories and models often revolve around visual forms and metaphors. Thus the visualisation of pattern is a cornerstone to the development of theories (Figure 1.1).

It is often the case that we need to view phenomena at varying levels of detail. At a fine scale of observation we discern many of the attributes that define individual phenomena (their form often revealing their function). At the broader scale we gain a more synoptic view – a sense of the regional context in which the phenomenon finds itself. At the global scale, we discern qualities such as patterns of distribution and association, of containment, connectivity, and contiguity between phenomena. By way of example the river engineer (in designing flood defences) would require detailed plans of the topography either side of a river. At a more general level they would require information on human habitation patterns and at a broader scale, they would also need to know a great deal about the river regime, such as the size of the catchment, the drainage capacity of the soil and the land cover within the catchment.

Very "granular" or synoptic models necessarily treat subcomponents of the model in a rather homogeneous way. An obvious example of this would be general circulation models – which model processes at the global scale. It is often the case that we need to move from the general to the specific, to understand the causal links between processes operating at fundamentally different scales (Wilbanks, 2002). Often it is important to observe phenomena at a range of scales in order to address problems of the sensitivity of scale effects (Meyer et al., 1992), to avoid excessive focus at any one scale (and by inference a particular set of patterns and processes), and as a way of understanding how information is transferred across scales. These ideas are mirrored in the cartographic literature (Monmonier, 1991a), highlighting the idea of operational scales at which processes operate – reflecting a linkage between scale and the persistence of a pattern through changing levels of detail. This idea chimes with the work of Muehrcke (1990) who presented the idea of "map stability" – defined as the extent to which the appearance of the map remains largely unchanged through changing levels of detail. This reflects the idea that dominant patterns among phenomena can be defined in terms of how much they persist over large changes in scale. Being able to represent geographic phenomena at various scales (or levels of detail) lies at the heart of the cartographic discipline. The rich history of cartography is testament to its importance to science and society.

1.2. A Brief History of Cartography

Maps have played a pivotal role throughout civilisation (Dorling and Fairbairn, 1997; Kraak and Ormeling, 2003). Some of the earliest records of maps (using clay tablets) date back to 2300 BC (Turnbull, 1989). Maps have been considered to be central to the discipline of geography leading Hartshorne (1939) to argue that if the problem "cannot be studied fundamentally by maps – usually by a comparison of several maps – then it is questionable whether or not it is within the field of geography" (Hartshorne, 1939, p. 249). Traditionally maps were used to define ownership; they were integral to discovery and exploration – a way of defining what we knew about the world. Turnbull (1989) and Tufte (1983, 1990) among others, beautifully illustrate a history of maps and graphics, revealing their critical roles in decision making through time. Sheppard and McMaster (2004) provide a more contemporary view, examining the different traditions of thinking about scale, including ideas of scale as a social or political construct (Marston, 2000; Delaney and Leitner, 1996) and illustrating how society and cultural values have fashioned the notion of the map through time. They go on to show how maps have afforded a more global perspective that has facilitated development of ideas of global governance (McMaster and Sheppard, 2004).

Traditionally, in many of these contexts, the cartographer was very much the custodian of the map – being responsible for selecting and symbolising information critical to the intended task, giving it meaning through the addition of appropriate contextual information. The map is an abstraction of reality, and the process of symbolisation attempts to make explicit what is implicit among the phenomenon being represented. Because of our capacity to interpret symbols, we are able to create linkages between referents (things represented) and signs (a language used to represent them); those signs or symbols, range from the mimetic (pictorial) to the highly abstract. The choice of symbol establishes a visual and conceptual logic, in which scale controls both geographic content, and the granularity of the feature being represented (MacEachren, 1995; Bertin, 1983). Various considerations act to constrain the choice of solution; the decisions the cartographer makes reflects their understanding of the phenomena being represented, as well as a sense of how the user will interpret the map – requiring a shared understanding of the symbols used. The design takes account of such things as the intended use (and the environment in which the map will be used), the map literacy of the intended audience, map styles, the medium and choice of cartographic tools, and the cost of production (Robinson et al., 1984; Keates, 1996). Specialist knowledge (and specialist training) is required to undertake what is clearly a complex decision making task. The challenges of automating cartography were apparent early on: "Imagine a gorge with a river and a road and a railway. First we plot the river, then we display the road. The railway is displaced further and finally the contours are moved. This presents a very difficult problem for the machine to solve" (Anon, 1965). But through time, what was a very manual process has slowly become, in all sorts of way, mechanised and more automated (Keates, 1989). Even following the advent of computer technology, the complexity of the task, combined with the need for aesthetic and interpretive skills, ensured that the cartographer remained "centre stage" to what was very much "computer assisted cartography" (McMaster and Shea, 1992). The result was that technology played a supporting role; the entire decision making process was undertaken by the human cartographer, with the emphasis on technology that would speed up what were tried and tested approaches.

1.3. Map and Model Generalisation

Many authors have written about historical developments in the field of automated cartography, and how ideas of map generalisation evolved into a research field (Sarjakoski, this volume (Chapter 2); Harrie and Weibel, this volume (Chapter 4)). Early research was somewhat stymied by a view that a map was nothing more than a set of points, lines, areas, and text, and that any solution should be viewed as manipulation of these geometric primitives, based on an analysis of a feature's geometry. Any deviation from this thinking was not helped by the "layered" approach to GIS (another cartographic anachronism) in which each theme is stored within a palimpsest of layers. This has not proved to be a fruitful starting point because map generalisation (as a geographical modelling problem) has much to do with modelling the behaviours and interactions *between* phenomena – whereas a layer based approach tends to keep things apart!

It is true that ultimately the results of generalisation do manifest themselves through the manipulation of geometric primitives. But current thinking argues that the reasoning behind those manipulations needs to be based on analysis of the context (Greenwood and Mackaness, 2002), taking account of the interdependencies, behaviours and characteristic forms of the phenomena being represented. Just as the interpretation of a written language depends on its context, the same is true of the interpretation of maps (Head, 1984; Pratt, 1993). Their value is embedded in the whole rather than the sum of its parts yet early attempts at automation adopted a stepwise refinement approach to the problem – breaking the problem down into manageable "chunks" – often dealing with manipulation of the geometric primitives used to represent a particular phenomenon. Unfortunately this approach ensured loss of context, and loss of associated meanings inherent among geographic phenomena. Consequently solutions of this type failed to take into account the changing representational form of other phenomena – thus generating design conflicts.

The need to make explicit and to model the interdependencies among phenomena now pervades much of the current research in map generalisation, and has led to an increasing focus on how the underlying data are stored. Current thinking is built around the idea that the geographic database is the first abstraction of reality, and via a process of generalisation, we distil and make explicit a subset of the relationships inherent among the phenomena being visualised (Harrie and Weibel, this volume (Chapter 4)). This is an approach being adopted by a number of national mapping agencies (Lecordix and Lemarié, this volume (Chapter 15)) – the idea being that from a single detailed database, it is possible to derive maps that vary in thematic and spatial resolution. The finest detail might be captured and maintained at 1 : 1250 scale, and a range of generalisation methods applied to create databases of notional scales of 1 : 10 000, 1 : 50 000, 1 : 250 000, up to 1 : 1 million (West-Nielsen and Meyer, this volume (Chapter 16)).

As new models have been developed to better understand the generalisation process, researchers have understood the importance of the underlying data model in supporting this process and sought better ways of structuring the data (Brewer and McMaster, 1999). The focus is much more about modelling relationships, and the changing levels of information at different granularities. At smaller scales, phenomena are transformed into higher order ones, and through varying levels of abstraction, reveal different characteristic properties and interdependencies inherent among those phenomena. The collective term used to describe this process is "map generalisation" and its goal is to reveal properties within and among geographic phenomena. Those operations concerned with the abstraction of the database come under the heading of "model generalisation", whilst the set of operations concerned with the optimal visualisation of the selected data are grouped under "cartographic generalisation". The two forms of generalisation are intimately connected with increasing emphasis now being given to data modelling and ways of reclassifying and aggregating geographic phenomena (Molenaar, 1998b; Mustière and van Smaalen, this volume (Chapter 6)). Though there is an intrinsic link between this process of model generalisation and cartographic generalisation, it is important to stress the fact that model generalisation is relevant to activities other than the visual. In particular it has relevance to data mining and pattern analysis techniques (MacEachren et al., 1999), and to data compression prior to the transmission and dissemination of information (Buttenfield, 1999) – something that is particularly relevant in the context of distributed systems and data rich worlds (Frawley et al., 1992). These ideas of generalisation now extend beyond conventional two-dimensional data to include three-dimensional data (Meng and Forberg, this volume (Chapter 11)), as well as spatio-temporal data.

1.4. A Changing Context

It remains the case today that a substantial amount of work in the field of map generalisation continues to focus on creating high quality cartographic products – tailored, unambiguous in their presentation, immediately interpretable and delivered via paper and digital media. Such work is of great interest to national mapping agencies, and map publishers. It has built on developments in analytical cartography (Kimerling, 1989) in which spatial analysis, content measures (Dutton, 1999) and cartometric measures have been used to develop evaluation techniques (Mackaness and Ruas, this volume (Chapter 5)). Developments in design methodologies, and geographical modelling techniques have grown in sophistication (Ruas and Duchêne, this volume (Chapter 14)), researchers have looked to more ambitious solutions with current research focused on ideas of autonomous design – systems capable of selecting optimum solutions among a variety of candidate solutions delivered over the web, in a variety of thematic forms, in anticipation of users who may have little or no cartographic skill.

More recently developments in the science of cartography, coupled with advances in technology have fundamentally altered how we author and interact with maps; the balance and process of decision making between the human and the machine has shifted significantly (Long, 1989). The database has become the knowledge store, and the map has become more a visual manifestation of some part of that database – a metaphorical window by which geographic information is dynamically explored. The idea of the long lived generalised paper map (expensive to produce) has been somewhat replaced by the idea of the instant, cheap, up-to-date, yet ephemeral specialised map – through which we interact and query geographic information. Geographic information systems (GIS) technology is just one part of this digital transition (Dodge and Kitchin, 2001) and now supports interactions between an ever broadening community of users; a community of users with high expectations of immediate access to all forms of information, combined from disparate sources, presented across a mix of media and environments. The term "cybercartography" has been coined to capture this idea. Cybercartography is defined as the analysis, presentation and communication of geographic information via the internet, across a broad spectrum of topics, in an interactive manner via a range of media and multimodal interfaces (Taylor, 2005). Cybercartography is seen as a unifying paradigm that goes beyond geographical information science. A paradigm, which argues for the centrality of the map as part of an integrated information, communication and analytical package (Taylor, 2005). Perhaps "Google Earth" exemplifies an early version of this paradigm (Google Earth, 2005) – the globe acting as a simple metaphor of interaction by which anyone can explore the earth, seamlessly travelling from the global view down to the street level and back again. These and other solutions begin to fulfil the idea of the "democratisation of cartography" (Morrison, 1997), in which everyone has simple and fast access to map making functionality and interaction.

In this and other paradigms of use the cartographer is no longer the gatekeeper and custodian of good design. In such contexts, the idea of consulting with a human cartographer is considered somewhat anachronistic – the assumption is that the human cartographer has been made redundant, replaced by symbol menus and colour defaults. The need for careful consideration of cartographic issues has apparently been further eroded by the emergence and development of scientific visualisation techniques (McCormick et al., 1987; MacEachren et al., 1999; Gahegan et al., 2001; Dykes et al., 2005) – visualisation techniques that enable dynamic representation and exploration of high-dimensional data in an interactive manner. But such notions are worthy of closer inspection. By analogy, providing a person with a car jack and an assortment of spanners, does not make them a car mechanic. The same is true of GIS and cartography. Ease of use is no substitute for cartographic skill (Fisher, 1998) and in the absence of an understanding of the art and science of cartography, it remains the case that GIS users are able to create "cartographic monstrosities with unprecedented ease" (Monmonier, 1984, p. 389).

MacEachren et al. (1992), following from DiBiase (1990) argue that visualisation can be defined as a four stage process of exploration, confirmation, synthesis and presentation. In the Geosciences, it is invariably the case that graphical imagery underpins all these activities – the process of model and cartographic generalisation facilitating both the identification, and the interpretation of patterns and relationships that typically exist between geographic phenomena. The process of model generalisation abstracts various characteristics and patterns of the phenomena, with cartographic generalisation seeking to optimally render those characteristics in a form that facilitates their interpretation. These ideas extend across media, from paper based series mapping, through to digital, interactive and mobile environments (Sarjakoski and Sarjakoski, this volume (Chapter 7)). The idea of abstracting space is as central to cartography as it is to all visualisation methodologies. The technology and the language might change, but the underlying principles do not.

What technology has done is to require the recasting and extension of cartographic theory to accommodate new models of interaction and to extend model generalisation techniques to include database abstraction that can support non-visual, spatial analysis. A more database centric view has encouraged us to view the map as a model of space. It has reminded us that cartography is very much about geographical modelling (Mustière and van Smaalen, this volume (Chapter 6)) – not just some "end" process in which we symbolise a set of data. These ideas have required extension to the art and science of cartography, requiring us to move well beyond the principles governing paper map design, to examine the links that exist between the modelling of geographic information, and how we model the process of abstraction and visualisation. These developments have even led us to challenge traditional understandings of what is meant by "scale". Scale is something that is precisely understood in the context of the paper map, but becomes much less clear in a digital context (Goodchild, 2004; Goodchild and Proctor, 1997). What might we surmise from all this? How does research in generalisation and multiple representation situate itself in such technologically evolving environments? In terms of the future context of this work, one can easily envisage environments comprising a set of distributed, fine scale databases. These databases would be continually updated using automatic feature extraction and change detection technologies, applied to imagery collected using remote sensing technologies. The process of model generalisation would be used to populate multiple representation databases (Bernier and Bedard, this volume (Chapter 9)), and in combination with cartographic generalisation be used to create multi themed, multi scaled products (at varying levels of detail) for a variety of tasks (interactive/non-interactive, spatial and non-spatial). Irrespective of the choice of scenario, it is important that research in this field responds to developing contexts of use lest it become an anachronistic irrelevance!

1.5. Why is Generalisation so Hard to Automate?

Decades of research have yielded some highly innovative solutions (Weibel and Jones, 1998), but for a variety of reasons, there exist few commercial solutions that really encapsulate the idea of cartography as a collaborative, decision making process. The ideas presented in this book clearly illustrate exciting advancements in response to developments in theory and changing contexts of use – themselves driven by technological developments. But it is an interesting question to ask: "why a task so effortlessly performed by a human, has proved so hard to automate?". I posit four reasons for this.

The first reason is that the process of design is complex – requiring assessment of multifarious factors, where multiple candidate solutions may exist, each reflecting a compromise among a sometimes competing set of constraints. The map is a complex mix of metric and topological patterns reflecting individual and interdependent gestaltic properties. Understanding these forms, and conveying salient characteristics requires both cartographic and geographic knowledge. Thus various attempts have, and continue to be made, at trying to engineer knowledge, to observe the cartographer at work, and to "capture" the rules of thumb and intuition that appear to underpin this subjective process. A review of attempts at knowledge engineering is given by Harrie and Weibel, this volume (Chapter 4), with current approaches reflected in work using multi agent systems (Ruas and Duchêne, this volume (Chapter 14)).

A second problem revolves around the morphosis of information through changes of scale. We look at phenomena at different levels of detail precisely because we wish to explore different characteristics and relationships. Minsky (1975) said "you cannot tell you are on an island by looking at the pebbles on a beach" but sail aloft and we see a different property of the pebbles – their connectedness as a perimeter ring providing the answer to our question. It is not the case that any one scale of observation contains less or more information, more that they contain different, albeit related information. Herein lies the conundrum – how to develop generalisation techniques that abstract out from the same data source, different properties of the phenomena? Being able to abstract out these characteristic forms is something that is explored by Heinzle and Anders, this volume (Chapter 12). Having derived these different characterisations and solutions, the question then becomes one of assessing the "correctness" of those solutions (Mackaness and Ruas, this volume (Chapter 5)). Just how good is any given solution, and with what do we compare?

A third challenge lies in seeing generalisation as a modelling problem rather than "something you do at the end". "Generalisation must be based on process rather than graphical appearance" (Muller, 1989, p. 203), and be based on the phenomenon being mapped (Ormsby and Mackaness, 1999) rather than the geometric primitive by which it is stored in the database. As a modelling problem it calls on a rich set of spatial analysis techniques to model many of the qualities inherent among surfaces, networks and point pattern distributions. Many of the chapters in this book illustrate the rich set of analytical tools now used to make explicit the qualities inherent among geographic phenomena.

A fourth area that has proved to be very challenging is in the area of interaction in the map generalisation modelling process. Is there a language of design or some sort of spatial syntax by which a user can interact, and effortlessly specify their needs? Krippendorff's (1995) comment that "design is all about making sense of things" appears to support the notion that this is a collaborative process. This begs the question, what is the optimal balance of decision making between the human and the machine? How are decisions of the system conveyed to the user in a way that is meaningful? How does the user retain a creative component in the task? Thus interfacing and interaction have proved to be difficult problems even to specify, let alone resolve.

But the chapters in this book are testimony to developments in all these areas. There is real optimism (and evidence) to suggest that innovative and meaningful solutions are being found to these problems – solutions that reveal a close linkage between generalisation and the understanding of geographic space.

1.6. The Chapters of This Book

The commissioned chapters of this book reflect three aims: (1) to explain the changing context of research in this field, (2) to lay out a framework that describes the theoretical underpinning of map generalisation (both from the database and the cartographic perspective), (3) to richly illustrate the broadening application domains, and (4) to illustrate the quality of solutions coming from commercial solutions. The early chapters have a theoretical focus, coupled with an emphasis on review. Sarjakoski, this volume (Chapter 2) reviews progression in conceptual models and the development of ideas of multiple representation; Regnauld and McMaster, this volume (Chapter 3) describe the various techniques used to manipulate geographic data and cartographic features; Harrie and Weibel, this volume (Chapter 4) present some of the frameworks in which the entire generalisation process can take place. The idea that systems might be more autonomous in their cartographic decision making has led to considerable effort being devoted to development of evaluation techniques (Mackaness and Ruas, this volume (Chapter 5)). But for all of the ideas presented in these chapters, it is clear that there are significant implications in terms of data modelling and data storage. Mustière and van Smaalen, this volume (Chapter 6) therefore examine various database issues, including the idea of storing multiple representations of phenomenon. This idea is developed further by Bernier and Bedard, this volume (Chapter 9) which explores ideas of "data warehouses" as a way of supporting fast delivery of map based information. Sarjakoski and Sarjakoski, this volume (Chapter 7) illustrate new application domains arising from developments in technology – in this case the role of generalisation in mobile mapping. Many research fields contribute to these endeavours – it seems that research in this field draws on developments in computational geometry, image analysis, computer graphics, interface design, database theory, informatics, cognitive science and geographical information science (to name but a few!). Edwardes et al., this volume (Chapter 8) argue for a need to develop "open generalisation systems" that could facilitate the sharing and integration of algorithms between these cognate disciplines – that compliance with ideas of interoperability could shorten development time, and facilitate development of commercial solutions.

The following chapters further illustrate the diversity of application domains. Sester and Elias, this volume (Chapter 10), examine how generalisation can be selectively applied to support the task of wayfinding (in the context of mobile applications) and how generalisation can be applied to three-dimensional abstractions of space (Meng and Forberg, this volume (Chapter 11)). Characterising phenomena is critical to their interpretation. Heinzle and Anders, this volume (Chapter 12) highlight the relevance of automatic pattern recognition techniques as a basis for identifying characteristic patterns inherent among geographic phenomena (in this case, applied in the context of road networks). Thomson and Brookes, this volume (Chapter 13) continues with this theme, showing how generalisation methodologies have been developed and applied to geographical networks of various types.

Ruas and Duchêne, this volume (Chapter 14) specifically detail the application of multi agent systems to the problem of map generalisation, reflecting the considerable interest and resources recently devoted to this particular methodology. There are many ways by which we might measure developments in this field. One of them is to examine the success with which solutions have migrated into commercial and business environments. Lecordix and Lemarié, this volume (Chapter 15) and West-Nielsen and Meyer, this volume (Chapter 16) are examples from two different national mapping agencies, and describe the success they have had in developing some of this research to the point that it is sufficiently reliable, and of sufficient quality, that it can be incorporated into map production environments. The final chapter (Mackaness et al., this volume (Chapter 17)) reflects on all of these developments, and attempts to distil a key set of research topics particularly in need of further endeavour.

1.7. Conclusion

Generalisation is about modelling geographic space – the scale at which we view the world profoundly affecting our ability to understand it. Both the scale of observation and of representation reflects a process of abstraction, an instantaneous momentary "slice" through a complex set of spatio-temporal, interdependent processes. Generalisation is about managing the complexity inherent in high-dimensional data (Bauer et al., 1999) and has much to do with reasoning about space and making sense of the emergent patterns. In essence meaning and interpretation can only be gained by viewing phenomena at a range of scales and in varying contexts (themes). It explains Monmonier's (1991a) argument that it is a travesty for GIS not to be able to support the viewing of phenomena at multiple scales.

Early research in the field of automated cartography focused on challenges associated with the design of paper based topographic mapping. Increasingly however, that research has had to adapt to (and take advantage of) the profound changes in how we handle and utilise geographic information – arising from developments in Information Technology (IT). MacEachren and Kraak (1997) argue that a fundamental shift in map use has taken place. To wit, an emphasis on map use that acknowledges the differences between *exploring unknowns* and *communicating knows with maps*. But generalisation can be viewed both as a process of exploration (through the process abstraction), and of communication (via the optimal design of maps). Consequently there is much overlap between the research agendas of those working in visualisation and map generalisation and multiple representation. As we will see in the following chapters, an excitingly diverse set of cognate disciplines contribute to the art and science of map generalisation and multiple representation.

The response has been to explore this research field within these changing technological contexts, and to extend and develop new methodologies by which we represent and interpret geographic information. The function of the map has expanded to become the "looking glass" by which we interactively search and retrieve information, both spatial and non-spatial (Skupin and Fabrikant, 2003). "The Map" has become but part of a continuum of visualisation methodologies used to contextualise and explore geographical information (GI). In this sense differentiating between internet mapping, geovisualisation, and map generalisation is a bit of an academic indulgence. What is becoming more apparent is a collective advancement in data capture techniques, data modelling, and visualisation methodologies, coupled with increasing Internet access. This process of democratisation of geographic information raises exciting challenges and opportunities for those working in this field. The interconnections among these cognate research fields would seem to suggest that greater collaborative effort between these research communities is the way forward.

Chapter 2

Conceptual Models of Generalisation and Multiple Representation

L. Tiina Sarjakoski

Dept. of Geoinformatics and Cartography, Finnish Geodetic Institute, P.O. 15, FIN-02431 Masala, Finland
e-mail: Tiina.Sarjakoski@fgi.fi

Abstract

The purpose of this chapter is to give an overview of developments in the conceptual modelling of generalisation and multiple representation. It shows how generalisation has evolved from a simple algorithmic approach towards comprehensive mathematical modelling. In parallel with this process, research on multiple representation databases has taken place. The background to research on the multiple representation approach lies in, and remains closely connected with, the generalisation problem and the underlying conceptual models for generalisation. The development of an object-oriented paradigm for modelling the real world as objects has played a central role in supporting the modelling and implementation of multiple representation solutions. In this chapter, the review steps through from cartographic generalisation to the era of model generalisation, and these steps are reviewed in relation to progress in computer science. Collectively these developments point to a need for future development of models and methods for generalisation and multiple representation that support a greater variety of tasks and applications.

Keywords: conceptual model, multiple representation, knowledge based methods, object-oriented, incremental generalisation, cartographic generalisation, model generalisation

2.1. Modelling Generalisation Processes

Over the last decade, considerable steps have been taken to improve methods for generalisation. However a deeper understanding of how to formalise and apply cartographic knowledge in the generalisation process is still required in order to support a greater variety of map users' tasks. Conceptual models are used as tools to describe and structure the geographic phenomena, and are prerequisite to communication in and between various disciplines (Grünreich, 1995). It is fundamental for a human being to model the space in the form of cognitive or mental maps, and in fact every modelling process can be primarily seen as a generalisation process. Cognitive maps in the minds of individuals can be translated into different languages of communication, such as gestures, sounds or words, tables, texts and graphics (Freitag, 1987). The cartographic communication model (Kolazny, 1969) describing the communication process related to graphic maps and spatial information has its roots in information and communication theory (Shannon and Weaver, 1949, 1998) and semiology (Bertin, 1967). Salichtchev (1983) argued conversely that a cartographic transmission of spatial-temporal information from a sender (cartographer) via the medium (map) to the receiver (map reader) is not linear, but is based on principles other than mathematical information theory. A map reading process was to be seen as a holistic spatial process since the relationships between map features provide additional information that is not determined by a specific continuous sequence. A map is a unique form of communication, and cartographic generalisation arises from the need to communicate. Knöpfli (1983) also emphasised the role of generalisation in the communication process and pointed out that the inefficiency in map communication is due to the amount of irrelevant or misleading information with respect to the user's question.

Manual generalisation was also regarded as a holistic process, where maps were products of art clarified by science (Eckert, 1908). The International Cartographic Association (ICA) defined generalisation in 1973 as: "The selection and simplified representation of detail appropriate to the scale and/or purpose of a map." The process of simplifying reality on a map has been referred to as cartographic generalisation by others (Robinson et al., 1995; Imhof, 1982; Bertin, 1983). Generalisation has always been one of the fundamental processes in cartography, since a map is an abstraction and can show only a subset of geographic reality. Bertin (1983) felt that understanding means simplifying and reducing a vast amount of data to the small number of categories of information that we are capable of taking into account in dealing with a given problem, and thus generalisation is fundamental to perceiving information. He argued that the necessity to generalise results from the contrast between the limitations of human perceptual constants and the infinite range of possible reductions of geographic order.

Generalisation has been directed by the choice of map scale, objective, graphic limits, and quality of data. Robinson et al. (1995) referred to these as the controls of cartographic generalisation and to scale as the dimensional relationship between reality and the map. They divided the generalisation process into selection, and into the elements and controls of generalisation. The elements of generalisation included simplification, classification, symbolisation, and induction. The authors noted that it is impossible for the cartographer in practice to separate these processes and model them individually without regard for the others. An interesting point regarding the elements of generalisation is that the authors did not include selection of information as being within the elements of cartographic generalisation. Robinson et al. (1995) stated that selection is regarded as the intellectual process of deciding which information will be necessary; it does not demand modification, and it can be undertaken irrespective of the map format or scale.

Generalisation methods and processes have been changed and improved alongside developments in the science and art of cartography and have been strongly influenced by progress in computer science. However, the development from manual to automated environments has not been without its critics. Keates (1996) pointed out that the practice of cartographic generalisation is essentially a visual skill and claimed that the tendency to concentrate on generalisation operations is not essential, because in reality, these play a subordinate role. Generalisation is affected by two sets of factors: (1) quantitative factors; i.e. the scale difference between source map and derived map, and the relative sizes of the symbols on the two maps, and (2) qualitative facts, i.e. the need to retain the characteristics of the topographic features and to judge their relative importance.

The cartographic research community is at the centre of rapid developments taking place in generalisation algorithms and methods. It is nevertheless important that the original purpose of what the cartographic generalisation process is trying to achieve is born in mind. Irrespective of the means, the map should always be a coherent, topologically correct, up-to-date, legible and visually pleasing abstraction of the world that is appropriate given the map's scale and task.

2.1.1. *From cartographic to digital generalisation*

The period of automated generalisation began during the 1960s. The term "digital generalisation" was often used to emphasise the transition from a manual to a computer based era (Shea and McMaster, 1989; Buttenfield and McMaster, 1991). One of the well-known outcomes from that period was the simple mathematical formula presented by Töpfer and Pillewizer (1966), in which the radical law of selection could be used to calculate the number of symbols that should be shown at different scales. The formula yielded the number of symbols to be displayed, but did not reveal which of the symbols should be selected and did not take account of local variation in the density of phenomena being portrayed. This selection law was not specifically intended for automated generalisation however, and additional efforts were made towards mathematical treatment of generalisation problems. Salichtchev (1976; Salistschew (1967)) was one of those who started to systematically study individual map generalisation operations, rules and quality criteria.

Ratajski (1967) presented one of the first formal models of generalisation and identified two fundamental types of generalisation process: quantitative and qualitative. "Quantitative generalisation involves a gradual reduction in map content which is dependent on scale change, while qualitative generalisation results from the transformation of elementary forms of symbolisation to more abstract forms" (after McMaster, 1991, p. 22). In Ratajski's model an important component was called "generalisation point" defining the level where a change in the cartographic method of representation was necessary (for example a change from line to area feature methods). Morrison (1974) presented a model of generalisation based on formal set theory and viewed the process in terms of four basic generalisation processes (simplification, classification, symbolisation and induction). He viewed the process in terms of probable transformation characteristics for a set of elements from the cartographer's reality that were to be portrayed on the map. His basic idea was to clarify the processes of generalisation through formal mathematics (Slocum et al., 2005).

The first period of digital generalisation (1960–1975) was focused on development of algorithms for line simplification, during which one of the most-used and still relevant algorithms for global line simplification was developed by Douglas and Peucker (1973). A detailed review of the linear feature based algorithms for local and global processing was given by Richardson (1993). During the second period of digital generalisation (the late 1970s and early 1980s), assessment of algorithm efficiency was emphasised (Buttenfield and McMaster, 1991). Scale dependency was modelled using parametric methods. During the early days of digital cartography, most research on generalisation was devoted to rather narrow aspects of the overall problem (Weibel, 1991), such as line simplification and selection according to a given threshold value. Examples of more complex approaches included the first studies on automated displacement. Christ (1978) studied automatic displacement of point and line symbols, and Lichtner (1978) developed algorithms for automatic displacement that regarded the relational positions of the objects at a target scale.

During the 1980s scientists began to direct their considerations more towards the conceptual modelling aspects of generalisation. Bertin (1983, p. 300) discussed two possible methods of generalising: conceptual and structural generalisation. In conceptual generalisation "one can change the implantation (for example, from a cluster of points to an area), which implies a new level of conceptualisation; thus, "mines" become a "coal field". It generally involves new information beyond that being processed. In structural generalisation one can maintain the level of conceptualisation, which implies maintaining the implantation and the planar structure of the phenomenon, but at the same time simplifying the distribution. Thus a cluster of points remains as a cluster of points representing mines, but as the scale is changed the number of symbols for mines will be reduced. Structural generalisation can be based on the information being processed, provided it is comprehensive." Although Bertin's model was not especially developed for the digital environment, his conceptualisation is particularly interesting in relation to conceptual models of generalisation, and furthermore to modelling the world as multiple representations.

McMaster and Shea (1988) proposed a framework for digital generalisation (Figure 2.1). The digital generalisation process was divided into three issues: (a) philosophical objectives in response to the question of *why* we generalise, (b) a cartometric evaluation of the conditions present *when* we generalise, and (c) the methods of generalisation, i.e. the generalisation operators arising in response to the question of *how* we generalise. A generalisation operator defines a type of spatial transformation that is to be achieved, while a generalisation algorithm implements that particular transformation (Weibel and Dutton, 1999). Instead of the term "operator", other terms such as "operation" or "process" have been variously used in the literature.

The decomposition of the 'how' aspect was divided by Shea and McMaster (1989) into twelve operators: simplification, smoothing, aggregation, amalgamation, merging, collapse, refinement, typification, exaggeration, enhancement, displacement and classification. These operators resulted in spatial and attribute transformations and are further discussed in detail by Regnauld and McMaster (this volume (Chapter 3)). The "when" aspect of the generalisation process consisted of three parts: conditions, measures, and controls (McMaster and Shea, 1992). The authors argued that organisation of the "when" and "how" processes are important if a complete solution for digital generalisation is to be achieved. Since the 1990s these considerations have been widely discussed and the why-when-how aspects have been referred to in a number of papers.

2.1.2. *Towards model generalisation*

A thorough summary of generalisation models proposed throughout the 1970s and 1980s was given by McMaster (1991); these models served as conceptual frameworks for digital generalisation. One of the models was developed by Nickerson and Freeman (1986) and was designed using an expert systems approach. Their model consisted of five tasks: (1) four distinct feature modification operations (feature deletion, simplification, combination, type conversion); (2) symbol scaling; (3) feature relocation; (4) scale reduction; and (5) name placement. Brassel and Weibel (1988) proposed a conceptual framework for automated map generalisation

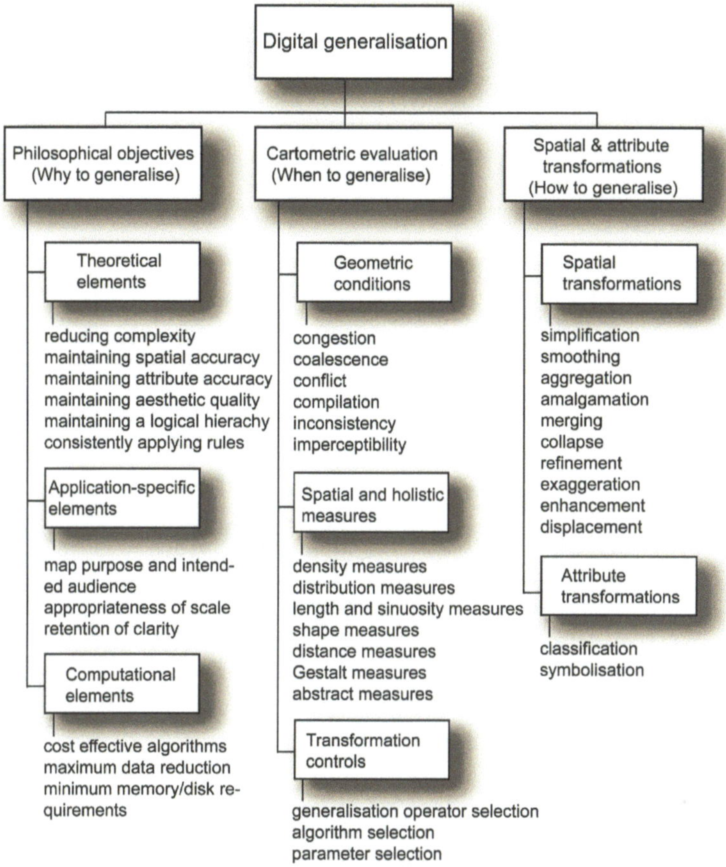

Fig. 2.1. McMaster and Shea's conceptual framework for digital generalisation (redrawn from McMaster and Shea, 1992, back cover). Reprinted by kind permission of the Association of American Geographers.

and distinguished between statistical and cartographic generalisation. Statistical generalisation was described as an analytical process that deals with information-content reduction in a database under statistical control. Cartographic generalisation was regarded as spatial modelling for visual communication aimed at modifying local structure. Generalisation in this framework consisted of five steps: structure recognition, process recognition, process modelling, process execution, and data display (Figure 2.2). What was earlier called statistical generalisation in Brassel and Weibel's (1988) framework was later rebranded as model generalisation.

During the late 1980s and early 1990s there was a shift in emphasis acknowledging developments in data modelling and database technologies. This period resulted in the two-fold concept of generalisation, namely cartographic generalisation and model generalisation. The motivation for this dichotomy arose from the idea that geographic databases have to be main-

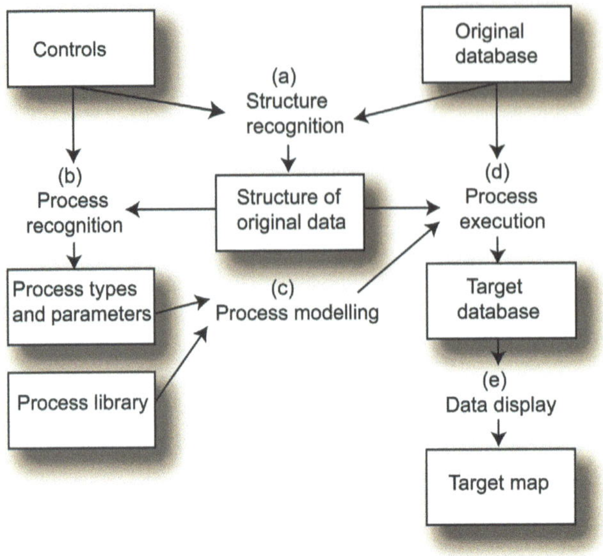

Fig. 2.2. Brassel and Weibel's conceptual model for generalisation (redrawn from Brassel and Weibel, 1988). Printed by permission.

tained at multiple levels of detail. It was felt necessary to make a distinction between, on the one hand, generalisation processes associated with geographic databases and, on the other, those needed for the final cartographic presentation of the output maps.

As stated previously (Kilpeläinen, 1997; Weibel and Dutton, 1999), the distinction between model generalisation and cartographic generalisation was already apparent in the early works of Grünreich (1985), in which the respective models were referred to as the Digital Landscape Model (DLM) and the Digital Cartographic Model (DCM). In the context of object generalisation, Grünreich (1992, 1995) referred to the process during data capture in which the basic primary models of the environment were set up. He referred to the derivation of primary models of lower semantic and geometric resolution from the basic digital object model as model generalisation. In this context the digital object model was comprised of the digital landscape model as a spatial reference system and the digital thematic models of all integrated disciplines, such as cadastral models. This terminology was used in the development of ATKIS (the Official Authoritative Topographical Cartographic Information System) in Germany during the late 1980s (AdV, 1989; Vickus, 1995). One of the essential contemplations was to construct topographic and cartographic landscape models separately. The DLMs were acquired by object generalisation from a state of the real world, and DCMs, resulting from cartographic generalisation, were applied to the DLMs (Figure 2.3).

In Kilpeläinen (1992, 1997) a model for conceptual generalisation within a database context was presented (Figure 2.4). Several researchers presented similar models prior to, and during this period (Grünreich, 1985). Model generalisation (i.e. conceptual generalisation) meant the simplification of the abstract and digital models represented by the geographic information (Kilpeläinen, 1997; Weibel and Dutton, 1999). The purpose of model generalisation

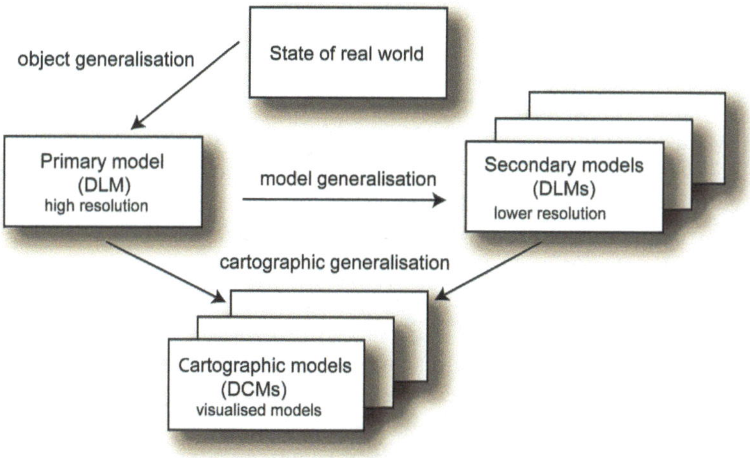

Fig. 2.3. The ATKIS model (after Grünreich, 1985). Printed by permission.

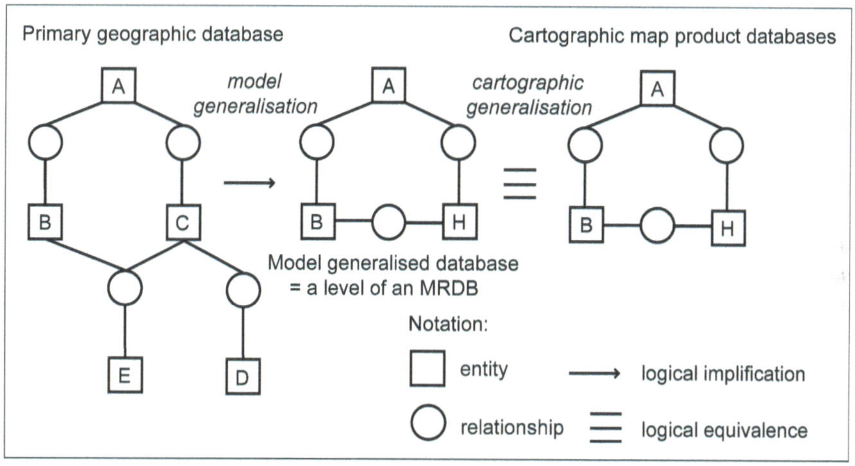

Fig. 2.4. A conceptual model for generalisation within the geographic databases. The semantics of spatial data may change in model generalisation (modified from Kilpeläinen, 1992, 2001).

was to create a modified database using the primary geographic database as an input source. During this phase the aspects related to graphical display were not considered. Model generalisation was seen as a preprocessing step prior to visualisation via cartographic generalisation and thus model generalisation involved no artistic, intuitive components. Model generalisation provided geographic data for analysis functions, whereas cartographic generalisation was performed for the visualisation of application-dependent products. In this context, cartographic generalisation involved the reduction, enlargement, and modification of the graphic symbolism on a map that must be performed in order to increase the effectiveness of cartographic communication.

Kilpeläinen (1992, 2001) emphasised that the semantics of spatial data may change in model generalisation (for example, the creation of city block areas by aggregation of buildings). This is illustrated in Figure 2.4 where the data in the primary geographic database is represented by the entity boxes A, B, C, D, E, and in the model generalised database by A, B, H. The entity H has now replaced C, D and E. The situation is different when deriving the cartographic map product databases from the model generalised database: the semantics of the data are not changed – which is illustrated by the unchanged entity boxes A, B, H in Figure 2.4. Model generalised database in the figure constitutes a level of a multiple representation database (MRDB) (see further § 2.2).

Weibel and Dutton (1999) made a distinction between conceptual models for process-oriented and representation-oriented generalisation. The process-oriented generalisation was understood as the process of transforming a detailed database into a database or map of reduced complexity at arbitrary scale. An alternative approach was called a "representation-oriented view" since it attempted to develop databases that integrated single representations at different fixed scales into a consistent multiple representation.

2.1.3. *Knowledge-based methods for generalisation*

Muller (1991) claimed that the definition of generalisation by the ICA from 1973 was misleading, as it confused the tools with the objectives. Muller wanted to emphasise that generalisation is an information-oriented process that is performed for map display, communication, and also for analytical purposes. Research on generalisation should be seen more as a comprehensive and holistic problem, and more intelligence should be applied to the process. In particular he pointed out the resolution of spatial conflicts that arise as a by-product of generalisation, as the competition for space increases with a decrease in scale. He stated that resolution of these conflicts required a simultaneous view of different cartographic features, priorities, and tools, and proposed, like many others at this time, the use of knowledge-based tools to support automated solutions.

Since the late 1980s, map generalisation research continued to progress with developments of knowledge-based methods, expert systems, machine learning techniques, neural networks, object-oriented techniques and models for formalising cartographic knowledge. One of the topics during the early 1990s was the establishment of rule bases for generalisation, a task that included data modelling and representation techniques (Buttenfield and McMaster, 1991). Cartographers' desire to utilise knowledge-based methods was driven by the development and apparent success in computer science in knowledge-based systems and artificial intelligence (AI) during the 1980s and 1990s. The cartographic process seemed to be an ideal candidate for representation as a set of rules, since manual generalisation was thought to be based on heuristics and experience borne of the cartographer's expertise and formal training.

By a knowledge-based system we refer here to a system in which the knowledge is stored in a knowledge base separate from the control and inference components. A knowledge base contains facts, procedures or rules expressing an expert's heuristics for the domain (Hayes-Roth et al., 1983). Knowledge engineering is the term used to describe the overall process of developing an expert system. Expert systems are composed of at least three basic entities:

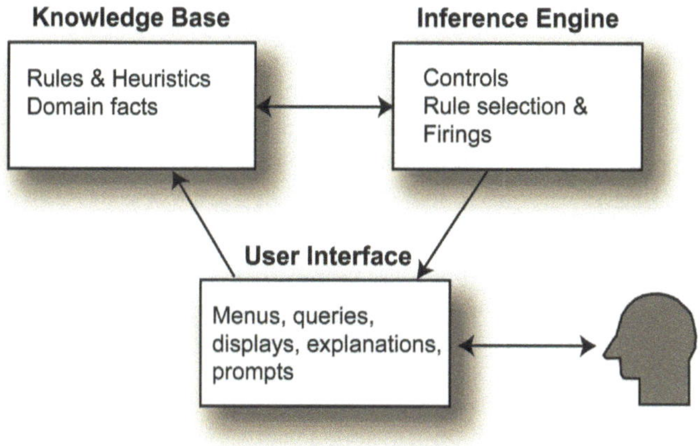

Fig. 2.5. Components of an expert system (McGraw and Harbison-Briggs, © 1989, p. 4). Adapted by permission of Pearson Education, Inc., Upper Saddle River, NJ.

the knowledge base, an inference engine, and a user interface (McGraw and Harbison-Briggs, 1989) (Figure 2.5). The textbook by Russell and Norvig (2003) gives a more recent review of knowledge-based systems and AI.

Knowledge-based approaches to generalisation were discussed for example by Muller (1990, 1991) and Buttenfield and McMaster (1991). Armstrong (1991) discussed the organisation of knowledge within knowledge-based generalisation and the use of procedural knowledge in selecting the appropriate generalisation operators and algorithms within a given map context. Mark (1991) discussed a certainty factor for determining the importance of rules. Shea (1991) proposed the establishment of meta-rules for controlling the priority of generalisation rules, and regarded forward-chaining reasoning as a more appropriate choice for generalisation, since the process is data-driven and situation specific. Laurini and Thomson (1992) regarded backward chaining as an interesting method for diagnosis and for discovering reasoning arising from a given situation.

Knowledge-based methods were partly used for generalisation of buildings from the scale of 1 : 5000 to the scale of 1 : 50 000, in the implementation of the GENEX prototype at the University of Hannover (Meyer, 1986). Mackaness and Fisher (1987) reported on a knowledge-based approach to displacement of point symbols. Schylberg (1993) developed a formalisation technique for gathering rules for small area replacement in topographic maps. His approach was implemented in raster-based area generalisation, and it appears that knowledge-based techniques were used quite successfully in other raster-based applications, for example in remote-sensing (Goodenough et al., 1987).

It should be noted that knowledge-based approaches were also applied to tasks that were, strictly speaking, not part of generalisation but essentially related to digital cartography. Examples of such studies included automatic cartographic name placement using rule-based systems (Cook, 1988; Doerschler and Freeman, 1989). Müller and Zeshen (1990) proposed a knowledge-based system for cartographic symbol design providing guidance in the production of cartographically acceptable products. Knowledge-based methods were used for the automatic recognition of names and numbers from maps, presented by Meng (1993). Mackaness et al. (1986) presented principles towards a cartographic expert system, and Forrest (1993) discussed cartographic expert systems, in particular for map design.

The research on knowledge-based systems for generalisation highlighted the complexity of the task and the need for generalisation tools. It reinforced the observation that whilst diagnostic tasks were relatively easy to model (for example the success of AI in medical diagnosis (Russell and Norvig, 2003)), tasks requiring synthesis and evaluation of solutions were harder to realise. Russell and Norvig (2003) argued that in areas such as game playing, logical inference, theorem proving and planning we have already seen systems based on rigorous theoretical principles that can perform as well as, or better than, human experts but in generalisation, this goal has been reached only in some specific tasks.

2.1.4. *Knowledge acquisition*

The lack of methods for exploiting and formalising cartographic rules for map generalisation was regarded as a major impediment to successful implementation of expert systems for generalisation (Buttenfield and McMaster, 1991). In response, a considerable amount of research was done on knowledge acquisition for generalisation during the 1990s. It was argued that to develop generalisation methods from a holistic standpoint, it might not be sufficient to use only parametric (metric/attribute) values for operator selection, but that more complex rules needed to be included in the model/cartographic generalisation process. McMaster and Shea (1992) and McMaster (1995) stated that since we have not been able to replace the human element of generalisation with computer algorithms or rules, more attention should be devoted to developing knowledge acquisition techniques for generalisation purposes in order to develop a more comprehensive system of generalisation. It was felt that it was difficult to formalise a cartographer's knowledge in such a way that it can be modelled and transformed into a form that could be processed by a computer. Parsaye and Chignell (1988) found that human experts are not explicitly aware of the structure of their knowledge, and though the intermediate steps in their reasoning appear obvious to them, they may not be able to provide an overall explanation of how their decisions are made at a level of detail necessary for implementation. The same difficulties were faced when trying to exploit the knowledge of cartographers (Nyerges, 1991; Rieger and Coulson, 1993). Lee (1996) was one of those who have studied rules for generalisation, and Richardson and Müller (1991) reported on rule selection for map generalisation.

Muller and Mouwes (1990) reported on a study using a topographic map series and stated that automation of the generalisation process requires great insight into the knowledge used in that process. They went on to discuss the different methods for knowledge acquisition and claimed that textbook descriptions are not sufficiently detailed for development of rules for automatic generalisation. Instead, they suggested that to obtain more specific rules the guidelines from national topographic map series production should be studied. One example of this was the study reported by Nickerson (1991), in which generalisation of the Canadian National Topographic Series of maps was studied. Cartographers' knowledge on generalisation in Finnish topographic map series was studied by Kilpeläinen (2000). Different methods potentially useful for knowledge acquisition in map generalisation were studied by Weibel (1995). The proposed methods included: conventional knowledge engineering techniques, analysis of text documents, comparison of map series (reverse engineering), machine learning, artificial neural networks, and interactive systems (amplified intelligence).

In addition to the lack of formalisation of appropriate knowledge for generalisation rule bases, application dependencies were also seen as a problem for rule formalisation (Beard, 1991a). Mapping approaches typically vary from country to country such that a different rule base would be required for almost every map series. Discussions on knowledge-based systems for generalisation have been a relatively minor topic of interest for researchers during the last decade but in other areas of geographic information science, progress has been made (Hui et al., 2004; Stefanakis and Tsoulos, 2005). In the field of map generalisation, the challenge of formalising knowledge as rules remains. The quality of the knowledge, and the efficiency of that formalisation are critical to their success (Parsaye and Chignell, 1988). New approaches to tackling this problem include constraint based problem solving and agent based methods. For further discussion see Ruas and Duchêne (this volume (Chapter 14)).

2.1.5. *Amplified intelligence*

To overcome problems connected with rule formalisation, Weibel (1991) proposed amplified intelligence as a way of supporting complex human tasks. The idea was that key decisions are normally made by the user, whose knowledge is amplified by a range of high-level tools for undertaking generalisation operations. Amplified intelligence was seen as a transitional approach; knowledge is gradually brought from the human operator into the system, thus leading to a full-scale expert system approach (for further discussion see Weibel and Harrie, this volume (Chapter 4)).

Similar ideas were reflected in semi-automatic approaches; the difference between them and amplified intelligence was that in the latter, the skill of the human operator was recoded into the system as it was detected. The idea of making partial use of the human operator throughout the process was proposed by Monmonier (1982). He suggested that the most practical approach would be one that combined the intelligent guidance of a cartographer with an interactive cathode ray tube (CRT) display and light pen. This approach was utilised in remote sensing for multispectral classification of land-cover information. For map generalisation, the map author may make certain primary decisions for a set of typical regions, and the computer would then replicate these decisions across the entire region. Mackaness (1995a) also proposed a constraint-based approach to human computer interaction from a user-centred perspective in automated cartography. He discussed the methods by which the user could feel more in control, and explore and navigate among alternative design solutions.

2.1.6. *Object-oriented methods for generalisation*

Object-oriented approaches have facilitated developments in generalisation methodologies since the early 1990s; Mark (1991), for instance, stated that an object-oriented database environment had the potential to support development of a prototype system for generalisation. He proposed that progress would be achieved by attempting to model and generalise real-world objects as features rather than their cartographic representations. Mark stated that a central concern in object-oriented approaches was to identify the specific object-classes to be represented, and in particular, to find classes of objects with common behaviour. Laurini and Thomson (1992) also argued that an object-oriented approach provided an improved way of modelling the real world. An object-oriented approach appeared to have promise given its roots in AI and knowledge-based systems.

Various object-oriented geographic data models have been studied (Egenhofer and Frank, 1989). In an object-oriented paradigm, real world entities are represented by objects, which have defined properties and behaviours. A geographical object can thus be described as a package of spatial information, attributes describing the characteristics of the objects and operations that are descriptions of their manipulations. The behaviour of the object can be realised by using methods, and the objects can communicate with each other by sending messages. Each object has a unique identifier. Rumbaugh et al. (1991) defined object identity as "a distinguishing characteristic of an object even though the object may have the same data values as another object." Each object belongs to an object class, object classes may form object hierarchies, and objects can inherit properties and characteristics in a parent-child class hierarchy. A class is a description of a group of objects with similar properties, common behaviour, common relationships, and common semantics (Rumbaugh et al., 1991).

In the past ten years, research on generalisation has undergone a period of transition and several works have attempted to implement systems based on object-oriented methods. Examples of approaches that have utilised object-oriented technologies in automated generalisation include work by Jones et al. (1996), Ruas (1998b), Hardy and Woodsford (1997), and Hardy (1999). Object-oriented methods have been developed for model generalisation (van Smaalen, 2003), and multiple representation databases (MRDBs) (Kilpeläinen, 1997; Harrie, 1998; Ormsby and Mackaness, 1999; Dunkars, 2004). Experiences from the work by Dunkars (2004) demonstrated the advantage of object-oriented modelling to share generalisation functionality among feature classes. A disadvantage with object-oriented modelling is that it encourages us to consider the use of generalisation operators sequentially rather than in parallel.

2.1.7. *Constraint-based modelling*

In recent years generalisation has started to be seen more and more as a holistic process, where actions applied to one object class may have repercussions for objects in other classes. Jones (1997) was among those who pointed out that successful generalisation requires holistic approach in which the interaction between cartographic objects can be monitored. One answer to holistic approaches is the constraint-based modelling in which the fundamental idea is to find a state in which a variety of constraints will be satisfied. Instead of focusing on individual actions, the focus is on how a goal state can be reached. Least-squares adjustment is one example of a constraint-based approach to generalisation. In least squares adjustment the generalisation is modelled as an equation system, itself defined in terms of constraints. The solution is treated as an optimisation task according to the least-squares principle. Sarjakoski and Kilpeläinen (1999) described how such methods have a long tradition in geodetic and photogrammetric applications, based on simultaneous adjustment of different kinds of observations. Such methods have been shown in many cases to be superior over methods based on a sequential solution of the problem at hand.

Højholt (1998) used finite-element methods for the solution of spatial conflicts related to displacement. His method was based on least squares techniques, as was Harrie's (1999,

2001), who presented a method for solving spatial conflicts. Harrie's work presented a comprehensive set of geometric constraints that can be applied to different practical situations. Sester (2005a) applied least squares adjustment for displacement and shape simplification (building ground plans). Ruas (1998b) used object-oriented constraint modelling to automate urban generalisation processes. Ruas and Plazanet (1997) and Ware and Jones (1998) used constraint-based Delauney triangulation for handling spatial conflicts. A constraint-based approach is discussed further in this volume (Weibel and Harrie, this volume (Chapter 4)).

2.1.8. *Agent-based technology*

Lamy et al. (1999) and Ruas (1999) suggested the use of multi-agent systems for providing a framework that enables the manipulation of map objects at different levels of detail. The concept of multi-agent systems has its roots in artificial intelligence. In a geographical database each object can be modelled as an agent (Figure 2.6). Each agent has as its goal the objective of satisfying a set of constraints which themselves reflect the requirements of the generalisation process. Agents can be seen as self-contained programs capable of controlling the decision-making process and acting on their perception of the environment, in pursuit of one or more objectives. For instance, an agent representing a building has the functionality to evaluate whether the building symbol is large enough or not. If not, the agent can utilise its functionality to improve the situation. Other constraints may involve several agents: for example among a group of overlapping symbols (Duchêne, 2003).

Duchêne (2003) described an implementation of an agent-based system in an EU-project called "AGENT", for performing generalisation in rural areas. Requirements for the target

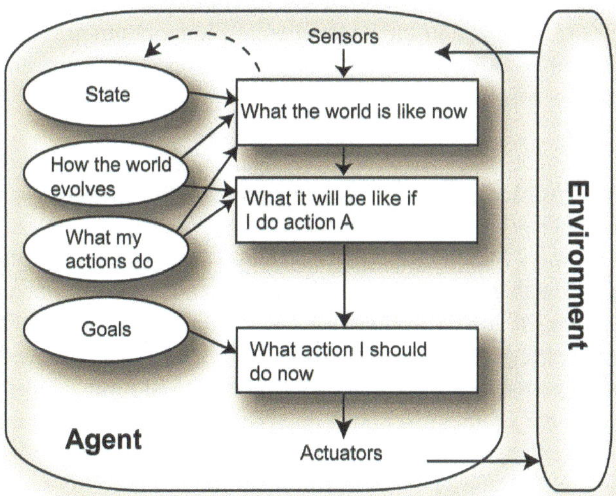

Fig. 2.6. An agent with explicit goals (Russell and Norvig, © 2003, p. 50). Reprinted by permission of Pearson Education, Inc., Upper Saddle River, NJ.

dataset were expressed as constraints at the micro-, meso- and macro-levels. Micro-agents corresponded to individual features such as buildings, while meso-agents corresponded to a natural group of objects such as the buildings in a city block. Macro-agents corresponded to the whole map. To generalise a dataset the different agents processed the information and interacted until a solution was found that satisfied the constraints. Galanda and Weibel (2002) presented an agent-based framework for generalising polygonal subdivisions. Agent-based technologies for developing generalisation methods in topographic map production are described by Ruas and Duchêne (this volume (Chapter 14)).

In a research proposal presented by Moore (2005) an agent-based generalisation system exploiting hierarchical ontologies of map features was outlined. In the work presented by Kulik et al. (2005) ontological information was also used for generalisation.

2.1.9. *Techniques in assessment and evaluation*

Critical to the success of any automated solution is the development of techniques for automatic assessment and evaluation of output from a system. Comprehensive generalisation models necessarily include analysis and evaluation techniques. Weibel (1995) discussed the three essential building blocks for automated generalisation – model generalisation, knowledge acquisition and the evaluation of generalisation alternatives. João (1995, 1998) also raised the question as to how to evaluate cartographic generalisation results. Up till this point the evaluation emphasis was on the quality of the final map, or in the data acquisition methods but she examined the impact of distortion of accuracy (arising from cartographic generalisation) on results from geographical analysis. In the framework proposed by Ruas and Plazanet (1997) evaluation was a part of the generalisation process. When a particular algorithm failed the system backtracked to an earlier stage in order to try alternate solutions. Ruas (2001) proposed principles and methods for evaluating generalisation. These were applied in evaluation undertaken by the European Organization for Experimental Photogrammetric Research (OEEPE). The importance and methods for assessment and evaluation in the generalisation process is extensively discussed by Mackaness and Ruas (this volume (Chapter 5)).

2.2. Modelling Geographic Data in Multiple Representation Databases

The central task of National Mapping Agencies (NMAs) has, in recent decades, been to establish digital geographic databases from which to produce maps. The maintenance and updating of databases has become an urgent problem for which there remains no uniform solution. At the same time, cartographic generalisation remains a key issue in map production, and efficient methods for automatic generalisation still need to be developed further. During the 1990s and up to the present time, the problem has been approached by introducing conceptual models for so-called multiple representation databases (MRDBs) (Kilpeläinen, 1997). The problems of (1) how to maintain and update topographic data and (2) how to organise map-product databases have been discussed in many NMAs, and some systems are under development (for example see Kreiter (2002)). The central question concerns the link between these two approaches and how to keep track of updates so that the user can be provided with a range of map-products that are up-to-date.

2.2.1. *What is an MRDB?*

The term "multiple representation database" refers to a database structure in which several representations of the same geographic entity or phenomenon, such as a building or a lake, are stored as different objects in a database and linked. An MRDB consists of various representation levels with different degrees of geometric or semantic abstraction providing a set of different views of the same object (Bruegger and Frank, 1989; Frank, 1990; Kilpeläinen, 1992; Kidner and Jones, 1994; Devogele et al., 1997, 1998; Weibel and Dutton, 1999). The different representations are stored at different levels of detail. They may consist of, for example, geometrical representations in 2D or 3D, representations from different time intervals, multimedia representations, as well as conceptual representations such as mathematical models (Kilpeläinen, 1997) (Figure 2.7).

An MRDB emphasises the utilisation of geographic databases for various spatial applications, not just for those that have been predefined for some specific map scales. Its flexibility lies in its ability to derive different types of maps from the representation levels of an MRDB, using generalisation methods. Both the geometric and semantic resolutions of objects vary at the different levels of an MRDB (Kilpeläinen, 1997). The term "resolution level" has often been used instead of "scale". Here the "resolution" refers to the smallest detail represented and is closely linked with the level of detail of the geographic data. The resolution of geographic data has been discussed in more detail by Stell and Worboys (1998), Vauglin (2002) and Skogan (2005). The ability to manage multiple representations relating to a single geographic phenomenon often requires us to be able to merge data from disparate sources with different resolution, without introducing data redundancy and inconsistencies between versions (Kidner and Jones, 1994).

Fig. 2.7. Several representations of the same real world entities are linked as objects in a multiple representation database. In this example the base level of an MRDB consists of a 3D city model of Helsinki. City model, © Fontus Ltd.

Characteristics of MRDBs include:

- Different levels of detail of the same geographical entities or phenomena are stored as different objects in the database;
- The corresponding objects at the different levels in an MRDB are linked. How these connectivities are implemented is a question of application;
- An MRDB is closely linked with the concept of model generalisation.

There are currently two main approaches to linking the different representations in an MRDB: bi-directional or one-directional links established between different representations. The conceptual model of an MRDB for maintaining topographic information presented by Kilpeläinen (1997) assumed that the connectivities between the representation levels are bi-directional. Recent implementations mostly include one-directional links (for example Harrie, 1998; Hampe et al., 2004; Dunkars, 2004).

In some research work, MRDBs have been referred to as multi-resolution databases or multi-scale databases. For example Jones and Abraham (1986) used the term "multi-resolution hierarchical database". The characteristics described above, where inter-level representations are linked, is seen here as the requirement for an MRDB.

2.2.2. *Creating an MRDB*

There are two main approaches to creation of an MRDB:

- Store independent datasets and link equivalent objects;
- Generate smaller scale representations via model generalisation.

Figure 2.8 shows the principle of creating the links for building objects at different levels of an MRDB. Using these bi-directional links it is possible to point at the city block area and zoom into the buildings that correspond to the block in the large-scale data set.

Dunkars (2004) emphasised that when designing an MRDB it is possible to either let one object have several geometrical attributes, or to let each object have one geometry and to connect objects that represent the same real-world entity. The first approach was utilised by Jones et al. (1996), and Hardy and Woodsford (1997), while Harrie and Hellström (1999), Hardy (1999), Kilpeläinen (2001), and Trévisan (2004) have used the second approach. Object-oriented approaches are often used in MRDB implementations. For modelling MRDBs traditional object-oriented modelling languages such as OMT (Rumbaugh et al., 1991) or UML (Booch et al., 1999) are used. In the EU project "MurMur" (Balley et al., 2004), the spatio-temporal modelling language MADS was extended to include capabilities for modelling multi-representations. In the work carried out by Dunkars (2004) UML was used.

In designing an MRDB, a key decision is in the choice of representation levels. This must in practice be based on in-depth analysis of the requirements for the respective geographic database. According to Kilpeläinen (2001) the solution should be based on the following criteria:

- What are the representations that are used most often?
- What are the update requirements? Levels that are supposed to be updated frequently should be included in their own representation level in the database.

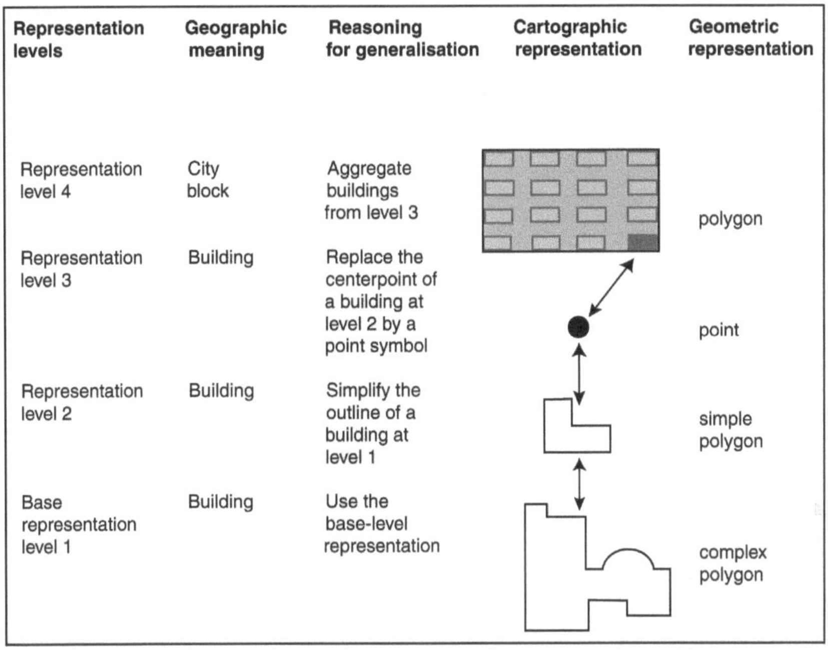

Fig. 2.8. Representation levels for an object "building" in an MRDB (modified from Kilpeläinen, 1997, 2001).

- What is the degree of automation that can be applied to derive the representation levels in an MRDB? Representations should be propagated from the base level by model generalisation. See Figure 2.4 for derivation of a level of an MRDB.

2.2.3. *MRDB versus cartographic map databases*

Kilpeläinen (1997, 2001) discussed the difference between an MRDB and a cartographic database (Figure 2.9). In her work the MRDBs, which were assumed to be present in model generalisation environments, consisted of real-world abstractions, whereas cartographic map product databases were cartographic representations for visualisation purposes. According to Kilpeläinen, the objects in an MRDB have no overlap conflicts in the model generalisation phase, since they have real geographic extent and do not appear in graphic space. Mark (1990) emphasised that since the objects in an MRDB have defined geographic extent, they do not have to compete for space for their representation. The database should be topologically correct and not contain consistency errors such as overshoots and undershoots of lines.

The following example after Kilpeläinen (1997) illustrates the differences between an MRDB and cartographic databases. In Figure 2.9, a point represents a building at a specific level of the MRDB. In a cartographic database the same building is represented by a rectangular symbol. To perform the cartographic representation satisfactorily, simplification, smoothing, symbolisation, exaggeration, and displacement would need to be applied. In Figure 2.9, to avoid the overlap between the road and the symbol for a building, the road and

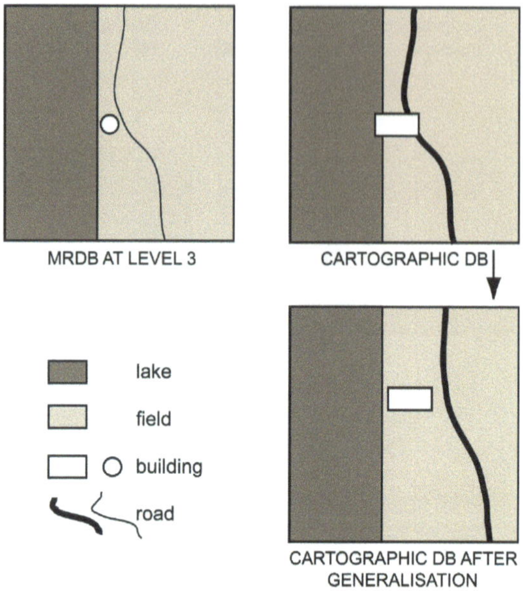

Fig. 2.9. Difference between an MRDB and a cartographic database (db). On the final cartographic data the following generalisation operators have been applied: simplification, symbolisation, exaggeration, and displacement (modified from Kilpeläinen, 1997, 2001).

the building should be displaced (taking into account the lake). The result of these generalisation operations is that whilst the topological relations have been preserved, there is a change in the positional accuracy of the objects. Whilst acknowledging their interdependence, the separation of model and cartographic generalisation helps to manage the complexity of the task.

A further difference between an MRDB and a cartographic database, reported by Kilpeläinen (2001), is that an MRDB does not include place names and map symbols associated with a specific map series (whereas the cartographic database does). Instead place names can be defined as attribute data of particular objects.

Kilpeläinen (1997) presented a model for a topographic MRDB system (Figure 2.10). In her model, the base level of the MRDB consists of the most accurate and highest level of detail of primary geographic data, and the other levels at lower resolution are derived from this. A knowledge base of generalisation rules and operators (Buttenfield and McMaster, 1991) is needed to support the two generalisation steps: a model generalisation step that is assumed to be fully automated and a cartographic generalisation step which seeks to produce cartographic map product databases that might support interactive functions.

2.2.4. *Research on MRDB*

Research on multiple representation was initiated in a research program at the National Center for Geographic Information and Analysis (NCGIA, 1989; Buttenfield and Delotto, 1989;

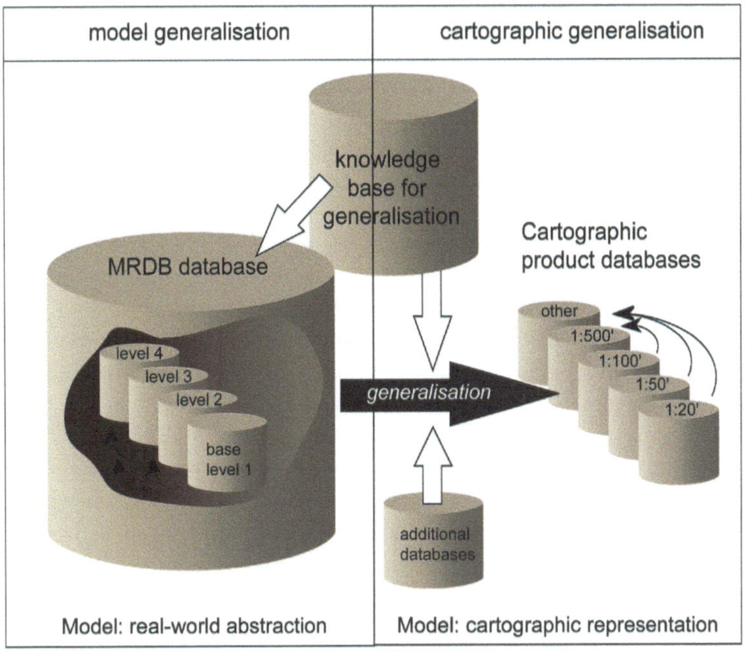

Fig. 2.10. A model for a topographic MRDB system (after Kilpeläinen, 1999, 2001).

Frank, 1990). In a closing report on this research initiative, Buttenfield (1993) stated that the main areas in multiple representation research that require further study were (1) database issues, especially the need to organise multiple topological and metric versions for efficient access and implementation of linkages between multiple representations, and (2) generalisation issues, such as flexible definitions of resolution for datasets, formalisation of digital feature description and categorisation models, and rules for generalisation.

The development of multi-resolution data structures was discussed by Bruegger and Frank (1989) and Bruegger (1994) in terms of achieving effective data maintenance. It was recognised that additional topological data structures are needed for the implementation of multipurpose GIS. To overcome the problems of insufficient topological data structures, Bruegger and Frank proposed a formal, dimension-independent approach for building multiple, hierarchically related representations of spatial objects. The same objects are presented in separate layers that support various spatial resolutions. The geometry of spatial objects is defined as aggregations of cells at a certain higher level. Hierarchical relations between the cells interconnect the layers. Richardson (1993) presented a model for spatial and thematic digital generalisation. The context transformation model presented could automatically decrease representation density through the application of "steering" parameters, classification and aggregation hierarchies, reduction factors, and topological structures for multiple abstractions.

Jones (1991) gave various reasons for storing multiple representations of the same objects in the database, one of which being the relatively limited capabilities of automatic generalisation. He presented the concept of a deductive knowledge-based system architecture that might provide a suitable basis for building multi-scale geographic databases. Jones was concerned that multi-resolution data structures should provide rapid access to generalised versions, which are geometric subsets and therefore prevent data duplication. Kidner and Jones (1993, 1994) described an object-oriented framework for handling multiple representations in a multi-scale GIS. In their framework each real-world entity within an object directory recorded references of all corresponding representations within the database. The approach facilitated the update capability, which could help in deciding whether new representations should be deleted or added to the database, or merged with an existing representation. The rules for recognising equivalences and differences between new data and stored representations were embedded within the object schema. The prototype presented by Kidner and Jones used Kappa, a C-based, object-oriented application development environment. An important component of this prototype system was the ability to incorporate deductive reasoning to maintain the integrity of the objects during input and update operations.

Laurini and Thomson (1992) discussed multiple representation in terms of devising different models for various spatial objects or instances of spatial objects. In multiple representations, particular classes of objects are stored in a database in various representations. These various representations have distinct inherent properties in terms of geometric computations, positional error and topological consistency. Bruegger and Müller (1992) suggested that with efficient human and machine reasoning, multiple levels of abstraction are necessary to systematically address problems without becoming lost in the detail. Kidner and Jones (1994) discussed how a multiple representation structure provides a good control mechanism for updating. Lagrange and Ruas (1994) discussed the suitability of applying an object model to multiple representation.

The resolution requirements of data collection have been mentioned as a possible reason for using a multiple representation structure (for example Frank and Timpf, 1995; Bruegger and Frank, 1989; and van Oosterom, 1990). Van Oosterom and Schenkelaars (1995, 1996) reported the development of the first GIS that could be used to manipulate a single dataset over a large range of scales. Timpf and Frank (1995, 1997) and Timpf (1998) applied the concept of multiple representation for zooming into geographic data at various levels of abstraction.

Martinez Casasnovas and Molenaar (1995) defined an aggregation hierarchy for the multiple representation of hydrographic data and implemented it in ArcInfo. The definition of the conceptual generalisation model was based on the use of a formal data structure (FDS). Molenaar (1996a) further developed the model for spatial database generalisation, which was formulated on the basis of the syntax of the FDS for single-valued, vector-structured maps (Molenaar, 1989). This syntax was used to formalise the database generalisation procedures. According to Molenaar, aggregation hierarchies for spatial objects can serve as basic tools for multiple representations of geo-data; these aggregation hierarchies can be based on an FDS. Molenaar (1996b) also presented the strategies for object generalisation. In the work by Kilpeläinen (1997), special emphasis was placed on the notion of bi-directional connectivities among the representations that enabled update-propagation.

Until recently, only a few implementations had been reported (exceptions being van Oosterom, 1990; Martinez Casasnovas and Molenaar, 1995; and Harrie, 1998). Friis-Christensen et al. (2002) addressed the multiple representation problem and proposed a system for maintaining consistency between independently developed databases. A case study was presented based on maintaining information about buildings in three different types of databases. Recent implementations have been presented by Hampe et al. (2003, 2004); Dunkars (2004) and Skogan (2005). Vangenot et al. (2002) studied modelling concepts needed to manage multiple representations of geographical data. Bernier and Bedard (2002) presented the Vuel concept, which is a viewing approach to multiple representations complemented with spatio-temporal and presentation features (see Bernier and Bédard, this volume (Chapter 9)).

Balley et al. (2004) presented a generic conceptual model for MRDBs in connection with the EU project "MurMur". They addressed multi-resolution in general and did not limit the representations to the geometry of objects, nor limit the approach to hierarchical structures (as proposed by Timpf and Frank (1995) and Zhou and Jones (2001)). Balley et al. (2004) proposed that the need to update topographic data at National Mapping Agencies (NMAs) could be managed if NMAs were willing to move from a map-sheet based update procedure to an object-based update procedure. The object-based update procedure can be realised if the different data sets maintained by NMAs were integrated into an MRDB. In the project a MADS-to-Oracle translation module was developed enabling implementation of an MRDB in the Oracle DBMS. The goal of the Institut Géographique National's COGIT laboratory was to merge the three source databases into a global database in order to improve global consistency of the data and to set up a structure that would be able to support automatic propagation of the updates. Braun (2004) proposed schema matching for finding correspondences between different database models, both user and producer schemas. A schema is a realisation of the data model, and includes definitions of the object classes, their attributes, and the consistency rules. According to Braun, schema matching can be used, not only for integration tasks, but also to control the consistency of MRDBs. The data management point of view is presented in more detail by Mustière and van Smaalen (this volume (Chapter 6)).

2.2.5. *What is an MRDB good for?*

Enabling automatic propagation of updates between topographic datasets has been regarded as the major advantage of MRDBs (Kilpeläinen, 1994, 1995; Kilpeläinen and Sarjakoski, 1995). Based on the findings presented by Egenhofer et al. (1994), Kidner and Jones (1994), Buttenfield (1995), Kilpeläinen (1994, 1997), Hampe et al. (2004) and Dunkars (2004) the benefits of an MRDB approach can be summarised as being:

- Maintenance of MRDBs is flexible: updates done at the most accurate level of the MRDB can be propagated to smaller resolution data levels automatically;
- Data redundancy is avoided since each object is stored only once;
- The links between objects at different representation levels provide a basis for automatic consistency and error checking;
- The speed of information access can be quicker (for example, in mobile applications with low bandwidth an MRDB can be used to quickly access relevant information as the user zooms);
- MRDBs can be used for multi-scale analysis of geographical information, to compare data at different resolution levels, for example in route finding;
- It is possible to derive application-dependent generalised outputs needed for a variety of media (such as printed maps or map series, screen maps, tables and diagrams, and on-the-fly maps for Internet and mobile use).

While these issues pertain to the data management side of things, it is important to consider the experience of using an MRDB from the user perspective. There are a number of affordances associated with an MRDB. Kilpeläinen (1997) pointed out that object representations at various levels do not only include geometric representations of the object, but may also include multimedia representations. For example a house that is stored at the base level might have a link to a photo of that house. It might also have a link to a sound representation that can demonstrate nearby traffic sounds at different times of day or videos showing the surroundings or the interior of the house (Kilpeläinen, 1997). It may also include richer, perhaps temporal attributes of the object such as the opening times of specific stores (to help plan and navigate a shopping trip) (Linturi and Simula, 2005). As noted by Balley et al. (2004) users often want to access available data for the same viewpoint but at different resolution levels. 3D-models could also be associated with objects (Sester and Klein, 1999) and could support activities such as architecture and planning. As an aside we might ask whether 3D-models should constitute the base level of an MRDB, as potentially they could be the most accurate representation of an object and would promote greater GIS use (Figure 2.7). The 3D modelling for building generalisation is discussed in detail in (Meng and Forberg, this volume (Chapter 11)).

Lehto et al. (1997) pointed out that an MRDB could also provide a suitable database structure for Internet users. The main benefit would be immediate access to updated information. The idea of MRDBs for on-line processing has been further developed by Jones et al. (2000), Cecconi et al. (2002), Cecconi (2003), Hampe et al. (2003, 2004) and Sester et al. (2004), and further examples are presented by Bernier and Bédard (this volume (Chapter 9)). Over the past decade there has been huge growth in provision of maps over the Internet, the latest developments include map services brought to users via mobile devices (Sarjakoski and Sarjakoski, this volume (Chapter 7)). It is possible to personalise maps to suit a specific purpose and integrate them into a variety of services, including route and "yellow pages" services. As a "quick and dirty" solution, a number of these services have been built on the top of the same maps as used in desktop and Internet applications (Nivala and Sarjakoski, 2005). The main problem however, is that mobile applications envisage a very different use scenario. Maps on mobile devices are used in field situations, and their visualisation and generalisation requirements are different from static indoor usage situations, for example depending on light conditions and size of displays (see Sarjakoski and Sarjakoski, this volume (Chapter 7); and Sester and Elias, this volume (Chapter 10)). Mobile applications certainly point to the need for generalisation methods and MRDBs that are flexible and able to support a range of tasks and environments.

2.2.6. *Incremental generalisation and updating*

Kilpeläinen and Sarjakoski (1995) proposed an approach called "incremental generalisation" for propagating updates through different abstraction levels in an MRDB. The principle for incremental generalisation was derived from software engineering, from a so-called modular approach (Ledgard, 1983). In a modular approach, a program is divided into modules, which are compiled incrementally. This idea was applied in the generalisation task, where a module was seen to be a subset of geographic data (Kilpeläinen, 1994, 1995).

The incremental generalisation proposed by Kilpeläinen and Sarjakoski (1995) is illustrated in Figures 2.11 and 2.12. In the figures, Structured Analysis is used to formalise the generalisation processes. Structured Analysis (SA) has been used for system design, analysis

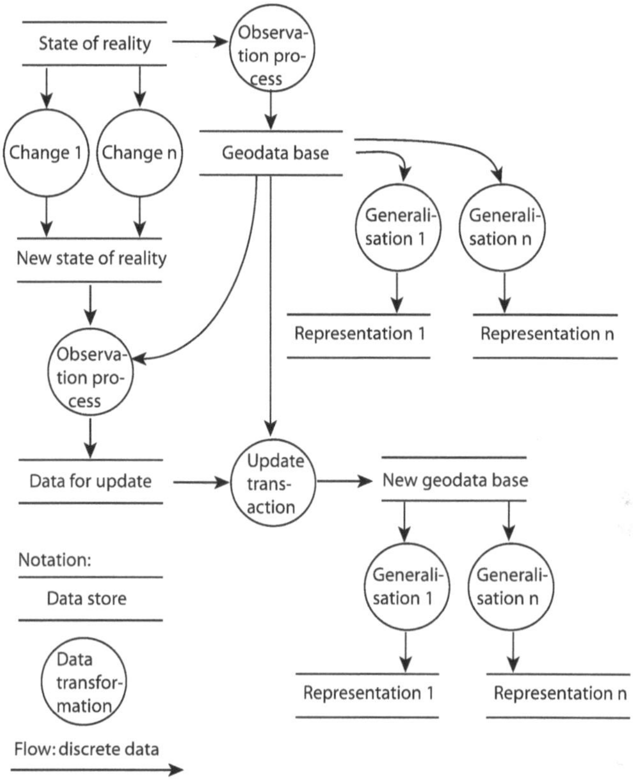

Fig. 2.11. Batch generalisation processes result in different generalised representations. The notation follows the SA/SD method (DeMarco, 1979). (After Kilpeläinen and Sarjakoski, 1995.)

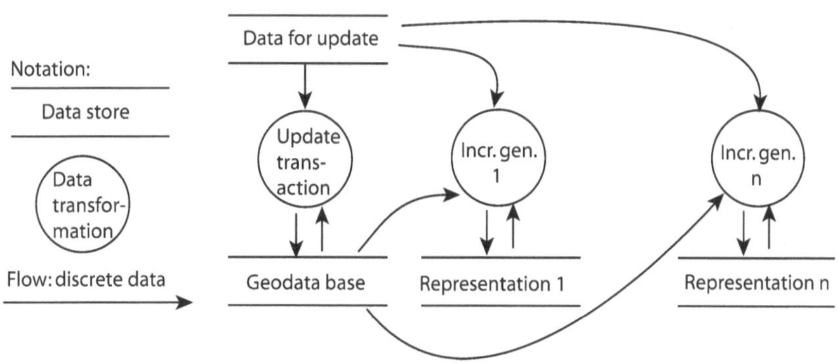

Fig. 2.12. The principle of incremental generalisation is shown schematically. The notation follows the SA/SD method (DeMarco, 1979). (After Kilpeläinen and Sarjakoski, 1995.)

and specification since the early 1970s (DeMarco, 1979). Usually, after updating even only a single object in a geographic database, all the data are completely reprocessed using batch generalisation (Figure 2.11). By contrast, when updating is done incrementally, the generalisation process needs to be performed completely for the entire geographic database only once, and the updates to the generalised versions are done incrementally, module by module (Figure 2.12). Only the modules that are influenced by the updates are processed. A prerequisite for an incremental approach is that the data can be divided into modules that can be generalised independently. The problem of how to divide the generalisation task into modules is similar to the problem of the displacement of generalised objects studied by Ruas and Plazanet (1997). Also the concept of conflict sets described by Aasgaard (1992) is closely related with the concept of the module. He divided generalisation operators into "conflict generating" and "non-conflict generating" operators.

Harrie and Hellström (1999) implemented a prototype system based on the ideas presented by Kilpeläinen and Sarjakoski (1995), and this was further improved by Dunkars (2004). His system performed automated propagation of updates at the scale range 1 : 10 000 to 1 : 50 000 for buildings and roads. He used cluster analysis to assist the generalisation process in updating an MRDB.

Hardy et al. (2003) reported on the incremental generalisation principles in Laser-Scan's object-oriented database. Here, the incremental approach is made possible by determining parentage information at the time of the original generalisation of the complete dataset, and stored on the derived features. When updates are needed, the update propagations can be quickly determined, and only the affected objects re-generalised. Skogan (2005) also introduced a method for incremental generalisation. The method used information about the changed objects to select only the source objects that needed re-generalisation after updating the source database. In order to test the method, a prototype was implemented utilising an Extensible Markup Language (XML) and Extensible Stylesheet Language Transformation (XSLT). He reported a case study where automated incremental generalisation was applied from 1 : 20 000 data to 1 : 50 000 data.

Badard (1999) presented a generic tool for automatic retrieval of the updates in geographic databases in order to make the integration easier. The mechanism was based on geographic data matching tools, which were implemented at the COGIT laboratory of IGN France. The tool allowed the extraction of the evolution of geographical entities that represented the same phenomenon in two versions of the same geographic database. The road network of the BDCARTO at 1 : 100 000 was used to test the implementation. The tool was an integral part of methods to further develop MRDBs at IGN France, as reported by Sheeren (2003). Sheeren et al. (2004) defined a rule-based system, based on a matching process, for assessment between multiple representations of geographic databases. The system was able to decide whether representations of matching pairs correspond to an equivalent real-world entity, or to an update between different databases, or a consequence of an error in the databases.

2.3. Future Research on Generalisation and MRDBs

A brief glimpse into the development of models and use of MRDBs has been given (§ 2.2). Some relevant research topics for generalisation and multiple representation approaches are briefly summarised in the following list:

- Further MRDB modelling towards more comprehensive real-world abstraction including not only urban but rural features, such as land-use features;
- Tools for creating MRDBs automatically;
- Management of inconsistencies and detection of conflicts (topological, semantic) between multiple representations;
- Tools for automatic update propagation;
- Methods for maintaining the links between an MRDB and cartographic product databases;
- Enrichment of GI databases to allow multiple views for the users;
- Distributed MRDBs: distributed according to different coverage areas, different applications, different object classes or different resolutions;
- Use of ontology-based methods;
- Real time processing and adaptive methods in support of new products or services (such as location-based services).

2.4. Concluding Remarks

The aim of this chapter was to describe the building blocks in the field of conceptual models for generalisation and multiple representation. Many of the topics were only briefly discussed, but should serve as an introduction to the more detailed discussion and presentation of applications in the following chapters. The role of generalisation is clear in the context of multiple representation, as the main task of model generalisation is to generate and maintain the various representations for an MRDB. Comprehensive models satisfying and supporting future user needs will be a key challenge within MRDB research. Many of the other challenges will relate not so much to technical issues, but to organisational and societal needs. Understanding the context of use will be key in understanding the importance of MRDB and associated generalisation methods. Comprehensive generalisation models are the building blocks in creating operational systems for generalisation and multiple representation. Conceptual models are also fundamental to finding pragmatic solutions to multi scale analysis and visualisation.

W.A. Mackaness, A. Ruas and L.T. Sarjakoski (Editors)

Chapter 3

A Synoptic View of Generalisation Operators

Nicolas Regnauld[a], Robert B. McMaster[b]

[a]*Ordnance Survey Research Labs, Ordnance Survey, Romsey Road, Southampton SO16 4GU, UK*
e-mail: Nicolas.Regnauld@ordnancesurvey.co.uk
[b]*Department of Geography, 414 Social Sciences Building, 267 19th Avenue South, University of Minnesota, Minneapolis, MN 55455, USA*
e-mail: mcmaster@umn.edu

Abstract

Most of the research in generalisation assumes that the process can be broken down into a series of logical operations that can be classified according to the type of geometry of the feature, into what we call generalisation operators. For instance, a smoothing operator is designed for linear features, while an amalgamation operator works on areal features. This chapter provides an overview of what has been achieved so far towards creating a comprehensive set of generalisation operators. It contains discussions related to the classification of these operators, and how different classifications have been defined to suite different contexts (such as raster vs. vector data, or 2D vs. Digital Elevation Model data); it proposes a generic list of generalisation operators, and a detailed list of implementation of these operators for different types of features. This provides a virtual toolbox that can be used when designing automatic generalisation solutions. The chapter concludes by discussing the changing nature of algorithms and operators in response to technological developments and changing contexts of use.

Keywords: classification, generalisation techniques, generalisation algorithms, virtual toolbox

3.1. Nature and History

Cartographers have written on the topic of cartographic generalisation since the early part of the twentieth century. Eckert, the seminal German cartographer and author of *Die Kartenwissenschaft*, wrote in 1908, "In generalising lies the difficulty of scientific map making, for it no longer allows the cartographer to rely merely on objective facts but requires him to interpret them subjectively" (Eckert, 1908, p. 347). Other cartographers have also struggled with the intrinsic subjectivity of the generalisation process as they have attempted to understand and define cartographic generalisation. For instance, Wright argued that, "Not all cartographers are above attempting to make their maps seem more accurate than they actually are by drawing rivers, coasts, form lines, and so on with an intricacy of detail derived largely from the imagination" (Wright, 1942, p. 528). Wright identified two major components of the generalisation process, simplification – the reduction of raw information that is too intricate – and amplification – the enhancement of information that is too sparse. This idea that generalisation could be broken down into a logical set of processes, such as simplification and amplification, has become a common theme in generalisation research. Raisz (1962) for example, identified three major components of generalisation, including combination, omission, and simplification, while Robinson and his colleagues (1978) identified four components: selection, simplification, classification, and symbolisation. In Robinson et al.'s model, selection was considered a preprocessing step to generalisation itself. Selection allowed for the identification of certain features, and feature classes, while generalisation applied the various operations, such as simplification. This is detailed in their model, as discussed below.

In specifically discussing generalisation in a digital environment, McMaster and Shea (1992, p. 3) noted that, "the generalisation process supports a variety of tasks, including: digital data storage reduction; scale manipulation; and statistical classification and symbolisation. *Digital generalisation* can be defined as the process of deriving, from a data source, a symbolically or digitally-encoded cartographic data set through the application of spatial and attribute transformations." They list the objectives of digital generalisation as: the reduction in scope and amount, type, and cartographic portrayal of mapped or encoded data consistent with the chosen map purpose and intended audience; and the maintenance of graphical clarity at the target scale.

The theoretical "problem" of generalisation in the digital domain is straightforward: the identification of areas to be generalised, together with the application of appropriate operations. In the first three sections we present a review of these algorithms. Section 3.1 provides some historic context, §3.2 reviews different generalisation frameworks, from which a list of common generalisation operators is extracted and described. Section 3.3 discusses the generalisation transformations that occur on the data schema, and the consequence of their application. The three following sections focus on existing specific implementations of generalisation operators, grouped by the type of features on which they operate (§3.4 for buildings, §3.5 for line features, §3.6 for networks, §3.7 for rural features and §3.8 for relief). Section 3.9 proposes a template for describing generalisation algorithms. Section 3.10 presents the algorithms available in the main commercial GIS platforms currently offering generalisation capabilities. Finally, the chapter concludes by discussing emerging research challenges.

3.2. Frameworks for Generalisation Operators

As discussed by Sarjakoski, this volume (Chapter 2), there have been several conceptual models, or frameworks, of the generalisation process created both in the manual and digital domain. Many of these models have attempted to define the specific operators of generalisation (the focus of this chapter). From American research, models by Robinson et al. (1978), in the many volumes of their classic textbook, *Elements of Cartography*, identified the fundamental operations of generalisation. Related to this work, Morrison developed a theoretic model describing the operators of selection, simplification, classification, and induction as mathematical functions. Brassel and Weibel (1988) also created a conceptual framework that focused on terrain generalisation, and discussed some of the basic operators including selection, simplification, combination, and displacement. In the early 1990s, McMaster and Shea (1992) designed a conceptual framework that identified the why, when, and how of the generalisation process, with emphasis on the special operators required for automated generalisation.

This section will briefly review the major frameworks of generalisation, focusing on the organisation of operators, and then provide an overview of many of the operators themselves. This is followed by a conclusion addressing the next step in the generalisation process, which is to combine the operators to achieve complex generalisation solutions.

3.2.1. *Robinson et al. model*

Robinson and his colleagues (1978) developed one of the first formal models or frameworks in order to better understand the generalisation process. They separated the process into two major steps: selection (a preprocessing step) and the actual process of generalisation, which involves the geometric and statistical manipulation of objects. *Selection* involves the identification of objects to retain in (or eliminate from) the database. *Generalisation* involves the processes of simplification, classification, and symbolisation. Simplification is the elimination of unnecessary detail in a feature, classification involves the categorisation of objects, and symbolisation is their graphic encoding.

3.2.2. *Brassel and Weibel model*

Brassel and Weibel (1988) at The University of Zurich have worked extensively in developing methods for terrain generalisation. Their research has two primary objectives: to design a strategy for terrain generalisation that is adaptive to different terrain types, scales, and map purposes, and to implement this strategy in an automated environment as fully as possible. Toward these ends, they have developed a model of terrain generalisation that consists of five major stages: structure recognition, process recognition, process modelling, process execution, and data display and evaluation of results. In structure recognition, the specific cartographic objects – as well as their spatial relations and measures of importance – are selected from the source data. Process recognition identifies the necessary generalisation operators and parameters by determining "what is to be done with the original database, which types of conflicts have to be identified and resolved, and which types of objects and structures are to be carried into the target database" (Brassel and Weibel, 1988, p. 232). Process modelling then compiles the rules and procedures – the exact algorithmic instructions – form a process library – a digital organisation of these rules. The final stages of Brassel and Weibel's model involves process execution, in which the rules and procedures are applied to create the generalisation, data display, and evaluation. The authors also describe the application of specific generalisation operators, including selection, simplification, combination, and displacement to illustrate the application of these operators to digital terrain models.

3.2.3. *McMaster and Shea model*

In an attempt to create a comprehensive conceptual model of the generalisation process, McMaster and Shea (1992) identified three significant components: the theoretical objectives, or *why to generalise*; the cartometric evaluation, *or when to generalise*; and the specific spatial and attribute transformations, or *how to generalise.* Most of the research in generalisation assumes that the process can be broken down into a series of logical operators that can be classified according to the type of geometry of the feature. For instance, a simplification operator is designed for linear features, while an amalgamation operator works on area features. McMaster and Monmonier (1989) proposed a framework for generalisation operators, dividing the process into those activities for vector- and raster-mode processing. The types of generalisation operators for vector and raster processing are fundamentally different. Vector-based operators require more complicated strategies since they operate on strings of x–y coordinate pairs and require complex searching strategies. In raster-based generalisation, it is much easier to determine the proximity relationships that are often the basis for determining conflict among features. Below, a more detailed discussion of individual vector-based operators is provided.

There are other classifications of generalisation operators; Beard (1991b) developed a classification that was based on the constraints of generalisation. The four constraints included: (1) graphic, (2) structural, (3) application, and (4) procedural. The fundamental classes identified by Beard were: (1) operators to reduce the number of objects (select, group, link, and aggregate), (2) operators for simplifying the spatial domain (link, aggregate, collapse, and simplify), (3) operators for simplifying the attribute domain (classification operator). Mackaness (1991) also developed a classification based on process and structure. His classification included the operators of change symbols, mask symbols, increase size difference, select, omit, simplify, combine (reclassify), displace, and exaggerate. McMaster and Barnett (1993) relate the operators of generalisation to the various levels of meaning such as the data structure or implementation level, conceptual vector or raster level, conceptual spatial object level, entity-object level, and real-world phenomena level. Based on this, many of the operators are classified into the categories of display, reduce, and fuse.

3.2.4. *Classification and symbolisation*

Two operators that are often included under the heading of generalisation are classification and symbolisation. In contrast to those operators that focus on the geometric, or geographical, component of the data, classification and symbolisation are considered to be attribute transformations, or those that transform the attribute component. Data classification involves reducing the raw data – at the nominal, ordinal, interval, or ratio levels – into a set of classes. Data classification has been a major research area in cartography, with a multitude of techniques developed and tested. Some of these include equal-area, nested means, standard deviation, and Jenks' optimal classification. A full discussion of many classification methods may be found in Slocum et al. (2005).

In a similar way, cartographic symbolisation is considered part of the generalisation process. Symbolisation involves the graphical encoding of the data, and can be applied to either the statistical or geographical component. For instance, the creation of a graduated circle map from classified data is a standard symbolisation technique. Likewise the symbolisation of boundary types (international, national, state) is an example of cartographic symbolisation applied to geometric data. Once again there is a rich literature on the development of symbolisation methods in cartography (see Slocum et al., 2005).

3.2.5. *The fundamental geometric generalisation operators*

Within each of the categories, one can find classifications of algorithms for each operator. As discussed by Sarjakoski, this volume (Chapter 2), and according to Weibel and Dutton (1999) a generalisation operator represents a type of spatial transformation that is to be achieved, while a generalisation algorithm is the implementation of a particular transformation. For each of these algorithms it is important to realise the fundamental difference between those approaches designed for raster-based data vs. those for processing vector data. Vector-based generalisation operators are often based on the geometry – traditionally thought of as points, lines, or polygons – or more complex geometries such as networks. As an example, one can simplify, or weed, unnecessary data from cartographic lines (strings of x–y coordinate pairs) by using a variety of approaches such as looking at the angular change between vertices. The complexity of vector-based approaches lies in the need to model various relationships, such as topology and connectivity. The generalisation of raster features is more straightforward, as the basic topology necessary for many generalisation operators is intrinsic in the data model (pixels next to pixels). Many of the operators for vector and raster processing carry similar names – simplification, smoothing, and enhancement for example.

Simplification

Simplification is the most commonly used generalisation operator. The concept is relatively straightforward, since it involves at its most basic level a "weeding" of unnecessary coordinate data. The goal is to retain as much of the shape of the feature as possible, while eliminating the maximum number of coordinates. Most simplification routines utilise geometrical criteria (distance and angular measurements) in selecting significant, or critical, points. A general classification of simplification methods consists of five approaches: independent point routines, local processing routines, constrained extended local processing routines, unconstrained extended local processing routines, and global methods. Independent point routines select coordinates based on their position along the line. For instance, a typical nth point routine might select every 3rd point. Although computationally efficient, these algorithms are crude in that they do not account for the true morphological significance of a feature. Local processing routines utilise immediate neighbouring points in assessing the significance of the point. Assuming a point to be simplified x_n, y_n, these routines evaluate its significance based on the relationship to the immediate neighbouring points, x_{n-1}, y_{n-1}, and x_{n+1}, y_{n+1}. This relationship is normally determined by a distance or angular criterion. Constrained extended local processing routines search beyond the immediate neighbours and evaluate larger sections of lines, comparing values of the 2, 3, or 4th neighbour in either direction. Unconstrained extended local processing routines also search around larger sections of a line, but the search is terminated by the morphological complexity of the line, not by algorithmic criterion. Finally, global algorithms process the entire line feature at once, and do not constrain the search to subsections. The most commonly used simplification algorithm – the Douglas–Peucker – takes a global approach, and processes a line "holistically." Details of the Douglas–Peucker algorithm (as well as classification and comparison of other algorithms for simplifying cartographic features) can be found in McMaster (1987) and McMaster and Shea (1992).

Smoothing

Although often assumed to be identical to simplification, smoothing is a very different process. The smoothing operator shifts the position of points in order to improve the appearance of the feature. Smoothing algorithms relocate points in an attempt to plane away small perturbations and capture only the most significant trends of the line (McMaster and Shea, 1992). As with simplification, there are many approaches to this process. A systematic classification and review of smoothing algorithms can be found in McMaster and Shea (1992). Research has shown that a careful integration of simplification and smoothing routines can produce a simplified, yet aesthetically acceptable result (McMaster, 1989).

Aggregation

Aggregation involves the joining together of multiple point features, such as a cluster of buildings. This process involves grouping point locations and representing them as areal units. The critical problem in this operator is determining both the density of points needed to identify a cluster to be aggregated, and the boundary around the cluster. The most common approach is to create a Delaunay triangulation of points, and use measures of distance along the Delaunay edges to calculate density and boundary (Ware et al., 1995).

Amalgamation

Amalgamation is the process of fusing together nearby polygons, and is needed for both continuous and noncontinuous areal data. A noncontinuous example is a series of small islands in close proximity whose size and detail cannot be depicted at the smaller scale. A continuous example is with census tract data, where several tracts with similar statistical attributes may be joined together. Amalgamation is a very difficult problem in urban environments where a series of complex buildings may need to be joined.

Collapse

The collapse operator involves the conversion of geometry. For instance, it may be that a complex urban area is collapsed to a point due to scale change, and resymbolised with a geometric form, such as a circle. Or a complex set of buildings is replaced with a simple rectangle – which might also involve amalgamation.

Merging

Merging is the operation of fusing together groups of line features, such as parallel railway lines, or edges of a river or stream. This is a form of collapse, where an area feature is converted to a line. A simple solution is to average the two or multiple sides of a feature, and use this average in the calculation of the new feature's position.

Refinement

Refinement is another form of resymbolisation, much like collapse. But refinement is an operator that involves reducing a multiple set of features such as roads, buildings, and other types of urban structures to a simplified representation. The concept with refinement is that such complex geometries are resymbolised to a simpler form – in essence a "typification" of the objects.

Exaggeration
Exaggeration is one of the more commonly applied generalisation operators. Often it is necessary to amplify a specific part of an object to maintain clarity during scale reduction.

Enhancement
Enhancement involves a symbolisation change to emphasise the importance of a particular object. For instance, the delineation of a bridge under an existing road is often portrayed as a series of cased lines that assist in emphasizing that feature over another.

Displacement
Displacement is perhaps the most difficult of the generalisation operators, as it requires complex measurement. The problem may be illustrated with a series of cultural features in close proximity to a complex coastline. Assume, for example, that a highway and railroad follow a coastline in close proximity, with a series of smaller islands offshore. In the process of scale reduction, all features would tend to coalesce. The operation of displacement would pull these features apart in order to prevent this coalescence. What is critical in the displacement operation is the calculation of a displacement hierarchy since one feature will likely have to be shifted away from another (Nickerson and Freeman, 1986; Monmonier and McMaster, 1991). A description of the mathematics involved in displacement may be found in McMaster and Shea (1992).

3.2.6. *Conclusion on the frameworks for generalisation operators*

This section has reviewed some frameworks that have been designed to formalise the context in which generalisation operators are applied. The challenge is then to implement these operators and see how much automation of the generalisation process can be achieved. A study from the OEEPE working group on generalisation has examined this issue (Ruas, 2001). The OEEPE working group on generalisation studied the process of generalisation and how to combine algorithms to achieve generalisation solutions. The classification was essential to allow a comparison between the sequences of algorithms used to generalise the same data on different platforms. This study highlighted the lack of contextual algorithms. It was noted that with the algorithms available at the time of the OEEPE study (1997), and the GIS used for the experiments, it was not possible to achieve good solutions without human interaction. Research indicates that each operator either requires many algorithms to implement it in different contexts, or much more complex algorithms able to take contextual information into account. Some specific studies presented in the next sections of this chapter address this contextual issue, but as yet, very few have found their way into commercial GIS.

3.3. Operations on the Data Schema

As discussed by Sarjakoski, this volume (Chapter 2), it has long been accepted that generalisation can be seen as a two stage process: The model generalisation and the cartographic generalisation (Müller et al., 1995b). The model generalisation transforms the classification and attribution of the data – in essence this is generalisation of the data schema. This is driven by the expectations of the user of the data (what kind of task do they want to perform). The cartographic generalisation modifies the representation of the features on the map, and is constrained ultimately by the choice made for displaying the data (resolution of the display device, scale, conditions of use). This section focuses on the generalisation of the data schema, which can be represented as a combination of information abstractions, as described by Nyerges (1991). Nyerges differentiates four types of information abstractions:

- *Classification* is the process of defining the categories of information (or classes in an object-oriented context) that should be included in a database;
- *Association* are relations that exist between two classes. They are used to model the possible influences between objects of different classes (such as the relationship between a bridge and a river);
- *Generalisation* is the process of making a class of objects less specific. It is the reverse operation of specialisation;
- *Aggregation* is an abstraction that allows the definition of a new class (composite class) where each instance is made up of a set of objects from other classes.

These abstractions require operations at two levels: at the schema level and/or at the instance level (object). The operations on the instance are used to adapt the data to the new schema. We describe below these basic operations on the schema and then on the instances (objects) and relate them to the abstraction (from the four specified in Nyerges's model) to which they belong. The new schema defines how the objects should be organised in the new database, and what the rules are for migrating the objects from the old schema to the new one.

- Class abstraction:
  A class abstraction allows for the regrouping of different classes into a super-class. For example, two classes "deciduous forest" and "coniferous forest" can be replaced by a single "forest" class ignoring the distinction between the types of trees. (This corresponds to Nyerges's "generalisation" abstraction.)
- Class elimination:
  When a class of features is not required in the target dataset, the class is simply removed (part of Nyerges's "classification" abstraction).
- Class composition:
  It may be necessary to create new classes represented by composite objects in the initial database. The new class is therefore a derived super-class of the classes containing the elementary object, and inherits their attributes (Molenaar, 1993). Van Smaalen (2003) calls such classes "composite classes", and gives the example of a new class "property" in which instances are made of an aggregation of the objects of class "building" and "lot". To relate this operation to Nyerges's abstractions, we can assume that the class composition is a "classification" abstraction, where a class is created that contains aggregated objects.
- Attribute elimination:
  When an attribute is not required, it is simply removed from the class. This again is part of Nyerges's "classification" abstraction.
- Attribute aggregation:
  Several attributes of a class can be aggregated to form a single new one in the new schema. The rule that defines how the new value is computed from the old ones needs to be specified.

- Modification of the class intension:
  The class intension is the set of conditions that determines which objects belong to a particular class (Richardson, 1993). So a class can change its intension in order for example to exclude objects that are too small and therefore not relevant to the target scale.

3.3.1. *Operations on objects*

Generalising the objects means migrating them from the old schema to the new one, following the rules specified by the new schema.

- *Elimination*
  When an object does not fulfil its class intention anymore, then it is simply eliminated.
- *Reclassification*
  The first reason for an object to change class is to reflect a change in the schema. Another case is in a space exhaustive tessellation of space where rather than remove objects, they are reclassified into the neighbouring class or classes.
- *Aggregation*
  Aggregation can be seen as the reclassification of all objects that constitute a composite object. Each individual object becomes reclassified to the target class. Their attributes are modified according to the intention of the new class.
- *Merging*
  After reclassification or aggregation, adjacent objects of the same class may be merged into one. Some rules must be set to determine how their attribute values become merged.
- *Attribute modification*
  Following the modification to the attribute definitions for the classes of the schema, the attribute values of their instances are modified. These modifications are made according to the rule specified during "attribute elimination" or "attribute aggregation" specified at the schema level.

The only type of abstraction from Nyerges's model that is not reflected in schema or object operations described above is the *association* abstraction. These associations are used to define relationships between classes that can be seen as integrity constraints on the database. These constraints are useful in controlling the cartographic side of the generalisation, when object geometries are modified to comply with cartographic rules.

3.4. Algorithms for Building Generalisation

Buildings have received lots of special attention in the context of generalisation. They occur frequently in maps, and have proved to be a useful basis for development of ideas that can then be applied to other classes of objects. They tend to be small (compared to natural features such as lakes, forests and fields), often requiring them to be enlarged except at the largest scales. They tend to be angular and orthogonal in form. They often occur in dense clusters – particularly in urban areas. The fact that they are often in dense clusters and need to be enlarged leads to situations in which not all can be represented individually. A wide choice of operators is therefore required to generalise buildings (discussed further under three categories). We have ordered these sections in the order in which they are usually used in the generalisation process. The first category contains operators that seek to reduce the number of features represented. Once the density of buildings is manageable, operators of the second category are used to make individual features comply with the readability constraints imposed by the map specifications. The operators in the last category are used to control the inter-feature relationships (such as distances, and relative orientation).

3.4.1. *Operators to reduce the number of buildings*

- Selection/Elimination:
  The challenge in selection or elimination is in deciding which features should be removed and which ones should be kept. The radical law, formulated by Töpfer, links the scale of the map to the amount of details it should contain. Töpfer and Pillewizer (1966) extended the study to take account of the exaggeration of the cartographic symbols. Ruas (1998a, 1999) proposed a measure of density in an urban block (area enclosed by streets) as a basis for triggering the reduction in the number of buildings. Buildings were eliminated until the density fell below 80%, which was considered to be a manageable density threshold for dealing with inter-feature conflicts. A cost function was used to combine a set of constraints in order to reach a decision on which to eliminate. The method tended to eliminate smaller buildings, further from roads, in highly congested areas and with no particular semantic importance.
- Typification:
  Typification reduces the number of buildings while preserving their distribution pattern. Regnauld (2001a) proposed a method for detecting groups of buildings based on gestalt theory, and represented these groups with a reduced number of buildings (Figure 3.1). Ruas and Holzapfel (2003) proposed another grouping method, based on detection and characterisation of building alignments, which could be used to define new typification algorithms. Sester (2001, 2005b) presented another approach based on neural networks, where a random set of buildings was selected, and then displaced to cover the initial distribution (Figure 3.2). Burghardt and Cecconi (2003) have used a mesh simplification technique to gradually reduce the number of buildings, and to use the mesh to position those remaining.

Fig. 3.1. Building typification (Regnauld, 2001a).

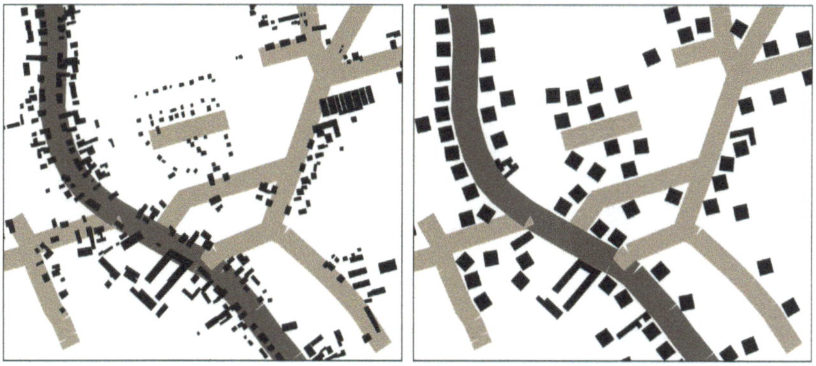

Fig. 3.2. Building typification (Sester, 2005b).

- Amalgamation:
  Several levels of amalgamation can be imagined. The first level would amalgamate adjacent buildings. Lichtner (1979a) proposed a method to displace and rotate small buildings to make them "adhere" to larger neighbours before unifying them. Ware et al. (1995) present several variations of this algorithm, using a Delaunay triangulation. More radical amalgamation may be required, especially as the target scale gets smaller. Regnauld (2003) proposed a method to represent a group of buildings along a road via a snake-shaped amalgam generated by buffering the road centre line. Li et al. (2004) proposed a method which selects the relevant algorithms to produce an adequate level of amalgamation of buildings in a city depending on the scale targeted. Research has also investigated amalgamation of raster data. For example Li (1994), Schylberg (1992) and Su et al. (1997) have used mathematical morphology operators (dilatation, erosion) to amalgamate buildings. Cámara and López (2000) also used similar principles to amalgamate city blocks.

3.4.2. *Operators to ensure readability of buildings*

- Enlargement:
  This operation can be achieved via simple geometric scaling, using an anchor point (usually the centre of gravity of the building). It preserves its shape and proportions. It is used to ensure that each feature complies with the minimum size stipulated by the map specification.
- Simplification:
  Detail in the outline of the buildings often needs to be simplified to comply with readability constraints. Lichtner (1979a) proposed a method to remove edges of a building's outline which were shorter than a specified threshold. Sester (2005a) used a similar approach where short edges were replaced according to a set of rules. Figure 3.3 shows the results obtained with an increasing threshold for the minimum length of an edge. Jones et al. (1995) used the "corner flipping" method to simplify square shaped buildings. Sester and Brenner (2004) proposed a continuous method to gradually simplify buildings, by applying a sequence of operators to their vertices. Four operators (remove vertex, displace vertex, duplicate vertex and insert vertex), were used in combination to perform different actions, such as elimination of extrusions and offsets or simplification of corners.

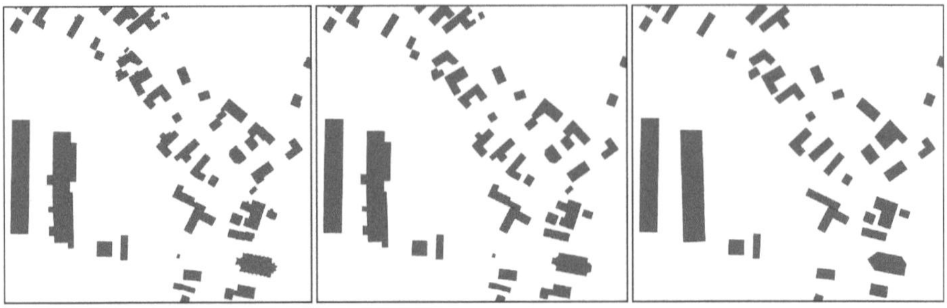

Fig. 3.3. Building simplification (Sester, 2005a).

- Local enlargement:
  Building shapes sometimes contain constrictions (such as corridors connecting two parts), which can be too narrow given the intended representation scale. A specific operator is therefore required to enlarge them.
- Squaring:
  This operator seeks to give a square look to a building and thus helps the reader to identify the building. This is often needed to overcome the accuracy limitations of the digitisation process. Airault (1996) proposed a holistic method that refines the position of the vertices of a set of buildings in order to improve the squareness of each building and retain the parallelism between walls of buildings in close proximity. This approach minimises a cost function, where cost reflects the displacement of each vertex and its incidence on corner angles and wall parallelism within and between buildings.
- Symbolisation:
  For small buildings, it is rarely worth trying to simplify their shape. The best solution is often to replace them with a predefined symbol (square or rectangle) of the appropriate size and orientation. Mackaness and Rainsford (2002) have developed a template matching algorithm that finds the best match for a group of buildings among a set of predefined template shapes. Once the best template has been selected, it is modified (scaled, rotated) to replace the original buildings.

3.4.3. *Operators to maintain relationships between buildings*

- Displacement:
  Where features are too close, one solution is to displace one or both of them. Displacement is a complex operator, difficult to apply in dense regions. It has attracted a great deal of research interest. The research can be divided into two types: incremental improvements and holistic approaches. Incremental methods evaluate proximity among a neighbourhood of objects, incrementally increasing one distance between a pair of objects and iteratively repeating this task for all neighbours until either the conflict is resolved, or until no further improvement can be found. Ruas (1998a) and Lonergan and Jones (2001) proposed two deterministic methods. In the first, at each step an evaluation is carried out to dynamically choose the next action to perform: either solving the side effects of the previous displacement, or identifying the current worse conflict and triggering a new displacement to reduce

it. The second method deals with the objects in no particular order. In both cases, the system can reach states where the displacement may not entirely solve the conflict. Ruas (1998a) used additional operators to solve the problem (amalgamation or elimination), while Lonergan and Jones (2001) identify clusters that require the use of these operators in the subsequent process. To overcome a common problem of being trapped in a local minima, Ware and Jones (1998) proposed a nondeterministic method based on simulated annealing. This method does not include elimination or amalgamation, and requires these operations to be carried out before processing, in order to reach a satisfactory solution.

Holistic approaches usually compute the displacement required for all the features at once. Mackaness (1994) proposes a radial displacement method. After identifying clusters of features in conflict, the features are moved away from the centre of the group. The displacement is proportional to the distance to the centre, to allow pattern preservation. A decay function is used at the periphery of the cluster to limit the expansion of the cluster to the rest of the map. More recent studies have modelled the problem as a system of equations that can be solved using an optimisation method. The equations are used to express the constraints (stiffness of the object, resistance to displacement, minimum clearance). Harrie (1999), Sarjakoski and Kilpeläinen (1999) and Sester (2001) used a least squares adjustment technique in their solution, while Højholt (2000) used a finite element method.

- Rotation:
  This operator is sometimes required when something changes in the environment surrounding a building. If the building was aligned to another feature that has changed (such as a road which has been simplified or smoothed), then rotating the building to preserve the relative orientation may be required. The difficulty lies in computing the orientation of a building (Duchêne et al., 2003), and deciding whether the absolute orientation or the relative orientation is more important.
- Clipping:
  This operator consists of truncating the geometry of a building using adjacent features. This is useful for big buildings or building amalgams, which are larger than the available space. This typically occurs in urban blocks, where the increase of the road symbology reduces the space left for buildings.

3.5. Algorithms for Line Generalisation

Some map features represent network structures (roads, railways, rivers, footpaths) and require special attention during the generalisation process. The generalisation of a network should not be confused with line simplification. Network generalisation refers to the attenuation of a network (akin to pruning a tree), while line simplification involves creating simpler, unambiguous forms of a linear feature – such as reducing the convolution of a line, smoothing a line, displacing a line, or reducing the number of points used to represent that line. Irrespective of the generalisation process, care must be taken to model the topology of the network to ensure connections during and after generalisation.

Most of the work in generalisation has focused on road and hydrology networks. Some research has focused on the roads themselves, such as how the road line can be simplified to stay readable through scale reduction whilst others have focused on the generalisation of the whole network, to find ways of reducing its density without loosing the connectivity that it provides between important places on the map. We therefore begin with a review of techniques for generalisation of lines, before discussing approaches to network generalisation.

3.5.1. *Generalisation of individual roads*

In order to keep the level of detail of roads adequate for the chosen scale and symbology, simplification of the road centre line is often necessary. Simplification is a collective term – embracing one or more actions: filtering: the action of removing some vertices from the geometry; smoothing: displacing some points on the line to attenuate shape details, and caricature: enlarging in order to make more explicit some detail in the line. These are combined in such a manner as to keep the characteristic form of the line.

3.5.1.1. *Filtering*

Filtering algorithms are critical to the reduction of space required to store the data, with consequent shortening of processing times. Lots of filtering algorithms have been developed. Some filter points depending on an analysis of their immediate neighbours (Jenks, 1989), others remove a point depending on local characteristics of a number of surrounding vertices (Lang, 1969; Reumann and Witkam, 1974; Thapa, 1989 and Visvalingam and Whyatt, 1993), and some consider either sections of, or the entire line (Douglas and Peucker, 1973; Jenks, 1989; Lang, 1969; Visvalingam and Whyatt, 1993). Distance between vertices, angular variation, and distance to the line between the previous and following vertices are some of the criteria used by these algorithms. McMaster (1987) compared several simplification algorithms and concluded that the Douglas and Peucker algorithm was one of the best. It has been extended to include variations such as prevention of self intersection via topological modelling of the line (Saafeld, 1999; Edwardes et al., 1998).

3.5.1.2. *Smoothing*

Smoothing algorithms seek to remove small line detail not relevant to the scale of representation. Similar to filtering algorithms, smoothing algorithms can be divided between those making modification via local inspection of the line, and those that apply to the whole line. Some local algorithms displace vertices based on a weighted average position computed using the neighbour vertices (Brophy, 1973; McMaster, 1989). Others are based on Perkal's ideas (Christensen, 1999) or the whirlpool principle, where vertices within a given radius are replaced by a weighted average (Beard, 1991b). Algorithms have also been based on Gaussian filtering (Babaud et al., 1986; Lowe, 1989). Lowe has adapted this method to avoid the flattening of bends in the line.

3.5.1.3. *Caricature*

Caricature algorithms seek to retain some specific characteristic of a line. They usually select relevant details and exaggerate them, while removing other ones. Wang and Müller (1998) proposed a simplification algorithm that also enlarges the most important bends. Several algorithms have been developed at the COGIT laboratory of IGN (French National Mapping Agency), and are presented in Lecordix et al. (1997). Amongst them, "accordion" increases the span of a series of consecutive bends, to avoid coalescence between them, and "schematisation" removes some bends from a series in order to clarify the continuity of the line.

3.5.1.4. *All-in-one algorithms*

Using these algorithms can be very difficult when the lines are not homogeneous. In such a case, no set of parameter values would ensure good generalisation of the whole line. Studies have therefore been done to automatically characterise a line (Buttenfield, 1991) and segment it based on homogeneity criteria (Plazanet, 1995). Fritsch (1997) proposes a method called "Plaster" whereby a line is smoothed and characterised at the same time. The line is split in sections of high or low curvature. High curvature sections are exaggerated and forced to remain in their original position, while low curvature sections are smoothed and absorb the exaggerations resulting from the high curvature sections. Similarly Mustière (1998) proposed an algorithm, called GALBE that automatically applies several algorithms to different parts of a line. The algorithm detects the sections of the line that show a self-coalescence problem (symbology of the road overlaps itself). Then, depending of the type of coalescence, it triggered different algorithms to solve the problem. Sections without coalescence problem are simply smoothed. Figure 3.4 shows some results produced using GALBE.

3.5.1.5. *Displacement of road sections*

One approach to resolving overlapping road symbology is through displacement. Jäger (1991) proposed a method to compute the displacement induced by the enlargement of a linear symbol to the surrounding features. It included a mechanism to propagate the displacement when several lines run parallel to each other. The method was developed for a raster context, but the principle could also be used in a vector context. Nickerson (1988) proposed a method to detect which lines interfere, displace them, and propagate the displacement to nearby features. However setting up the parameters for the algorithm is problematic. Other approaches have represented the problem using a deformable model on which forces can be applied. These forces are used to represent the repulsion between features, and internal resistance to deformation. Burghardt and Meier (1997) proposed a model "the snakes model" – to represent these forces. It has since been improved by Bader (2001) and Bader and Barrault (2000) to better respect cartographic constraints. Bader (2001) and Bader and Barrault (2001) investigate the use of elastic beams, which unlike the snakes method, allows lines to be represented in a true two dimensional model. This improves the flexibility of the algorithm, allowing control over the stretching and bending of individual line sections (Figure 3.5).

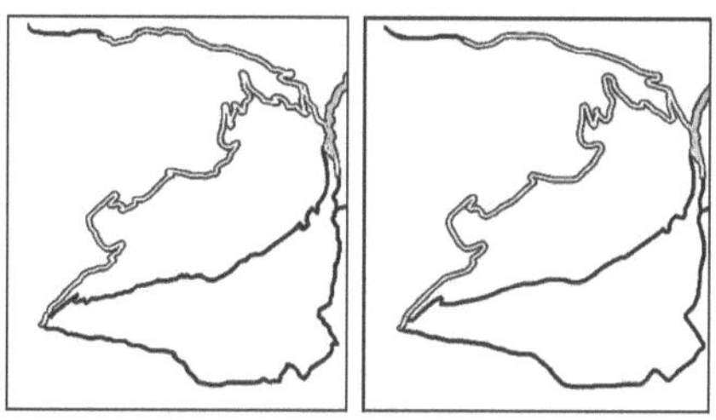

Fig. 3.4. Road generalisation using GALBE (before and after) (Mustière, 1998).

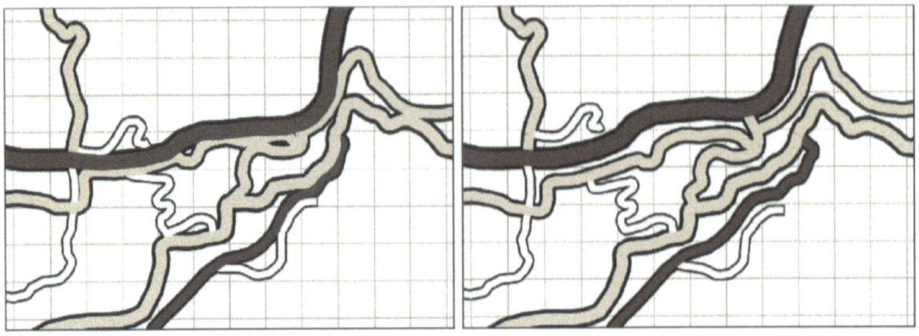

Fig. 3.5. Displacement using the "beams" approach (Bader, 2001).

3.5.1.6. *Algebraic representation of lines*

Instead of specifically designing algorithms to generalise a line, some research have proposed methods for representing lines using algebraic curves, such as splines, Bezier curves, or cubic curves (Akima, 1974; McConalogue, 1970). Affholder (1993) proposed a method to split a very sinuous line at its main inflexion points (such as roads in mountainous area), and approximate each bend via a couple of cubic curves, joined at the top of the bend using circular sections.

3.5.2. *Generalisation of individual rivers*

Both rivers and roads can be represented either in polygonal or linear form. Rivers are more complex to model since their width can vary enormously – extending to form features such as lakes and estuaries. A number of the road generalisation algorithms presented in §3.5.1 are equally applicable to rivers.

Often the same database will contain rivers in linear and polygonal form, with narrow rivers represented by linear features and wider ones by polygons. An operator is therefore needed to collapse polygons into lines, to allow for the representation at small scales of rivers stored as polygons. The challenge with this operator is to decide which part of a polygon needs to be collapsed, and which ones should remain as a polygon. Attribution or algorithms to detect river/lake break points are therefore needed (Regnauld and Mackaness, 2006). A large number of skeletonisation algorithms have been developed for finding the centre line of a river. Particularly relevant is the work of Gold and Snoeyink (2001); and Ogniewicz and Ilg (1992).

3.5.3. *Summary*

Other network structures (such as railways) have received less attention, though it is argued that much of the work on road and river generalisation are equally applicable. Railway tracks are traditionally displayed on maps using a single line. Due to the nature of the railway, the line is smooth and curves have large radius, which makes operations such as simplification and smoothing largely unnecessary. Simplification of a track is only required at very small scale, and does not present significant problems. When several tracks are running parallel, they can either be collapsed (in the case of mainline tracks) or typified (for sidings or marshalling yards). Other networks include cables, pipes and overhead powerlines. Typically these features are not shown on maps, or are represented at a large scale sufficient not to require generalisation.

The other theme that has received specific attention from the generalisation community is representation of the coastline (Christensen, 1999; Wang and Müller, 1993). Though not a network, it is sufficiently complex to warrant attention in its own right. By nature, they are very different to all other types of lines. They can have a very complex and angular shape in rocky areas, while having a smooth shape in areas made of long sandy beaches. Characterisation of the line is therefore necessary, in order to choose the appropriate operators for each section.

3.6. Algorithms for Network Simplification

3.6.1. *Road network simplification*

Despite their importance, it is not possible to show every road at smaller scales. It is necessary to simplify junctions and remove sections of road from the network whilst maintaining connectivity of the network overall, as well as its characteristic form.

Mackaness and Beard (1993) showed the potential of graph theory, and in particular how the use of weighted graphs to govern the attenuation of a road network. Thomson and Richardson (1999) also used graph theory plus the "good continuation" principle, coming from Gestalt theory, to ensure that roads did not break abruptly. Their ideas were based on the idea that roads having long smooth continuity were more important than short sections of roads. This was used to create a hierarchy of roads, thus providing a basis for generalisation. Morisset and Ruas (1997) use a simulation mechanism based on agent methodologies to evaluate the importance of each road in the network. Again, the less important roads were removed first. A more complete discussion of this topic is given by Thomson and Brooks, this volume (Chapter 13).

The road network simplification problem has also been studied from a different perspective. Instead of analysing the network to decide which edges could be removed, we can analyse the urban blocks adjacent to the road network, and amalgamate some of these blocks. The effect of this amalgamation process is to remove the intervening road section. This approach ensures that the generalised network will leave urban blocks of sufficient size given the anticipated target scale. Peng and Müller (1996) used such a method to aggregate urban blocks that were too small. The choice of which neighbours to amalgamate depends on the semantic of the road separating them. Ruas (1999) extended this idea by putting additional constraints on the method. Urban blocks that needed to be amalgamated were selected depending upon their size and the density of buildings within a block. The amalgamation did not take place where the main network would be broken (use of road semantic). The algorithm ensured that urban blocks had a similar building density, and that the shape of the amalgam was typical of an urban block (reasonably compact). Edwardes and Mackaness (2000) proposed a mixed approach that used the characteristics of the urban block (its size) to decide which ones should be amalgamated, together with the properties of the network to decide which streets to remove to perform the aggregation. They reused the concept of "strokes" (Thomson and Brooks, 2000) to assign resistance to elimination of those roads which were important axes in the network. Figure 3.6 shows some progressive network simplification achieved by this method as the minimum size for an urban block is gradually increased.

Fig. 3.6. Gradual attenuation of the network (Edwardes and Mackaness, 2000).

In addition to the simplification of the network, it is also necessary to simplify the form of junctions that connect roads. Multiple road intersections and slip roads may need to be reduced in complexity down to a single point intersection. The simplification of road junctions was undertaken by Mackaness and Mackechnie (1999). Their approach involved a combination of graph theory and cluster analysis to identify junctions of various complexity and reduce them by varying degrees to simpler forms whilst still retaining topology inherent among the roads. There is great variation in the complexity of junctions – at some conceptual level, cities can be viewed as road junctions. Each solution often requires intermediate generalisation solutions across a range of scales.

3.6.2. *River network simplification*

The properties of a hydrologic network are quite different from the properties of a road network. The hydrologic network usually looks like a tree, where the leaves correspond to springs, and the streams join together to form bigger and bigger rivers, before ending in a lake or at the mouth of the sea. Horton (1945) proposed a method, known as the Horton order, to weight each section of the hydrologic tree, depending how far in terms of the number of confluences, it was from a spring. Strahler (1952) attempted something similar. From this a hierarchy is set between the different branches of the tree. This can be used to select which branches should be retained during generalisation. In the absence of topological modelling, this approach may result in the isolation of lake features from the rest of the river network. Thomson and Brooks (2000) have also applied their good continuation principle to the simplification of a hydrologic network, with impressive results – a sample of which is shown in Figure 3.7. The dark lines are the rivers that have been kept; the light grey ones have been suppressed.

There is a growing understanding of the requirements for network generalisation, and the need to take account of the characteristic form of the features, whilst maintaining the topology of the network. Networks often connect features of different data types, making it necessary to develop techniques that determine the connectivity of the network (Regnauld and Mackaness, 2006).

3.7. Algorithms for Rural Features

Generalisation has tended to focus on anthropogenic forms in an urban context. Increasingly attention is turning towards the modelling of features found in rural areas. Further still, attention is turning to the generalisation of categorical maps – thematic maps such as soil or

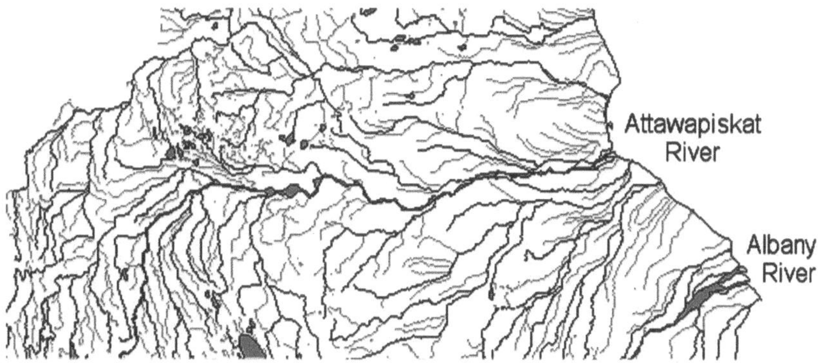

Fig. 3.7. River network simplification (Thomson and Brooks, 2000).

geological maps. Categorical maps are often formed of a space exhaustive tessellation of space – a mosaic in which each part of the surface is categorised. Finally there are a whole set of algorithms developed for the modelling and generalisation of terrain.

3.7.1. *Algorithms for rural features*

Generalisation of rural features such as forests and agricultural fields, has recently received more attention. An early piece of work examining the generalisation of a collection of water bodies was undertaken by Müller and Wang (1992). They proposed a method to generalise a multitude of lakes across a region. The method in effect achieved a typification of the area, retaining the overall distribution of the lakes as well as their size differences, through a wide range of target scales. The method is based on the following sequence of operators: enlargement/reduction in size of lakes (the small are reduced and the big are enlarged), elimination of the smallest ones (using a threshold depending on the target scale), reselection of eliminated lakes in depleted areas, fusion of overlapping lakes, displacement (for the lakes too close to each other), and simplification.

It is considered that much of the research done on anthropogenic and urban contexts could equally be applied to the rural domain, though clearly more work is required. It is thought that simplification of boundary polygons could be achieved via line simplification algorithms and algorithms such as the ones proposed by Jones et al. (1995) could be used to amalgamate polygons based on a triangulation. Displacement algorithms based on optimisation techniques, such as those proposed by Bader and Barrault (2001), Harrie (1999), Højholt (2000), Sarjakoski and Kilpeläinen (1999) and Sester (2001), could also be adapted to handle rural features.

3.7.2. *Algorithms to generalise categorical data*

Categorical data can be represented in both raster and vector form. In a raster data model every pixel has a single value, specifying its category (theme). Groups of adjacent pixels of the same category form area objects. Such maps have quite a lot of similarities with digital images. As a result, some tools for processing digital images have been adopted and adapted to suit the purpose of map generalisation. McMaster and Monmonier (1989) proposed a classification of generalisation operators for raster imagery. Under the heading of "categorical generalisation" they defined merging, aggregation and attribute change – all involved filtering of data using kernel processing. Cost functions have also been used to control the reclassification process. For example, Le Men (1996) proposed a method that processes small gaps and areas by either enlarging, aggregating, dissolving or bridging between them. In each case, the operator that generates a minimal thematic transition cost is chosen. Fuller and Brown (1996) also used a cost function to model separation of classes during generalisation.

Categorical data can also be represented in a vector data model. We call this type of map a "polygonal subdivision" or a "polygonal map" (Galanda, 2003). The conceptual generalisation operators are similar to those in raster, even though their implementation is very different. In any solution it is important to ensure that all the space is fully covered by a mosaic of polygons, whilst satisfying various constraints (such as minimum size, minimum width). For example, Downs and Mackaness (2002) developed a technique for generalising geological maps. Bader and Weibel (1997) developed a technique for dissolving polygons among neighbouring polygons by splitting the polygon, and reassigning each portion to the closest neighbouring polygon. They also proposed other operators, such as enlargement and displacement, to solve size and proximity conflicts. Galanda and Weibel (2003) present an optimisation-based approach using snakes to generalise polygon maps. They used the snake based model to perform displacement, enlargement and exaggeration (local enlargement). These operators have been used in a Multi-Agent system environment to automate the generalisation of polygonal datasets to produce maps at different scales (Galanda, 2003).

3.8. Algorithms to Generalise Relief

There is a growing number of techniques for capturing, storing and manipulating information on relief (Bjørke and Nilsen, 2002; Wood, 1996), in response to a growing number of application domains (Raper, 1989; Wilson and Gallant, 2000). The most common ones are the representation by contour lines and by Digital Terrain Models (DTMs). Nowadays, the tendency is to build and maintain DTMs, from which the contour lines can be automatically derived. However, contour lines are still more widely available than DTMs. Gold and Thibault (2001) propose a method to smooth contour lines, using a skeleton based approach. The skeletons are computed between consecutive contour lines, simplified and used to adjust the position of the original contour vertices. Contour lines, with or without smoothing, can be used to automatically generate DTMs using the method described by Gold and Dakowicz (2003).

Generalising directly a DTM can be done in two ways, according to Weibel (1992). The grid constituting the DTM can be generalised directly using filters similar to those used for raster images. Otherwise, structuring lines can be extracted from the DTM and generalised (Wood, 1996), in some cases using techniques developed for line generalisation. Two types of structuring can be considered: the drainage network (see O'Callaghan and Mark (1984)), and the ridge network. Weibel (1992) proposes a strategy that would use the different approaches in different situations. Generalising the DTM directly works well in relatively flat areas, while generalising the structuring lines works better in irregular (mountainous) regions. It is therefore important to have techniques available to automatically characterise the terrain and identify these different types of regions. Monier (1996) proposes some tools to characterise a DTM to support the choice of generalisation tools.

Table 3.1. List of factors to select appropriate algorithms.

Raster-mode generalisation	Vector-mode generalisation
Structural generalisation – simple structural reduction – resampling	Point feature generalisation – aggregation – displacement
Numerical generalisation – low-pass filters – high-pass filters – compass gradient masks – vegetation indices	Line feature generalisation – simplification – smoothing – displacement – merging – enhancement
Numerical categorisation – minimum-distance to means – parallel piped – maximum-likelihood classification	Areal feature generalisation – amalgamation – collapse – displacement
Categorical generalisation – merging (of categories) – aggregation (of cells) o non-weighted o category-weighted o neighbourhood-weighted – attribute change	Volume feature generalisation – smoothing – enhancement – simplification
	Holistic generalisation – refinement

3.9. Generic Description and Evaluation Criteria for Algorithms

This chapter has illustrated the breadth of approaches and implementations. There are good reasons for this. Each algorithm can be applied in different contexts – different themes, different scale variation, different regions (for example urban area vs. rural area) and to achieve different types of representation (depending on the specifications of the target map or dataset). Each algorithm has its own limits, in terms of their conditions of use, efficiency, the quality of output and consistency. Though we have a large choice of algorithms, it is not always clear which is the most appropriate for a given task, the order in which they should be combined and how much they should be applied. Table 3.1 is a list of factors that can help in that process of selecting appropriate algorithms.

3.10. Operators and Algorithms Within Existing GI Systems

This section presents an overview of the generalisation operators and algorithms that have been implemented in commercial GIS software. As discussed in §3.2 a generalisation operator defines the spatial transformation, while a generalisation algorithm is an implementation of that particular transformation (Weibel and Dutton, 1999).

3.10.1. *Map Generalizer*

Map Generalizer is a generalisation software produced by Intergraph. Although still in use in numerous places, Intergraph has now replaced it with DynaGen. Map Generalizer is a platform for performing interactive generalisation, and contains the following operators: elimination, collapse, simplification, smoothing, aggregation, typification, displacement, point scaling, boundary extend, point orientation and area squaring. Essentially Map Generalizer has been superceeded by DynaGen, whose functionality we now describe in greater detail.

3.10.2. *DynaGen 3.0*

DynaGen is the current generalisation product of Intergraph, based on Dynamo. The software, described in (Intergraph, 2003), allows the user to either perform manual, interactive or automatic generalisation. The following list of operators is included in the software, often with a choice of different algorithms to perform them:

- Simplification. DynaGen contains a wide range of simplification algorithms:
  Line simplification:
  - *Douglas Global Tolerance Band Thinning*, based on (Douglas and Peucker, 1973)
  - *Lang Local Tolerance Band Thinning*, based on (Lang, 1969)
  - *Point Relaxation Thinning*
    This algorithm works by constructing a relaxation circle around each point on the feature. Then it draws straight-line segments as long as possible that touch or pass through each circle.
  - *Reuman–Witkam Corridor Search Thinning*, based on (Reumann and Witkam, 1974)
  - *VectGen Thinning*
    This algorithm appears to be based on Jenks's algorithm (Jenks, 1989).
  - *Nth Point Thinning*
  - *Thinning Elevation Process*
    This algorithm simplifies a line by taking into account the height of its vertices, and keeping the significant height variation. The algorithm reuses principles from the Douglas and Peucker algorithm.

  Area simplification:
  The line simplification algorithms listed above can be used on boundary lines or area features. In addition, the following algorithms are available specifically for areas.
  - *Area Preservation Thinning*
    The algorithm removes details but keeps the size of the feature. It is adapted for nonregular shape areas.
  - *Building Clarification*
    This is a simplification algorithm for building-like shape features.
- Smoothing. DynaGen proposes three different algorithms:
  - *Brophy smoothing*, based on (Brophy, 1973)
  - *Simple average smoothing*
  - *Weighted average smoothing*
- Aggregation.
  - *Intersecting Area Merge*: combines areas which are overlapping
  - *Area Merge Irregular*: Amalgamates area features which are close together, by bridging the gap between them.
  - *Area Merge Orthogonal*: same action as above, but adapted for buildings (produces a squared shape).

- *Aggregate points.* This algorithm merges a set of point features that are within a specified distance of one another into a single area and/or line feature.

- Collapse. This operator reduces the dimension of the input data. Different types are available: area to line (total or partial), area to point, line to point, point set to point and dual line to single line.
- Typification. This operator reduces the number of features in a group while preserving their distribution characteristics. It is applied to a line or set of points, and different algorithms are used to perform the typification on a drainage network or on a road network.
- Network merge. This operator merges multiple touching features into one. In effect this constructs intelligent feature networks out of a set of linear features. Two different algorithms can be applied to rivers and roads.
- Squaring. This operator forces the angles of a polygon to be 90 degrees. This is intended to be used on buildings, or other similar shapes.
- Boundary extent. This operator extends the boundaries between areas and lines, to remove small gaps between them. In effect it allows the features involved to locally share part of their geometry.

3.10.3. *CHANGE*

The system CHANGE (Institute of **C**artography – University of **HAN**over – **GE**neralisation Software) has been designed in the nineties to perform batch generalisation for buildings and roads. It is intended to generate maps at scales from 1 : 10 000 to 1 : 25 000, from a detailed vector database. CHANGE contains two modules: one for road generalisation, one for buildings (CHANGE, 2004). More details about these algorithms can be found in Powitz (1992) and Staufenbiel (1973). The following operators are available for road generalisation. They allow the production of a topologically structured generalised road network:

- Collapse: generating centre lines for roads and construction of a network;
- The module GALINO in CHANGE contains several algorithms for simplifying and smoothing roads:

LIRE:	simplification based on linear regression.
HIGHPASS:	simplification based on a high pass filter.
DOUGLAS:	simplification based on the Douglas and Peucker algorithm (Douglas and Peucker, 1973).
GLEITM:	simplification based on the moving average principle.
COSINUS:	simplification shifting each vertex of the line based on angular properties of the triangle made with its direct preceding and following vertices.
AKIMA:	smoothing based on the AKIMA interpolation method (Akima, 1974).
SPLINE_INT:	smoothing using a spline interpolation.
POLYNOM:	smoothing using an interpolation using a polynomial of degree 5.

The following operators are available for generalising buildings:

- Selection
- Simplification. The simplification algorithm processes all the edges of the boundary line that are too short. A set of rules triggers the process depending on the neighbouring edges of the shortest. The algorithm preserves circular shape portions if they are large enough.

- Amalgamation. Buildings that are close to each other (less than a distance provided by a parameter of the method) are amalgamated. Amalgamation can be constrained by the type of the buildings (for example, only industrial buildings are amalgamated together).

All the parameters are values set by the user in advance, and the system performs the generalisation in batch mode.

The University of Hanover has other operators commercially available:

- PUSH is a displacement method based on least squares adjustments that allows the user to specify different constraints on each class of features and also on individual features. These constraints include the minimum distance to neighbours, the resistance to displacement, orientation change, and shape deformation. The operator also provides a quality indicator with the result. PUSH works on all types of geometric features (line, points and polygons).
- TYPIFY is a typification operator using neural network techniques to select a subset of features, and adjusts their position such that the initial distribution pattern is retained in the result. Density variations are retained implicitly by the algorithm.

These operators and algorithms are described in more detail by Sester (2005a).

3.10.4. *Workstation ArcInfo 9.0*

Workstation ArcInfo has been developed by ESRI. It contains a library of generalisation tools available via a command language. Some are invoked interactively, by selecting the features on which to apply an algorithm, specifying the parameter values and invoking the algorithm required. Others are available in batch mode. The following operators are proposed (Lee, 2000), with different algorithms available for each of them:

- Simplification
  - *PointRemove*: based on Douglas and Peucker (1973);
  - *BendSimplify*: line simplification that removes full bends from the line, giving a result less angular than *PointRemove* (Wang, 1996);
  - *BuildingSimplify*: simplification for buildings (Wang and Lee, 2000);
  - *FindConflicts* is an algorithm that detects proximity conflicts occurring after running the *BuildingSimplify*.
- Smoothing
  - *McConalogue*: smoothing algorithm based on the interpolation method described in McConalogue (1970);
  - *Cubic Polynomial.*
- Collapse
  - *CentreLine*: collapses regular dual line features into one centre line;
  - *CreateLabels* followed by *CentroidLabels*: collapses polygons to points at either the centroid or the nearest location inside the polygon if the centroid is outside the polygon;
  - *Arclabels*: collapses lines to points at a specified side of the line and a distance perpendicular to the middle segment of each line.
- Amalgamation
  - *AreaAggregate-Orthogonal*: for amalgamating close buildings or other features with orthogonal angles;

- *AreaAggregate-NonOrthogonal*: for amalgamating features with natural boundaries;
- *ConstrainedAreaAggregation*: amalgamates close buildings, but takes into account obstructing features (ESRI, 2000);
- *Dissolve*: merges adjacent polygons or lines which have the same value for a specified attribute;
- *Eliminate*: merges the selected polygons with the most appropriate neighbouring polygon (based on the area or the length of the shared border).

- Refinement
  - *AreaExtend*: this operator extends polygon boundaries between closely located polygons, and creates shared boundaries wherever they meet (ESRI, 2000).

Note that *ConstrainedAreaAggregation* and *AreaExtend* are not released in Workstation ArcInfo, but delivered as benchmark functions. The steps are described in (ESRI, 2000).

3.10.5. *ArcGIS 9.0 (ESRI)*

Some of the generalisation tools available in ArcInfo 9.0 are already present in ArcGIS 9.0. Others will be migrated in future releases. In addition, ArcGIS 9.0 also contains new tools. We only describe here the new additions present in ArcGIS, while the tools imported from ArcInfo can be identified in the table at the end of this section.

- Simplification
  - The line simplification algorithms are the same as in ArcInfo, but they have been enhanced to manage topological errors. Topological errors are detected and resolved by performing local corrections, where the portions of line incorrectly simplified are resimplified with gradually reduced tolerance;
  - *Area to rectangle*: creates a rectangle that covers the original area.
- Smoothing
  - *Polynomial Approximation with External Kernel (PAEK)*: PAEK smoothes lines using a parametric continuous averaging technique (Bodansky et al., 2002);
  - *Bezier_Interpolation*: the algorithm fits Bezier curves through every line segment along an input line. The Bessel Tangent is used to connect the curves smoothly at vertices (Farin, 2001). The resulting lines pass through input vertices.
- Collapse
  - *Feature To Point – Centroid or Inside*: Creates points based on input polygons, lines, or multipoints. The centroid or inside option allows the user to specify whether the centroid should be on the initial geometry or if it is allowed to lie outside.

3.10.6. *Lamps2 v5-2 (Laser-Scan)*

The GIS Lamps2 has been developed by Laser-Scan, and contains a list of operators accessible interactively through a user interface. These operators can also be called using the internal language Lull or C, to perform generalisation in batch mode. The following operators are available:

- Simplification
  - *Manmade simplification* (for simplifying buildings);

  - *Topographic* (version of the Douglas and Peucker algorithm);
  - *MBR:* Minimum bounding rectangle.
- Squaring
  - *Angle squaring*. Modify the orientation of sides so that they match the base orientation or its perpendicular. Only modifies the orientation of sides when it is within a given tolerance of the base orientation or its perpendicular;
  - *Side squaring*. Similar algorithm, but checks that a side is suitable for squaring by evaluating the displacement of the vertices which is induced by the modification of orientation.
- Typification

  The operator works on a set of points. It detects clusters of points and replaces each one by a single representative point. This is more of a collapse operator than a typification operator.
- Aggregation
  - Convex hull (Computes the convex hull of the selected features);
  - Shrink wrap hull (Wraps the features into a polygon that is a closer approximation to the original feature than the convex hull).
- Collapsing
  - Area to Area/Line (Performs a selective collapse of area or linear objects);
  - Line to Line (Performs the collapse of a dual line into a single one);
  - Area/Line to Point (Replaces the input geometry with a point).
- Exaggeration
  - *Area* (scaling algorithm that changes the size of an object using a seeding point and a coefficient);
  - *Line exaggeration.*
- Smoothing
  - *Akima Interpolation*, is based on a bicubic spline (Akima, 1974);
  - *McConalogue interpolation* (McConalogue, 1970);
  - *Other Cubic Spline Interpolations*;
  - *Linear interpolation.*
- Area merging

  Amalgamates two polygons close to each other into one.
- Displacement

  For each object intended for displacement, the set of objects in proximity are identified. Displacement vectors are computed, amalgamated, and used to perform displacement. Feature classes can have different resistance to displacement. Line and polygon are treated as set of points, with additional constraints on endpoints or intersections.

3.10.7. *Clarity 2.0 (Laser-Scan)*

Laser-Scan has also developed a platform dedicated to generalisation, called Clarity, which is based on the technology developed during the AGENT project (Lamy et al., 1999). The platform proposes a generic and extensible framework for setting up automatic generalisation. The generalisation operators and algorithms available in Lamps2 are called using Java, C or LULL. Clarity does not contain an interface for triggering generalisation algorithms interactively. A subset of the operators and algorithms developed for the AGENT prototype have been ported to the first release of Clarity. They are:

- Bend caricature
  - *Minimum break*: resolves self-coalescence on a single bend of the line, by finding the skeleton of the bend and using it as the shared boundary of the two sides of the bend;
  - *Maximum Break*: resolves the same type of problem by moving the two sides of the bend away from each other, until the coalescence disappears;
  - *Accordion*: resolves coalescence occurring between bends or a series of bends. This is done by stretching the series of bends until the coalescence is resolved;
  - *Bend removal*: resolves the same problem as above (coalescence in a series of bends), by removing a pair of bends (in opposite directions) and stretching the others to reconnect them.
- All-in-one line generalisation
  - *Plaster*: this is a combination of smoothing and bend exaggeration to smooth the line without loosing its main shape characteristics;
  - *Generalise-by-parts*: This algorithm splits a line to isolate coalescence conflicts. Then different algorithms are used on each section before being reconnected to provide the generalised line.
- Area enlargement
  - *Enlarge to rectangle*: replaces a building with its minimum bounding rectangle, scaled to match a given size and orientated to match the original building;
  - *Polygon elongation*: enlarge or reduce a polygon along its main axis;
  - *Polygon local width*: enlarge bottlenecks in buildings, until they match a specified minimum width.
- Amalgamation
  - *Amalgamate:* amalgamates buildings which overlap each other, following an enlargement.

3.10.8. *Summary*

Tables 3.2a and 3.2b summarise the operators and algorithms that are available on the different commercial GIS platforms.

3.11. The Changing Classification of Generalisation Operators

To date, most generalisation algorithms have been designed to be used in an interactive environment. Development of operators usually concentrates on the algorithmic side, at best trying to extract parameters that can be used to control the algorithm. This approach is sufficient to prove the concept used by the algorithm, and allow its deployment in an interactive environment. However, in order to use operators and algorithms efficiently in an automated context, they need to come with additional information. They need to provide metadata that enables an automated system to use operators and algorithms in a much more dynamic way than is possible today. Such tools would be much more interoperable, and would allow generalisation systems to be more quickly developed.

Table 3.2a. A summary of the operators and algorithms available on the different commercial GIS platforms (part 1).

Operators/ algorithms	ArcInfo	ArcGIS 9	DynaGen 3.0	Change	Lamps2	Clarity
Line simplification						
*N*th point			YES			
Douglas	YES	YES	YES	YES	YES	YES
Lang			YES			
Reuman–Witkam			YES			
House algorithms	Bend simplify	Bend simplify	1) Point relaxation 2) VectGen 3) Elevation processing	GALINO		
Line smoothing						
Brophy			YES			
Averaging			YES	YES		
McConalogue interpolation	YES				YES	YES
Bezier interpolation		YES				
Akima interpolation				YES	YES	YES
Other cubic splines	YES				YES	YES
House algorithm		PAEK		GALINO		
Bend caricature						
Accordion						YES
Min break						YES
Max break						YES
Bend removal						YES
Exaggeration					YES	YES
All-in-one line generalisation						
Plaster						YES
Generalise-by-parts						YES
Line merging						
Blend line	YES	YES	YES			
Hierarchical Blend line			YES			

In early work in the field of generalisation, operators and algorithms were identified from observation of traditional approaches to mapping. Technology has hugely influenced the context of design and offered environments in which users can request tailor made maps. Consequently it is important to link the algorithm (and its associated parameters) with the user's

Table 3.2b. A summary of the operators and algorithms available on the different commercial GIS platforms (part 2).

Operators/algorithms	**ArcInfo**	**ArcGIS 9**	**DynaGen 3.0**	**Change**	**Lamps2**	**Clarity**
Area enlargement						
Scaling					YES	YES
Enlarge to rectangle						YES
Enlarge bottleneck						YES
Area simplification						
Irregular shape			YES			
Orthogonal shape	YES	YES	YES	YES	YES	YES
Turn to rectangle		YES			YES	YES
Area enhancement						
Area extend	YES		YES			
Squaring			YES		YES	YES
Amalgamation/ Aggregation						
Merging	YES	YES	YES		YES	YES
Irregular amalgamation	YES	YES	YES			YES
Orthogonal amalgamation	YES	YES	YES	YES		
Point aggregation			YES		YES	YES
Collapse						
Area > Point	YES	YES	YES		YES	YES
Line > Point	YES	YES	YES		YES	YES
Points > Point		YES	YES		YES	YES
Area > Line			YES		YES	YES
Area > Edge			YES			
2 Lines > Line	YES	YES	YES	YES	YES	YES
Typification						
Points			YES			
Network simplification						
Street network			YES			
River network			YES			
Neighbourhood detection						
Conflict detection	YES					
Clustering					YES	YES
Displacement						
Vertex displacement					YES	YES

requirements or a particular map specification. Given the context of use, it is important to model the quality of the solution, and to record the processing steps undertaken in generating that solution. It is therefore important for each algorithm to have its own associated metadata as part of the quality control process. As new algorithms are developed for each type of operation, it is important to discern what differentiates each algorithm and to specify the appropriateness of its application (the conditions of use). This information would include the types of features to which the algorithm could be applied, the scale range over which the algorithm works, and constraints associated with the input data. The completeness of this information will depend on rigorous testing of the algorithm in different contexts.

Advances in hardware and modelling tools provide new opportunities to develop new types of algorithms. The list of different types of phenomenon (Ormsby and Mackaness, 1999) that need to be processed during the generalisation is very large, and the list of representations that we may want to derive for each of them is larger still. Thus there is always a need to develop new algorithms, particularly those that can take into account the context in which they operate. This lack of contextual algorithms has already been reported by Ruas (2001) and it remains the case that such algorithms are critical to the development of generalisation systems that have a high degree of automation.

Acknowledgement

We would like to thank ESRI, Intergraph, Laser-Scan and the University of Hanover for providing us with information about algorithms available on their different GIS platforms.

Generalisation of Geographic Information: Cartographic Modelling and Applications
W.A. Mackaness, A. Ruas and L.T. Sarjakoski (Editors)

Chapter 4

Modelling the Overall Process of Generalisation

Lars Harrie[a], Robert Weibel[b]

[a]*Lund University, GIS Centre, Sölvegatan 12, SE-223 62 Lund, Sweden*
e-mail: lars.harrie@nateko.lu.se
[b]*University of Zurich, Department of Geography, Winterthurerstrasse 190, 8057 Zurich, Switzerland*
e-mail: weibel@geo.unizh.ch

Abstract

Research on the automation of cartographic generalisation has led to the development of a large number of generalisation algorithms. This chapter describes modelling techniques for using these algorithms to form a comprehensive generalisation process. Important issues include: when to use the generalisation algorithms and how to trigger and control them. Three main modelling techniques are described: condition-action modelling, human interaction modelling and constraint based modelling. In a condition-action modelling process an identification of objects and relationships between objects is first performed. Then, based on the identified conditions, generalisation algorithms are triggered. Human interaction modelling is based on the principle that the cognitive workload can be shared between computer and human. The computer typically carries out those tasks which can be sufficiently formalised to be cast into algorithms, while the human assumes responsibility for guiding and controlling the computer software. Finally, in constraint based modelling the starting point is the requirements (constraints) of the generalised map. An optimisation process then finds a generalisation solution that satisfies as many of the constraints as is possible. The chapter describes the historical evolution of these modelling techniques as well as their strengths and weaknesses.

Keywords: agents, amplified intelligence, batch process modelling, combinatorial optimisation, computational efficiency, condition-action modelling, constraint based modelling, constraints (typology of), continuous optimisation, finite element method, human interaction modelling, interactive systems, least-squares adjustment, optimisation, snakes (elastic beams), structural recognition

4.1. Introduction

The automated generalisation process can be divided into several sub-processes (Brassel and Weibel, 1988; SSC, 2005). These sub-processes, or generalisation operators, are performed by generalisation algorithms (further described by McMaster and Regnauld, this volume (Chapter 3)). The subject of this chapter is on ways of modelling the overall process of map generalisation using these algorithms as building blocks.

Early software systems with generalisation functionality relied on batch processing to sequentially chain together several generalisation algorithms and feed the necessary control parameters. In a batch process, all the necessary algorithms, control parameters and input data have to be made known before the process starts. In such a process, no interaction with the system is possible at program run-time, and the entire batch process has to be re-run with new parameter settings in instances where the results are not satisfactory. In the "batch processing era" (up till the early 1980s) this strictly sequential approach was the only possible way to model the overall generalisation process. It was only later that important advances in computing technologies prepared the stage for more sophisticated approaches to process modelling, including personal computing, graphical user interfaces, and advances in computational intelligence. This chapter describes three successors of batch processing for modelling the overall generalisation process:

1) Condition-action modelling:
   A condition-action process consists of two phases: structural recognition (condition) and execution (action). In the structural recognition phase identification of objects and relationships between objects is performed. Based on the identified conditions, algorithms for generalisation are triggered in the execution phase.
2) Human interaction modelling:
   Human interaction modelling is based on the principle that the cognitive workload can be shared between computer and human. The computer typically carries out those tasks which can be sufficiently formalised to be cast into algorithms, while the human assumes responsibility for guiding and controlling the computer software.
3) Constraint based modelling:
   Several requirements must be fulfilled in a generalised map; these requirements can act as constraints to the generalisation process. A constraint based approach seeks a generalisation solution that satisfies as many of the constraints as possible.

4.2. Condition-Action Modelling

4.2.1. *Basic philosophy*

Condition-action modelling is based on rules saved in a rule base and is conducted in two phases. To perform the generalisation the cartographic data are analysed in a structural recognition phase. The result from this phase is "structural knowledge"; that is, information about single objects or groups of objects. For example, the information could be that a building object (represented with polygon geometry) has an area of 120 m^2. The structural knowledge is then used, together with the rules in the rule base, to determine which generalisation action to trigger in the execution phase. In the example above, the action could be to represent the building as a square point object.

To generalise the left map in Figure 4.1 several condition-action rules are required. Examples of such rules for a target scale of 1 : 50 000 are:

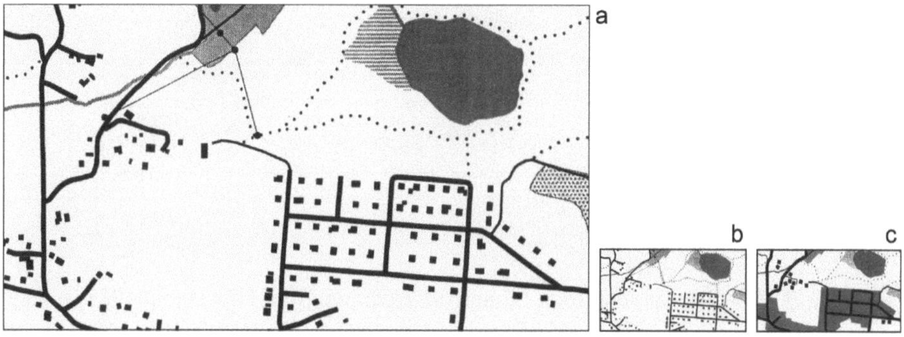

Fig. 4.1. Three maps of the same area; (a) at a scale of 1 : 10 000, (b) same map scaled down to 1 : 50 000, and (c) generalised from (a) for presentation at a scale of 1 : 50 000. ©Lantmäteriverket, Sweden. Printed by permission (2005/4413).

if (objectType == *major road*)
then simplify the road using Douglas-Peucker's algorithm with a threshold value of 10 m.
if (objectType == *minor road*)
then remove the road object.
if (objectType == *building* and area > 80 m^2)
then represent the building as a point object (using the gravity point of the original corner points).
if (objectType == *building* and area < 80 m^2)
then remove the building object.
if (distance between object *A* (objectType == *building*) and object *B* (any object represented with a line) < 10 m)
then move object *A* in a perpendicular direction from the line object *B* so that the distance is ⩾10 m.

The first four rules are defined for single objects and are all quite simple. But the last rule, that considers a group of objects, is more complex. The last rule also requires that the structural phase can identify proximity conflicts between objects.

4.2.2. *Implementations*

Approaches based on condition-action modelling were mainly pursued in the late 1980s, in a period when rule based systems (or expert systems) were generally favoured as a methodology to solve complex and ill-structured problems requiring iterative decision making. Inspired by successful implementations of rule based systems in other areas such as medical diagnosis or engineering (Hayes-Roth et al., 1983), a number of researchers also attempted to exploit this approach in map generalisation and related tasks of cartographic design (Mackaness et al., 1986; Nickerson and Freeman, 1986; Mackaness and Fisher, 1987; Fisher and Mackaness, 1987; Gill and Trigg, 1988; Nickerson, 1988; Robinson and Zaltash, 1989; Doerschler and Freeman, 1989; Jankowski and Nyerges, 1989; Buttenfield and Mark, 1991; Schylberg,

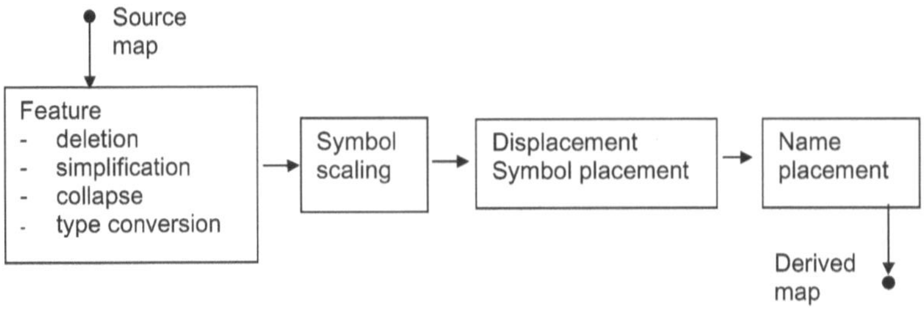

Fig. 4.2. The Nickerson–Freeman framework (redrawn from Nickerson, 1988, p. 18).

1993). In these implementations, important progress was made over the previous state of the art. Most importantly, these systems, for the first time, offered an avenue to move away from a simple mechanistic modelling of the generalisation process as atomic "hard wired" algorithms that were applied in batch processes, applying the sort of condition-action modelling that can indeed be found in human reasoning processes. Some of these systems, of which Nickerson (1988) and Schylberg (1993) are probably the most elaborate examples, generated very encouraging results for particular generalisation tasks.

Despite the initial success of the cartographic rule based systems, the implementations also revealed some limitations of condition-action modelling. The first problem is that it relies on the ability to formalise rules that define the condition-action pairs. Formalising cartographic knowledge, however, is extremely hard and cumbersome as generalisation often involves a great deal of intuitive judgment and graphical operations are inherently difficult to describe in a formal language. This situation contrasts with the classical application domains of rule based systems such as medical diagnosis or fault diagnosis in computer systems, which usually rely on relatively well structured reasoning chains. In cartography, this has rapidly led to a situation that might be termed a *knowledge acquisition bottleneck* (Weibel et al., 1995).

A second problem of condition-action modelling is that generalisation of a map requires a great many rules to be defined in the rule base, especially if rules are set up for the relationships between map objects (Beard, 1991a). Maps are often complex and there exists almost an infinite number of possible relationships between objects. To categorise all these relationships into different classes and to pre-define suitable actions for each class is a difficult if not an intractable task. To keep the number of rules used for a complete generalisation process at a tractable level, researchers employing the condition-action modelling paradigm (Nickerson and Freeman, 1986; Nickerson, 1988; Schylberg, 1993) tended to use pre-defined sequences of generalisation operators. These sequences were inspired by the literature on manual generalisation as well as experimentation and attempted to optimise the choice of generalisation operators. A typical example of such an operator sequence is the Nickerson–Freeman framework shown in Figure 4.2. Generally, selection and classification are applied first, followed by other operators associated with model generalisation (e.g., aggregation and collapse), with operators of graphical generalisation (simplification, smoothing, exaggeration, and displacement) applied last, where displacement is usually the very last operator.

Operator sequencing in the condition-action model, however, leads to a third problem. Operators have different goals and effects. Hence, when an algorithm is applied to solve a particular cartographic conflict, the algorithm may create other conflicts that have to be solved by subsequent algorithms. Since condition-action modelling always chains a condition (i.e., conflict) to a particular action (i.e., generalisation algorithm with associated parameterisation) it is generally inflexible and cannot model all situations in map generalisation. In particular, this approach lacks the capability of evaluating the results of its generalisation actions (Mackaness and Ruas, this volume (Chapter 5)).

4.3. Human Interaction Modelling

4.3.1. *Basic philosophy*

In the early 1990s, the problems with cartographic knowledge acquisition and implementations of approaches based on the condition-action modelling paradigm had become apparent. The focus of academic research and commercial software development alike was now on the development of strategies that relied mainly on human interaction. In the commercial world, the focus was placed on interactive generalisation systems (e.g., MGE MapGeneralizer by Intergraph, and LAMPS2 Generaliser by Laser-Scan) that could be deployed rapidly. These systems implemented toolboxes containing a wide range of generalisation algorithms that were available from academic research at the time. On the other hand, academic research still had not been able to deliver the necessary procedural knowledge; that is, the knowledge needed to decide when and how to apply the various algorithms of the interactive generalisation toolbox. That meant that the interactive systems lacked automated process modelling to drive the algorithms.

In academic research, an attempt was made to integrate some of the more successful elements of condition-action modelling methods (such as rule based systems) with human interaction to form a new paradigm coined “amplified intelligence” (Weibel, 1991). The concept of amplified intelligence is based on the assumption that the cognitive workload can be shared between the computer and the human. The computer (i.e., the generalisation software) typically carries out those tasks which can be sufficiently formalised to be cast into algorithms, while the human, given their ability of holistic judgment, assumes responsibility for guiding and controlling the computer software. That is, the human intelligence is amplified by the processing power of the computer, while the limited capabilities of the computer for holistic reasoning as well as visual perception and quality evaluation of the generalisation results are amplified by the presence of human intelligence. This arrangement is illustrated in Figure 4.3.

In the concept of amplified intelligence, the functionality of the generalisation system is not limited to a mere toolbox of generalisation algorithms that can be interactively triggered and parameterised. Instead, the system may also offer complex spatial data models (e.g., triangulated data models) capable of supporting complex generalisation operations, algorithms for structure recognition (e.g., shape analysis), and knowledge based methods. This compares with a simple interactive system, where the user is offered a series of tools that allow editing operations such as moving, deleting, or modifying map objects, as well as some generalisation operators (e.g., simplification of building outlines, aggregation of adjacent buildings) that can be interactively triggered on the map objects. The user will choose the appropriate operations one-by-one under continuous visual inspection of the original map and the resulting map. In an amplified intelligence system, the system should be able to *propose* suitable operations and

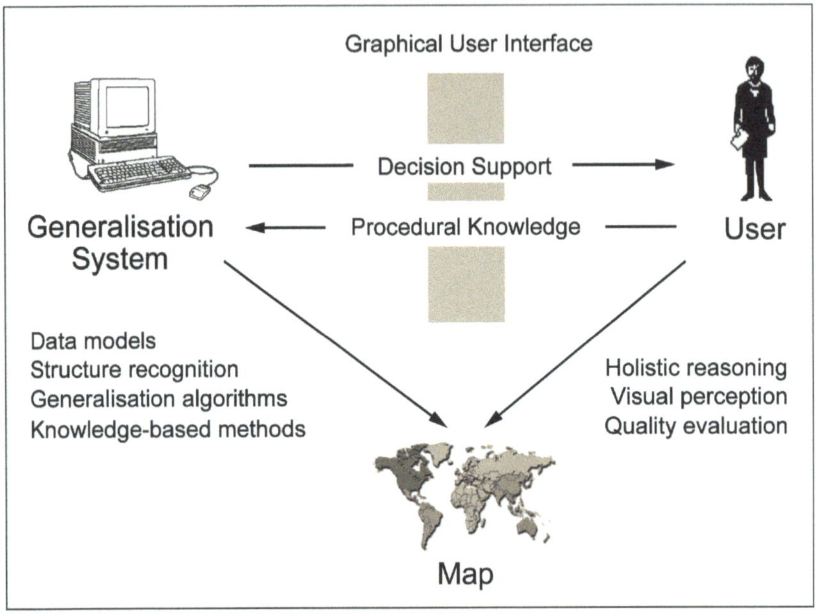

Fig. 4.3. The concept of amplified intelligence as an example of human interaction modelling (after Weibel, 1991).

is possibly even capable of executing entire rule based sequences of operations. For instance, buildings that fall under the size threshold may be highlighted for deletion or exaggeration, and buildings that are too close are highlighted for displacement, together with a possible displacement solution. These are the key differences between an interactive generalisation system and one that supports amplified intelligence (Figure 4.3).

4.3.2. *Implementations*

The software industry has produced a number of interactive software solutions for map generalisation. Software vendors with a focus on general GIS applications (e.g., ESRI, MapInfo and others) usually confined themselves to implementing selected generalisation operators in a toolbox approach (Lee, 1999, 2003). Vendors with a particular interest in the cartography market developed specific interactive generalisation systems. Intergraph started this evolution in the early 1990s with a product called MGE MapGeneralizer, followed in the mid-1990s by Laser-Scan's LAMPS2 Generaliser. Both systems offered a variety of algorithms to implement several generalisation operators, with a focus on operators that consider only single objects. Parameter settings for these algorithms could be controlled via interactive slider bars, allowing visual assessment of the resulting generalisation effect before actually executing the algorithm. Both systems offer the capability of recording appropriate algorithm sequences and parameter settings after tuning them on a small map sample, providing the option to subsequently run them in "batch mode" on entire map data sets.

In the mid-1990s, the Working Group on Generalisation of the OEEPE (Organisation Européenne d'Études Photogrammétriques Expérimentales) conducted a series of tests to evaluate the cartographic performance of generalisation software systems available at that time (Ruas, 2001). The conclusions drawn from the tests, with respect to interactive generalisation systems can be summarised as follows: The systems that were tested worked fairly well for generalising individual objects; results that could be obtained were highly dependent on the skills and experience of the user; the productivity gain was rather low; the amount of interaction in retouching was cumbersome and remained high; important algorithms (exaggeration, line displacement, typification) were missing; maintenance of shape characteristics was often poor; and tools for structure recognition and decision support were generally missing. Although commercial systems existed that would indeed allow generalisation, at least semi-automatically, the OEEPE test revealed problems with these tools. These observations led Ruas (2001, p. 21) to comment "Many algorithms have already been developed to tackle individual generalisation tasks. While this complex task should continue, the most urgent questions address their use: when, on which objects, with which parameter values, in which order?"

The concept of *amplified intelligence* might have offered an interesting alternative to interactive generalisation systems and rule based systems. However, *amplified intelligence* has remained a concept thus far and has not yet been implemented in any production generalisation system. Instead, the limitations of interactive and rule based systems gave rise to the development of a new paradigm in the mid-1990s, namely, constraint based modelling.

4.4. Constraint-Based Modelling

4.4.1. *Basic philosophy*

A generalised map should satisfy several conditions. The main idea behind constraint based modelling is to let these conditions act as constraints in the generalisation process. For example, the generalised map in Figure 4.1 should satisfy constraints such as:

- Major road objects should be smooth and have a level of detail of about 10 meters;
- Minor road should not be represented in the map;
- Building objects with an area larger than 80 m^2 should be represented with a point symbol;
- The building point should not be further away than 10 m from the gravity point of the original corner points;
- Building with an area less than 80 m^2 should not be represented in the map;
- The closest distance between a building object and any other object should be at least 10 meters.

The constraints above are similar to the rules given in §4.2.1. Generally, the distinction between rules and constraints is: rules state what is to be done in a process, while constraints stress what results should be obtained. However, this distinction is not always that clear. The main difference between condition-action and constraint based modelling lies in the way the rules/constraints are used to perform the generalisation. In condition-action modelling each condition is connected to a specific action, but in constraint based modelling the generalisation is ruled by a *synthesis* of conditions (Ruas and Plazanet, 1997), leaving more flexibility in defining the method by which a satisfaction of these conditions (i.e., constraints) is reached.

To use constraints in map generalisation they must have a measure; most often these measures are of interval or ratio scale, but they could also be of nominal or ordinal scale. In order for the generalised map to satisfy all constraints, all the measures must have satisfying values. Most often, however, there exists no generalisation solution that completely satisfies the constraints and the solution must be a compromise. This is due to the fact that many constraints are inherently in conflict with each other. For instance, the constraint that a building symbol must be large enough to be legible may conflict with the constraints that a building symbol should not overlap other symbols and that a building object's position should not change. To find the best compromise there must be a cost connected with the violations of the constraints (computed by means of a set of measures). The *optimal* generalised map is then the solution with the lowest total cost. In this section three methods for finding the *optimal* solution are described: agent modelling, combinatorial optimisation and continuous optimisation.

One should be careful with the use of the word *optimal* here. That a generalised map is an optimal solution, in a constraint based modelling sense, does not mean anything more than that the constraints used are violated as little as possible. This implies that a solution will never be better than the constraints used. To compute a *perfect* generalised map would in addition require that we have a complete set of constraints for the generalised map. Currently, the constraints used for generalisation are limited since they cannot describe all aspects of a generalised map; hence, the generalised map using constraint based modelling is not optimal in a more general sense. In §4.4.2, we take a closer look at some constraints that have been used in map generalisation.

4.4.2. *Typology of constraints*

There are two main requirements that a generalised map must satisfy. Firstly, it must be a correct, though abstracted, representation of a physical reality. Secondly, the map must be readable by the user. If both these two requirements are not taken into account the map will not be of high quality; hence, both *representation constraints* and *map readability constraints* are required.

The whole process of constraint based modelling can be viewed as follows. The original map is assumed to be correct which implies that the representation constraints are satisfied. But the map is too detailed and needs to be generalised; that is, the map readability constraints are severely violated. In the generalisation process the map is transformed in order to decrease the violations of the map readability constraints. This transformation will lead to minor violations of the representation constraints. The final generalised map will eventually be a compromise between good readability and good representation. This can be seen in Figure 4.1, where the generalised map (Figure 4.1c) is an abstracted representation of the original map (Figure 4.1a); yet distinctively more readable than if it was simply scaled down (Figure 4.1b).

Table 4.1. Six categories of constraints and their properties. The constraints consider either single objects, groups of objects or both. The constraints govern representation quality or map readability.

	Consider individual objects	**Consider groups of objects**	**Consider the requirement of**
Position	X (absolute)	X (relative)	Representation
Topology		X	Representation
Shape	X		Representation
Structural		X	Representation
Functional	X	X	Representation
Legibility	X	X	Map readability

A number of researchers have contributed to development of constraint based modelling. Beard (1991a) was probably the first author to explicitly propose the use of constraint based modelling in map generalisation. She argued that it is difficult to formulate generic condition-action rules for all map types and that these rules are inflexible for the diverse relationships between cartographic objects. The advantage of using constraints instead is that they are not bound to a certain action. Instead any action can be applied to resolve the constraints. Furthermore, Beard (1991a) proposed an initial typology of constraints. Also Mackaness (1995a) argued that it is difficult to find deterministic rules for the generalisation process. Instead he proposed a modelling approach based on testing several possible solutions of the generalisation process and an evaluation mechanism for choosing the best alternative. This evaluation process should be based on constraints of the generalised map. Thus, a major difference compared with a pure condition-action modelling approach is the possibility of evaluating different strategies and navigating back in the generalisation process if the result is not satisfactory. Later, Ruas and Plazanet (1997) proposed a model that formalises Mackaness' evaluation strategy. In their approach constraints would not only be used for evaluation of the generalisation result, but also for triggering different algorithms. Ruas and Plazanet (1997) also proposed their own typology of constraints (partly based on Beard's earlier work). Weibel (1997) and Weibel and Dutton (1998) proposed typologies of constraints that are influenced by thematic mapping (e.g., the generalisation of polygonal subdivisions that can be found in geological or land use maps), Harrie (2003) proposed a typology especially for graphic generalisation, while Ruas (1999) defined a comprehensive set of constraints for use in medium scale generalisation of urban settlements in topographic maps. The following discussion of constraints is mainly based on the typology set forth by Ruas and Plazanet (1997) (Table 4.1).

Position constraints

Position constraints are concerned with the movement of objects in the generalisation process. There are two types of position constraints: *absolute* and *relative*. Absolute position constraints state that individual objects should not move in relation to the geodetic reference system, while relative position constraints dictate that distances between objects must be maintained.

Topology constraints

The aim of these constraints is to ensure that the topological relationships in the map should be correct. In many cases these rules are quite simple (e.g. that the connection of the roads in a network must not change in the generalisation process) but in other cases the formulation of the constraints are more complex (e.g. to ensure correct topological relationships after an aggregation transformation).

Shape constraints
It is essential that the characteristics of single objects are maintained in the generalisation process. Several constraints have been proposed for single objects, such as preservation of area and angularity.

Structural constraints
Structural constraints seek to preserve object patterns such as road networks, built up areas, etc. Examples of constraints on groups of objects are: alignment of objects, mean distance between objects and size distribution.

Functional constraints
The purpose of these constraints is to ensure that the generalised map can be used for a certain purpose. For example, if the generalised map is intended to be used for car navigation it is especially important to have a correct representation of the road network.

Legibility constraints
The visual representation of cartographic objects is important. The data must not contain any spatial conflicts, objects (and features within objects) must be large enough and not too detailed, and the chosen symbolisation must conform to graphic limits.

4.4.3. *Agent modelling*

Agent modelling techniques started to be used for map generalisation in the late 1990s. In this section we will only summarise the main points of this approach necessary to compare agent modelling with other process modelling techniques. More details of the technique are given in Ruas and Duchéne, this volume (Chapter 14). The principle idea behind using agent modelling in generalisation is that map objects can be understood as active and autonomous objects, or *agents,* that attempt to satisfy *goals* in order to achieve optimal rendering at the target scale. For instance, if a building object is modelled as an agent, then it may have goals such that it maintains: (1) a minimum size in order to remain legible, (2) a minimum length of its sides, and (3) its position. These goals can then be equated with constraints of the types listed above in §4.4.2. As explained in §4.4.1 each constraint needs a measure to evaluate whether the appropriate threshold value for the constraint is met. To use one of the examples of §4.2.1 a building needs a minimum area of 80 m^2 to be clearly legible at a scale of 1 : 50 000. If this value is not met, then the agent has to have a *plan* to trigger an appropriate generalisation algorithm that improves the situation. For instance, the plan of the building agent could be to first try to enlarge itself, and if that does not help (e.g., because it starts to overlap with a neighbouring object) it could then remove itself (the most drastic action an agent could take).

An agent needs to have *perception* of its *environment*. The perceptive capabilities of cartographic agents can be modelled in several ways. The AGENT project (Barrault et al., 2001) distinguished two types of agents following the approach outlined in Ruas (1999): *micro-agents* which are only capable of evaluating themselves but not perceiving their spatial context (e.g. to detect an overlap with another map object), and *meso-agents* which are responsible for a group of micro-agents and hence capable of analysing the spatial context related to that group. Examples of micro-agents are individual buildings or roads. Examples of meso-agents are city blocks (responsible for the buildings contained within) or road networks. Recently, Duchêne (2004) proposed a new approach called CartACom that still uses micro- and meso-agents, but where even the micro-agents are capable of analysing their spatial environment, able both to perceive potential conflicts such as overlaps and capable of engaging in a communication with other agents to resolve these conflicts.

A multi-agent system requires an engine that controls the evolution of agents in order to let them reach their goals, both individually as well as jointly as a system of agents. Figure 4.4 shows the lifecycle of an agent. Once activated, the agent goes through an iterative process, evaluating its associated constraints for possible violations, trying out plans to remedy potential conflicts, until all constraints have been evaluated, and the best possible solution has been found. For each constraint that is in conflict, available plans are tried out until either the conflict has been resolved or no more plans are available. In the agent engine of the AGENT project this search process follows a depth first strategy which is described in Ruas (1999) and Regnauld (2001b). In this search for an optimal solution, backtracking is possible and plans that deteriorate the agent's state can be cancelled. Invoking this lifecycle on all the micro- and meso-agents eventually leads to a generalisation of the entire map modelled as a system of

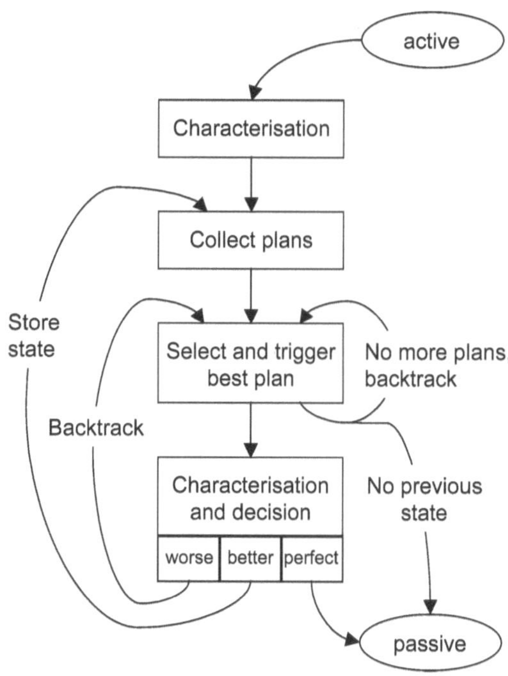

Fig. 4.4. The lifecycle of an agent's evolution (after Ruas, 1999 and Regnauld, 2001b).

agents. Each agent reports its "happiness", that is, the degree to which its goals (constraints) have been met (see Ruas and Duchêne, this volume (Chapter 14)).

An important property of the multi-agent modelling approach is that it is generic. The reasoning mechanism is simple and allows evaluation and evolution of any type of constraint. Also, any type of map object can be handled, individual map objects as well as groups of objects (in hierarchical or non-hierarchical, overlapping or non-overlapping organisations). And finally, any type of generalisation algorithm can be integrated into this strategy via plans. A plan can be a simple geometric operation (e.g. enlargement of a building about its centroid) or a sophisticated and complex generalisation procedure. Due to their versatility, agent modelling has been used in very different application domains such as the generalisation of urban areas and overland road networks. Examples include the use of AGENT (Barrault et al., 2001) in the generalisation of rural areas Duchêne (2004), and generalisation of polygonal maps (Galanda, 2003).

4.4.4. *Combinatorial optimisation modelling*

In combinatorial optimisation an optimal solution, according to a certain criterion, is found in a discrete search space. In generalisation this search space could consist of several possible maps. The combinatorial optimisation process then searches for the best map according to the constraints defined. There exists a trade-off between the computational cost of the search process and the cartographic quality that can be obtained. The more trial solutions that are generated the better the cartographic quality can potentially be – but the more time the search process will require. Clever search strategies are therefore required to converge efficiently towards an optimal solution, allowing a fine-grained discretisation of the search space.

Combinatorial optimisation techniques were introduced to cartography quite some time ago for solving parts of the map-making process. The first use was for map label placement (Zoraster, 1986, 1997); today combinatorial methods are well-established techniques in commercial map labelling packages. In the early 1990s, Cromley and Campbell (1992) used integer programming to formulate line simplification as a combinatorial optimisation problem. Combinatorial optimisation techniques have also been used for object displacement (Ware and Jones, 1998). In Figure 4.5, each object that is found to be in spatial conflict with another object can be displaced to a total of 19 trial positions within the radius of the maximum displacement distance. The trial positions are constrained by a user defined maximum displacement distance. In assessing these trial positions, any position that avoids spatial conflicts with other map objects is permissible, whereby the displacement cost increases with increasing distance from the original position. Hence, positions with small displacements are preferred.

In order to solve the displacement problem shown in Figure 4.5 using discrete trial positions one could use a brute force approach that simply tests each trial position against each other, for each object. However, given n objects on the map, and k trial positions per map object, this would lead to k^n possible solutions, or $19^7 = 893\,871\,739$ different generalisation results even in this trivial case. Fortunately, various standard techniques are available from the literature to implement an efficient search for an optimal solution in a discrete search space. Ware and Jones (1998) used *simple gradient descent search* (SGD) and *simulated annealing* (SA) to implement their feature displacement method. Their experiments showed that SA

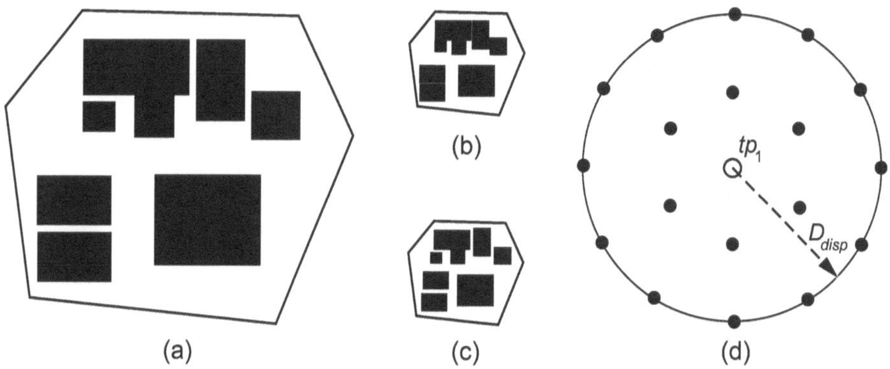

Fig. 4.5. (a) Original map. (b) Map at reduced scale; buildings are too close and congested. (c) Spatial conflicts have been resolved by displacement. (d) Trial positions; tp_1 is the original position of the object in question; D_{disp} is the maximum displacement distance. (After Ware and Jones, 1998.)

clearly outperformed SGD in terms of the cartographic quality of the results as SGD has a tendency to get trapped in local minima during the search process, precluding it from finding better solutions that are globally optimal. More recently, Ware et al. (2003b) and Wilson et al. (2003) have used a different combinatorial optimisation strategy using genetic algorithms, to implement their spatial conflict resolution procedure. Results are similar as for simulated annealing.

Of particular interest in the context of this chapter is the recent extension of Ware et al. (2003a) to use simulated annealing to model a generalisation process consisting of multiple operators rather than a single operator. In their paper, Ware et al. (2003a) generated and tested not only different candidate positions but also different candidate operators, including displacement, enlargement, deletion, and shrinking of map objects (shrinking is used as a last resort if other operators fail). The selection of generalisation operators can be controlled by assigning a cost to each operator. Thus, preference can be given to a particular operator by assigning a low cost compared to other operators. As the results reported in Ware et al. (2003a) show, this approach can indeed be used to model multiple generalisation operators. However, due to the fact that a fixed, discrete number of map realisations are used in the combinatorial optimisation process, the approach appears to be limited to treating discrete map objects, that is, point objects or small area objects. If objects with continuous geometry such as lines or polygons are involved, or if continuous shape deformation (such as in smoothing or displacement of lines) is sought, alternative methods are needed that adopt a continuous approach.

4.4.5. *Continuous optimisation modelling*

A continuous optimisation technique aims at finding the minimum (or maximum) value of an objective function defined on a continuous space. In generalisation this technique has been used as follows (Figure 4.6). First, the set of constraints are chosen and the constraints are formulated as analytical measures. The measures are functions of point locations, where the

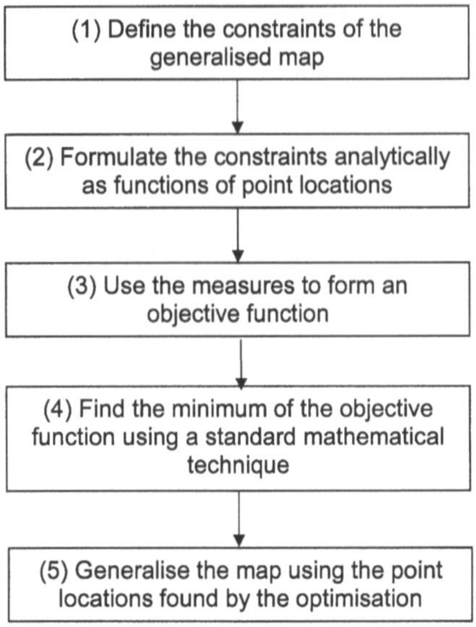

Fig. 4.6. The steps used in continuous optimisation in map generalisation.

points are the vertices that define all the cartographic objects. Then an objective function is formed by the measures. The minimum of the objective function is the best possible generalisation solution. By solving the optimisation problem, using a standard mathematical technique, the optimal solution of the generalisation is found relative to the given constraints.

A number of computational techniques used for continuous optimisation in generalisation have been proposed. Common to all these techniques is that they are based on the steps in Figure 4.6. Some of the techniques are: finite element method (Højholt, 2000), snakes or elastic beams (Burghardt and Meier, 1997; Galanda and Weibel, 2003; Bader et al., 2005; Figure 4.7), or least-squares adjustment (Harrie, 1999; Sester, 2000; Bader, 2001; Figure 4.8). At first continuous generalisation was used for solving only spatial conflicts (displacement). To perform this, constraints such as *the minimum distance between objects* (legibility constraint) and *the allowed distortion of objects* (shape constraint) are used in the optimisation process. Recently, the optimisation process has been extended with more constraints for example including simplification and smoothing operators (Harrie and Sarjakoski, 2002). In general, continuous optimisation methods are particularly suited for generalisation operators that require continuous shape deformation or continuous shifting of map objects, such as feature displacement, smoothing, shape simplification, or exaggeration of objects.

Example of using a continuous optimisation method for generalisation

In this section a brief example of how continuous optimisation can be applied in map generalisation is given. Assume that the visual representation of the building objects in Figure 4.1 has been changed to point symbols (main buildings) or removed (annex buildings). This could be a first step in the generalisation process. A second step would then be to solve the spatial conflicts between the new building symbols and other symbols. In this example the least-squares

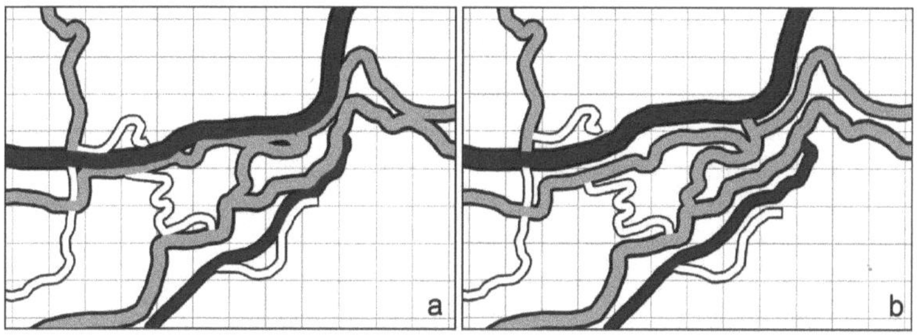

Fig. 4.7. Road displacement using elastic beams, (a) before and (b) after displacement (Source: Bader, 2001, p. 98).

Fig. 4.8. The effect of generalisation when performed by the least-squares technique. Constraints that correspond to the cartographic operators displacement, simplification and smoothing were used. The original objects are shown in half-tone and the generalised in full-tone (Harrie and Sarjakoski, 2002, p. 255). ©Kluwer Academic Publisher, with kind permission of Springer Science and Business Media.

adjustment technique is used for solving these spatial conflicts (other continuous optimisation techniques could be used in similar ways). In this example we only use three constraints (movement, internal geometry and spatial conflict). These constraints are somewhat restricted in what they actually can model, but the aim here is only to present the general idea.

One key issue is to find measures for the constraints. In least-squares adjustment the number of points is invariant, which enables the formulation of the measures on point movements. For the sake of computational simplicity, we restrict ourselves to linear equations; that is, all the measures are of the analytical form:

$$\text{const}_{x1} \cdot \Delta x_1 + \text{const}_{y1} \cdot \Delta y_1 + \cdots + \text{const}_{xn} \cdot \Delta x_n + \text{const}_{yn} \cdot \Delta y_n = \text{const}_{obs}, \tag{4.1}$$

where

$\Delta x_i, \Delta y_i$ are point movements,
const_{xx} are constant values, and
n is the total number of points.

All three types of constraints should be formulated using this general form.

Movement (a position constraint of subtype absolute)
These constraints simply imply that each point should not move. This is enforced by requiring that the movement of each point (i) is zero, or in analytical form:

$$\begin{aligned} \Delta x_i &= 0, \\ \Delta y_i &= 0. \end{aligned} \tag{4.2}$$

Internal geometry (a shape constraint)
The shape of the objects should not change, but it can be translated and/or rotated. To simplify matters here the objects are only allowed to be translated. If two points (i and j) have a common edge in one of the cartographic objects the following relationship should be kept:

$$\begin{aligned} \Delta x_i - \Delta x_j &= 0, \\ \Delta y_i - \Delta y_j &= 0. \end{aligned} \tag{4.3}$$

Spatial conflicts (a legibility constraint)
Objects should not be closer than a predefined threshold (this threshold is dependent on the symbol width and the minimum separation of symbols on the map). If the objects are close then there should be a relationship defined on the distance between the objects. A major difficulty here is to identify the spatial conflicts in the map. One of the most common methods to perform this is to rely on (constrained) Delaunay triangulation. In Figure 4.9, constrained Delaunay triangulation is used to identify the spatial conflicts marked on a map – the dotted line marks edges in the constrained Delaunay triangulation. Figure 4.9 contains a point i that has a proximity conflict with a line segment between points j and $j+1$. It is assumed that the shortest distance (h) must increase by a distance equal to *distance_value* to solve the conflict. In this example only the point i is moved (to the point marked in grey) to solve the proximity conflict; in a normal case all the points would have been moved in order to resolve the conflict.

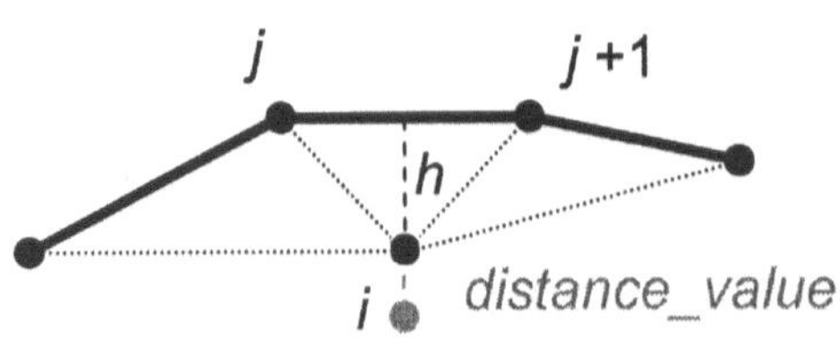

Fig. 4.9. A point i has a proximity conflict with a line segment between points j and $j+1$ requiring movement of *distance_value* to resolve the conflict.

The shortest distance between the point object and the line segment can be formulated as a function (h) of the point locations of the point i and the points j and $j+1$ on the line segment. Then, a measure of the spatial conflict constraint can be formulated as:

$$\left.\frac{\partial h_k}{\partial x_i}\right|_{\bar{\mathbf{x}}} \cdot \Delta x_i + \left.\frac{\partial h_k}{\partial y_i}\right|_{\bar{\mathbf{x}}} \cdot \Delta y_i + \left.\frac{\partial h_k}{\partial x_j}\right|_{\bar{\mathbf{x}}} \cdot \Delta x_j + \left.\frac{\partial h_k}{\partial y_j}\right|_{\bar{\mathbf{x}}} \cdot \Delta y_j + \left.\frac{\partial h_k}{\partial x_{j+1}}\right|_{\bar{\mathbf{x}}} \cdot \Delta x_{j+1}$$
$$+ \left.\frac{\partial h_k}{\partial y_{j+1}}\right|_{\bar{\mathbf{x}}} \cdot \Delta y_{j+1} = \mathit{distance\_value}, \tag{4.4}$$

where $\partial h/\partial x$ denote a partial derivative (computed at the original point locations, denoted by $\bar{x}$), and *distance_value* is equal to how much the distance h must be increased by in order to solve the spatial conflict (Figure 4.9).

The constraints in Equation (4.2) are set up for each point; the constraints in Equation (4.3) are set up for point pairs in rigid objects; and, finally, the constraint in Equation (4.4) is set up for all proximity conflicts. This implies that there are more constraints than unknowns (since the number of unknowns is equal to twice the number of points). All the constraints together constitute an equation system in which the point movements are the unknowns. In matrix form this equation system can be written as:

$$\mathbf{Ax} = \mathbf{l} + \mathbf{v}, \tag{4.5}$$

where

A is the design matrix (containing constants from the left-hand side of Equations (4.2), (4.3) and (4.4)),
x is a vector containing the unknown point movements (Δx_i, Δy_i),
l is the observation vector (containing the right-hand side of Equations (4.2), (4.3) and (4.4)), and
v is the residual vector.

Equation system (4.5) is over-determined (i.e., more constraints than unknowns). Generally an over-determined equation system does not have a unique solution and therefore a residual vector has to be introduced (as in Equation system (4.5)). The size of an element in the residual vector states the degree to which the corresponding constraint is violated. This implies that we could use the residuals to form the objective function (step 3 in Figure 4.6). Normally, we would not consider all constraints to be of equal importance. The internal geometry of an object is, for example, often more interesting than its absolute position on the map. The different degrees of importance can be modelled when forming the objective function by setting a weight for each constraint (p). If weights are used, then the least-squares technique gives the following objective function (g):

$$g(\mathbf{x}) = \sum_{i=1}^{n} p_i \cdot \left[v_i(\mathbf{x})\right]^2, \tag{4.6}$$

where n is the number of constraints. The size of the objective function is a function of the point movements (Δx_i and Δy_i stored in vector **x**). By using a standard mathematical approach the values of the point movements that minimise the objective function are found. And, finally, these *optimal* point movements are added to the points in the original map to generalise the map.

The least-squares method to generalisation, as well as other continuous optimisation methods, can be viewed as follows. Before we start the generalisation all the point movements are equal to zero ($\Delta x_i = \Delta y_i = 0$ for all points i). Considering Equations (4.2)–(4.4) we see that the movement and internal geometry constraints are satisfied at this stage, but that the spatial conflict constraints are violated. The residuals for the latter constraints are equal to the distances the objects must be additionally separated by in order to resolve the spatial conflicts. In general terms the following holds true: the representation constraints (such as movement and internal geometry) are satisfied in the original map, while the map readability constraints (such as spatial conflicts) are violated. By minimising the objective function (Equation (4.6)) the violation of the map readability constraints are decreased at the expense of the constraints for representation.

4.4.6. *Comparison of methods for constraint based modelling*

Agent modelling, combinatorial optimisation modelling and continuous optimisation modelling are equal in the sense that they all are based on constraints. Apart from that, however, the nature of the modelling methods and their application domains are quite different.

Agent modelling is versatile and can, theoretically, handle all types of generalisation processes. If needed other optimisation techniques can simply be built into an agent process as a particular plan. This is demonstrated by Galanda and Weibel (2003) who integrated a snakes-based procedure (continuous optimisation technique) for displacement, as well as enlargement and exaggeration of polygons into a multi-agent system used to generalise polygonal maps. To generalise from Figure 4.1a to Figure 4.1c, a diverse set of generalisation tasks must be performed in a complex overall process. Some building objects must be aggregated into built-up areas, roads must be simplified, dead-ends removed, individual buildings displaced, and so on. Agent modelling can handle such tasks by, for instance, creating meso-agents of building clusters contained in city blocks. However, the success of such a strategy depends on a number of factors. First of all, a large number of constraints need to be accurately defined and measures for their evaluation must be available. Second, meaningful groups of objects that form meso-agents must be determined. And finally, appropriate plans (i.e., procedural knowledge) must be formalised to guide the selection of appropriate generalisation algorithms and control parameters. So, while agent modelling has a high potential to be used for overall generalisation process modelling, there are still a few parts missing from a complete solution.

The success of combinatorial techniques in cartographic label placement stems from the fact that it is natural to determine a fixed number of trial positions for each label. For map generalisation tasks that are amenable to discretisation of the search space (such as those requiring selection, displacement and simplification), combinatorial optimisation modelling has proven to be effective, and relatively computationally efficient compared with other approaches. But there are generalisation tasks where it is difficult to define all possible solutions. In theory, it is possible to define a search space, for instance, for the required building aggregations shown in Figure 4.1, but this would result in an excessively large search space.

To date, constraints associated with continuous optimisation have been defined on point locations. This implies quite strict limitations as to what kinds of generalisation can be performed. Theoretically, the continuous optimisation technique (as used so far in generalisation) is restricted to a rubber sheet transformation (a homeomorphism). Imagine that your original map is drawn on a rubber sheet. The continuous optimisation techniques are then capable of stretching and distorting the rubber sheet map to be more readable according to the requirements (constraints). This implies that (pure) continuous optimisation techniques cannot handle operators such as aggregation, selection and typification. It is, for example, not possible to perform all required generalisation operators in Figure 4.1 by only using continuous optimisation techniques. Yet, for all generalisation operators that are of a continuous nature (as in the rubber sheet metaphor), such as feature displacement, smoothing, simplification, enlargement, or exaggeration, continuous optimisation techniques are capable of delivering convincing results.

Overall, in terms of applicability, agent modelling is the most powerful modelling method followed by combinatorial optimisation and finally continuous optimisation. Of the three approaches, only agent modelling has the potential to model the complete set of generalisation operators. Furthermore, it can integrate other constraint based modelling techniques such as sub-processes, or plans. However, to rank these three modelling methods, it is important to note that the value of the modelling methods is inherently dependent on the type of application. If the required generalisation can be solved by a rubber sheet transformation continuous optimisation is likely to be the best choice both from map quality and computational complexity perspectives. And if it is possible to define combinations of generalisation solutions for each map object (or group of map objects), then combinatorial optimisation techniques is most likely to be the best modelling method. To conclude, the choice of modelling method for constraint based modelling requires that the cartographer has knowledge about both the nature of the modelling methods as well as a firm knowledge of the generalisation situation.

4.5. Conclusions and Outlook

This chapter described approaches for modelling the overall process of generalisation: batch processing, condition-action modelling, human interaction modelling and constraint based modelling. In doing so, we attempted to show that there is a historical evolution between these modelling methods and their predecessors. Condition-action modelling was an improvement over the earlier batch process modelling. However, condition-action modelling, as exemplified by rule based systems, suffers from a severe knowledge acquisition problem (especially regarding procedural knowledge). That was the main reason for the evolution of the human interaction modelling approach. In this model the human is the main contributor of procedural knowledge. However, it turned out that human interaction modelling, or at least the available interactive generalisation systems, did not significantly improve productivity (given the high degree of human interaction), highlighting the need for a new process modelling method. Presently the main focus of research is therefore on constraint based modelling. The potential for this modelling method, in a production environment, has partially been established (Lecordix and Lemarié, this volume (Chapter 15); West-Nielsen and Meyer, this volume (Chapter 16)) but remains to be explored further.

There is still no perfect solution for modelling the overall process of map generalisation. It is conceivable that a new modelling method might evolve that may replace or complement the current constraint based modelling paradigm. However, there are still several improvements that could be made to increase the usefulness and performance of constraint based modelling. The following are some of the areas were we believe further research will be necessary to extend the approach.

Improvements of constraints

To date, the main limitation of constraint based techniques is the limitations of the constraints themselves. Important cartographic constraints still remain to be defined, others are poorly defined. More work is required in formulating constraints and associated measures for groups of objects. Work on pattern recognition and on constraints for preserving cartographic patterns will be particularly important in this context.

Enriching databases

Several national mapping agencies are currently working on establishing links between their cartographic core datasets and external datasets such as building registries, cadastral registries and road information databases. This development will provide more semantic information that could be used to inform in the orchestration of the generalisation process. Additionally, cartographic datasets could be further enriched by integrating information that can be extracted in pattern and structure recognition processes.

Improving agent modelling

The agent modelling approaches used so far in generalisation are still limited. Various improvements could be explored. For instance, the formation of meaningful groups of map objects to build meso-agents (e.g., from clusters, alignments, or topological neighbours) will need to be developed further. Also, models of agent organisation will require further study. Strategies built into the plans of agents will need to be equipped with better procedural knowledge. And the reasoning mechanisms used in agent engines will need to be made more efficient.

Computational efficiency

Many future applications, such as web mapping or location based services, will require real-time generalisation solutions (Sarjakoski and Sarjakoski, this volume (Chapter 7)). The real-time generalisation techniques today only consider quite simple generalisation tasks. Studies should be carried out in the use of more powerful constraint based techniques for real-time applications (cf. Harrie, 2005). Studies into improving the computational efficiency of these techniques will be essential. Possible departure points could be to use more efficient search and optimisation algorithms in agent based systems or in combinatorial optimisation. Such methods can be found in artificial intelligence (for example see Russell and Norvig, 2003), and in mathematics (cf. Papadimitriou and Steiglitz, 1982). For continuous optimisation more studies should be performed in sparse matrix techniques for improving computational efficiency; such methods are common in geodesy and photogrammetry (for example, see Bjerhammar, 1973; Zhizhuo, 1990; Sarjakoski and Kilpeläinen, 1999).

Multiple representations

This chapter concentrated on modelling applications of generalising one dataset into another dataset. There are also other types of generalisation modelling applications such as those needed for updating multiple representation databases. Using constraint based approaches for updating processes of multiple representations would seem like a logical choice, as constraints cannot only be defined between map objects of a single scale but also between multiple levels of scale (Mustière and van Smaalen, this volume (Chapter 6)). Furthermore, constraints could help to ensure database integrity in the updating process.

Web applications

In this chapter we have treated process modelling only from a single system perspective. But there is clearly a trend in which an increasing number of mapping applications are performed in a distributed environment. The Open Geospatial Consortium (OGC) has, in cooperation with the International Organization for Standardization (ISO), recently defined a geospatial service taxonomy (ISO/DIS19119). In this taxonomy a *feature generalisation interface* is listed. To the authors' knowledge, all implementations of this interface published to date are restricted to the level of individual algorithms (see for example Burghardt et al., 2005; Sarjakoski et al., 2005b). This implies, however, that the modelling of the generalisation process is the responsibility of the client. In the future, new interfaces might be developed that specify the generalisation tasks at a higher level (e.g. using constraints) and then the modelling will be a responsibility of the server side.

Quality evaluation

Evaluation of quality is an important aspect of map generalisation and has also been a driving force in the evolution of modelling methods. One weakness of the earliest modelling methods, batch-mode modelling and condition-action modelling, were that they lacked an evaluation mechanism. In the human-interaction modelling approach the human could perform the evaluation and guarantee the quality of the generalised map. The constraint based modelling approaches perform, to a certain degree, a self-evaluation. That is, rather than blindly performing generalisation actions (as in condition-action modelling), the generalisation process is monitored by monitoring the satisfaction of constraints. The level of constraint satisfaction in the final generalised map can be used as an indicator of success. This approach is quite appealing, but it also has some drawbacks. The constraints are used both for controlling and evaluating the generalisation process. Since the constraints do not describe all the aspects of the generalised map, a constraint based approach to evaluation might give an overly optimistic view of the final map quality. Mackaness and Ruas, this volume (Chapter 5), discuss in more detail the issues concerned in evaluating the quality of generalisation solutions.

W.A. Mackaness, A. Ruas and L.T. Sarjakoski (Editors)

Chapter 5

Evaluation in the Map Generalisation Process

William A. Mackaness[a], Anne Ruas[b]

[a]*Institute of Geography, The University of Edinburgh, Drummond St, Edinburgh, Scotland, EH8 9XP, UK*
e-mail: William.mackaness@ed.ac.uk
[b]*Institut Géographique National, COGIT Laboratory, 2-4 av. Pasteur, 94165 Saint-Mandé, France*
e-mail: Anne.ruas@ign.fr

Abstract

In this chapter we discuss the nature of evaluation in the context of automated solutions to map generalisation and highlight the challenges of defining evaluation criteria. Considerable progress has been made in defining and incorporating evaluation criteria in systems that generalise maps over moderately small changes in scale. We seek to give an idealised view of how evaluation methodologies might be incorporated into generalisation systems. Examples are given that illustrate these ideas. For larger changes in scale, it will be necessary to evolve these techniques, and incorporate evaluation techniques linked to pattern analysis and a more synoptic evaluation of the success of the design solution overall. The chapter highlights the challenges of developing and prioritising evaluation criteria, and where in the process they can be applied. Given the complexity of the task it is important to consider the role of the human in the evaluation process.

Keywords: cartographic evaluation, spatial analysis, qualitative reasoning, evaluation criteria, types of evaluation, contextual generalisation

5.1. Evaluation in the context of Map Generalisation

Evaluation is the process of examining a system to determine the extent to which specified properties are present (Foldoc, 2005); it means we can determine the value or worth of something. For example, we might ask: In reducing the number of coordinates in a file, have we managed to compress the file whilst minimising the loss of information? We might then ask: What amount of information loss is tolerable? What processing effort was required to achieve this? Does the algorithm work in all circumstances? Can it be applied to very large datasets? When we think of evaluation from a map generalisation point of view it tends to take a more graphical perspective. We might ask: have we reduced the detail sufficient that more regional patterns are discernable? Is the map aesthetically pleasing? Have we conveyed the maximum number of ideas in the shortest space of time with the least amount of ink (a goal of design suggested by Tufte (1983))? Does it reveal or conceal information inherent among a set of abstracted data? Most importantly does it enable the user to succeed in their given task (such as exploring, route following, observing)? These many aspects of evaluation can be variously grouped under three headings: algorithmics (focusing on measuring the efficiency and robustness of generalisation algorithms), informatics (information handling including user specification and integration in the evaluation phase) and cartographic evaluation (optimising the presentational form by modelling symbolisation, abstraction and aesthetics).

There are numerous benefits to incorporating evaluation methodologies throughout the spatial data handling cycle (from capture through to presentation). In particular:

- developing of autonomous systems: the creation of systems capable of self evaluation, and comparison between alternate solutions;
- directing human intervention: directing the user in specific tasks in order to make effective use of their time (for example, directing the user to regions of the map known to contain poor generalisation solutions);
- enabling quality control: as part of an audit trail revealing the reasoning behind different decisions made throughout the process;
- supporting the educational process: users being able to discern when and where decisions were made and the basis for those decisions (for example, illustrating when and where different types of evaluation were applied in the map generalisation process);
- enabling consistency in batch processing: consistency in application of processing techniques (for example, being able to verify consistency in the application of generalisation techniques in map series production).

For each of these tasks, we can define many criteria against which we could evaluate any given activity; however the greater challenges are in (1) formalising and prioritising those criteria, (2) knowing when to apply evaluation techniques, and (3) how to record and convey, at the appropriate moment, evaluation parameters and decisions to the user. It is important to acknowledge the close linkage between evaluation methodologies and issues of data quality (Goodchild and Jeansoulin, 1998), specifically relating to a map's fitness for use (Guptill and Morrison, 1995). In this chapter we discuss these matters in following manner: §5.2 starts with an overview of research work on evaluation techniques for generalisation. Section 5.3 discusses the importance of data structures able to support analysis and evaluation, and §5.4 explains the difficulty of calibrating evaluation criteria. Section 5.5 proposes a formal mathematical framework of evaluation, incorporating ideas of levels of detail for evaluation and a synthetic classification of measurements. Finally §5.6 describes the different types of evaluation in the context of recent research work.

5.2. An Overview of Developments in Evaluation Techniques

5.2.1. *Analysis, synthesis and evaluation*

A review of evaluation techniques illustrates the spectrum that have been variously developed – the lion's share falling under the heading of "cartographic evaluation" techniques. They range from simple metric and topological measures, through to more complex ideas relating to the measurement of changing information content with scale. Automated evaluation, closely linked to cartometric analysis techniques, is central to the successful development of

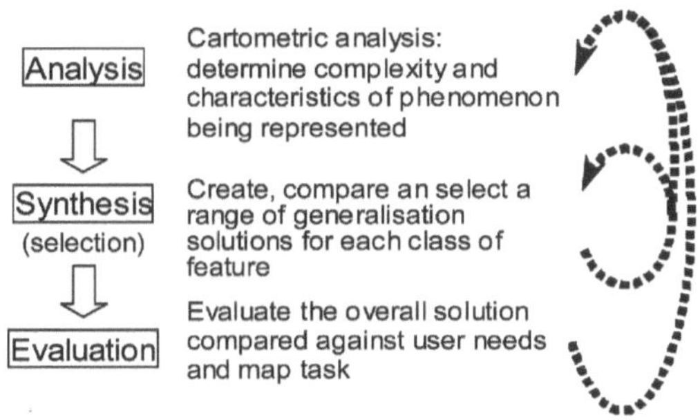

Fig. 5.1. The role of analysis and evaluation in the map generalisation process. The feedback arrows reflecting the iterative nature of the whole process.

autonomous systems. Generally speaking, evaluation begins with an analysis phase. This is a precursor to deciding what actions should be taken to improve the map in anticipation of a change in the level of detail or scale of representation. We need to know about the complexities and interdependencies inherent among the phenomenon being represented. This analysis provides a basis from which the most appropriate solution can be selected. This is synonymous with Brassel and Weibel's (1988) idea of structure recognition and process recognition. The second stage might be to then choose a solution from a variety of candidate solutions. The final stage is to evaluate the success of those changes that arise from the generalisation process, both at the fine scale, and more broadly across the map as a whole. The solution could be compared with paper maps of an equivalent scale, or from a task defined perspective. This idea is summarised in Figure 5.1. The intrinsic link between analysis and evaluation is reflected in the fact that the analysis phase often provides the benchmarks against which evaluation criteria can be compared.

A great deal of information is inherent among the set of objects that comprise a map. Generalisation technique make changes to the map within a competing set of design goals. Consequently a very broad set of techniques are required to measure and evaluate each of these changes to ensure optimal compromise within those design constraints. It is not surprising therefore to note that a large and valuable collection of work on evaluation techniques exists under the headings of spatial analysis, cartometric analysis, graph theory, network modelling, pattern analysis and description, spatial tessellations, metric and topological measures. Here we summarise some key research in the field of evaluation techniques.

5.2.2. *Research in the field of evaluation techniques*

Evaluation for assessing the appropriateness of content

Some of the earliest work relating to modelling content was done by Töpfer and Pillewizer (1966). They proposed an equation that enabled content thresholds to be calculated based on scale (Dutton, 1999), though their approach did not enable the identification of which features should be generalised in order to achieve that threshold. Another methodology for determining content levels was proposed by Mark (1990) who suggested that competition for space be used as a paradigm for controlling map content. The minimalist design of Beck's Map of the London Underground (Garland, 1994) demonstrates the weakness of this approach however.

Other researchers have proposed information theory/entropy modelling as a basis for modelling the changing levels of content that arise as a result of map generalisation (Bjørke, 1997). It is extremely difficult to measure information content in a map; it varies with symbology and varies depending on the interpretation skills of the user. A symbol is imbued with meaning (for example, a simple dot with the word "London" next to it captures the notion of city) but may require additional context to support the process of interpretation. Knowing how much or how little geographical information to include in the map will depend on the task, and the user's knowledge, experience and expectations of what constitutes a "good map" (João, 1998; Monmonier, 1991b).

Evaluation as a basis for triggering generalisation

McMaster and Shea (1992) defined and listed a set of "Spatial and Holistic measures" as part of the "cartometric evaluation" process. The list is composed of: density, distance and distribution measures, length and sinuosity measures, shape and gestalt measures. In this case the evaluation process begins with symbolisation of the features followed by scale reduction without any changes to the map objects. The scale reduction process results in a number of design problems and acts as a "trigger" for map generalisation. The need for map generalisation arises from a number of consequences of that scale reduction – namely congestion, coalescence, complication, inconsistency and imperceptibility (McMaster and Shea, 1992). Analysis techniques are required to identify the severity of these five consequences – evaluation techniques are required to assess the degree to which application of generalisation techniques have resolved them. We might call this a bottom up approach, one that reflects a cartographic driven approach to generalisation. An alternate approach might be to select the intended scale and theme, and on this basis select the phenomena, and their representational form appropriate to that scale. In effect, this alternate approach which is a top down approach acknowledges the consequences of scale reduction and attempts to bypass them by choosing representational forms appropriate to the scale (level of detail) and task. In general terms, the bottom up approach works well for relatively small changes in scale, and the top down approach works better over large changes in scale, where the changes in content are much more fundamental. Irrespective of the approach taken, evaluation requires the system (1) to gain a clear picture of the user's requirements (e.g., how will the map be used, what level of precision is required), (2) convey back to the user the methods used and their degree of application in deriving the generalised solution.

Interfacing issues and evaluation

Thus there are important interfacing issues relating to evaluation – identifying what needs to be done, and conveying the strategy by which a solution is reached. Therefore complex sets of quantitative parameters need to be collectively conveyed in easy to assimilate forms. This was the objective behind work undertaken by Mackaness (1991) who proposed the use of rose diagrams in order to compare generalised solutions, not in terms of changes in content, but in terms of the quantity and the degree of generalisation operators applied.

Evaluation for parameter setting in generalisation algorithms

Evaluation techniques continue to be developed; in some cases evaluation takes place after generalisation, while in other situations, evaluation criteria are intrinsic to the incremental application of a particular method. For example, Cheung and Shi (2004) developed a model for assessing positional uncertainty in a simplified line, and used this to determine parameters to a line simplification algorithm. Edwardes et al. (1998) measured changes in polygon topology during generalisation and used this measure as a "cut off" in an algorithm that iteratively increased the parameter value of the Douglas–Peucker algorithm applied to polygon simplification. Similar methods to choose a parameter value for Gaussian filtering for generalisation can be found in Plazanet (1995, 1996).

A template for describing and sharing evaluation criteria

Given the breadth of evaluation techniques that continue to be developed, it seems appropriate to devise a generic template by which evaluation techniques can be described, ascribed, and shared thus avoiding duplication of effort among the many disciplines contributing to work in this field. This was the task undertaken by the OEEPE working group on generalisation during 1993–1996 (Ruas, 2001). The working group defined two templates: one to describe the generalisation algorithms, and one to describe the measuring algorithms. In this last template, a classification based on the work of McMaster and Shea (1992) was used. Each measuring algorithm could then be classified according to the criteria that it evaluated: remoteness, size, orientation, convexity, concavity, sinuosity, orthogonality, elongation, compactness, density, proximity, singularity, regularity, repetition, symmetry, distribution, complexity, and homogeneity. The measuring algorithm template is composed of a set of fields for describing each measure. The first field was a reference, followed by a description of the quality being measured (brief description, a classification), a mathematical description, the input parameters (pre-processing required, range of value), the output (the properties, specific cases where the measure fails to detect the quality it is intended for), the purpose (application and usage, the point of use during the process of generalisation), the algorithm (its complexity, possible improvements), and illustrations of the output. An example of how the template was used is given in Table 5.1 (in this case a method to measure the congestion of a building surrounded by a set of buildings (Ruas, 1999, pp. 172–175)). During 1997–2001, the OEEPE working group on generalisation was able to perform a series of tests in order to review and compare generalisation solutions among existing GIS platforms (Ruas, 2001). The first step was to define a list of conflicts. A list of 62 types of conflicts were chosen (Ruas, 2001, p. 27), structured in the following way: Conflicts of granularity, size, shape, proximity, overlapping, relative position, absolute position, and density. These conflicts could be in or between one line, one polygon, a set of lines, one line and one polygon, one line and one symbol, a set of polygons or a set of symbols.

Evaluation within agent modelling

More recently the AGENT project provided considerable impetus to the further development and integration of these and other measuring algorithms (AGENT, 2000). A consortium of organisations was led by the French National Geographic Institute (Institut Géographique National, IGN) with funding from the European Union (EU) under the ESPRIT programme. The AGENT project sought to develop autonomous solutions to map generalisation based on agent-based methodologies (AGENT, 2000; Ruas and Duchene, this volume (Chapter 14)). Sophisticated evaluation techniques were critical to the success of the project. A summary of those algorithms is published online in a number of reports (AGENT, 2000) AgentDC1,

Table 5.1. Example of the description of a measurement of congestion (Agent DC4, 1999).

Tool type	Measure/Instantiated structure
Tool name	Building-Congestion
Level	“Micro” level – among a group of buildings
Location	Initial analysis phase, and post application of generalisation methods
Preprocessing	Proximity computed by means of Delaunay triangulation
Input data types	Micro within a “city block”
	Building
Concept	The aim is to quantify the proximity of an object in different directions and to evaluate the degree of congestion – prior perhaps, to removal or displacement.
Short description	1 – The proximity between the building and its neighbours is computed from Delaunay triangulation. 2 – The smaller the proximity value, the greater the congestion. Congestion is calculated radially from each building, using a rose structure (*a disk shared into n equal sectors*). 3 – A propagation function allows us to model the knock-on effect on the neighbouring rose sectors. 4 – If a part inherits from different congestion vectors, the largest value is preserved. A congestion value within a part is between 0 and 1. As a consequence each orientation sector of the rose has a value (from 0 to 1) that represents the congestion in this direction. The figure below shows the congestion of four buildings. 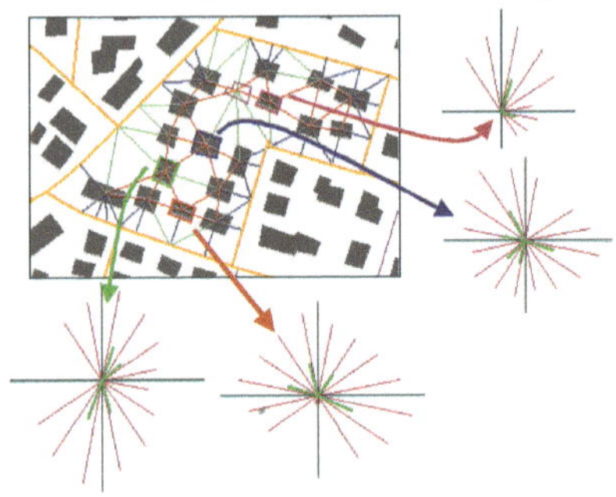
References	Ruas (1999, pp. 172–175)
Output data types	Two possible outputs: 1 – a rose structure: each part of the rose has a congestion value [0, 1], 2 – two global values: average congestion = sum (part-congestion)/total number of parts, freedom ratio = number of free parts/total number of part.
Parameters’ significance	Three parameters are used: – dist-max: to compare congestion either within a block or between blocks, – number of radials of a rose: default value = 16 parts, – maximum angle for knock-on effect – Default value = 45°.
Present state	Implemented on the Stratège Platform (LISP + SMECI).
Current use?	Used for Building removal in case of high density.
Drawbacks	IF the dist-max is too large OR the number of part is too small OR the propagation angle is too large, THEN all directions are congested.
Possible improvements	1 – The angle of propagation could depend on the distance (the greater the distance, the smaller the angle). 2 – A more accurate congestion could be computed from angles between buildings, while integrating distance variation.
Similar tools	Voronoi gardens (Hangouet, 1998) could provide similar results.
Remarks	Could be used for displacement purposes.

AgentDC4 reports (1999). This project made extensive use of this generic template reflecting the endeavours of the IGN COGIT research laboratory. The measuring algorithms were classified according to measures of internal structure, and "between object" measures (Barillot, 2002). Internal measures included position, orientation, shape, and internal topology. Inter object measures included relative alignment, topological measures (such as containment, connectivity, contiguity) and gestalt properties such as patterns of dispersion and distribution. Here again the measuring algorithm template (as illustrated in Table 5.1) facilitated the sharing and comparison of these evaluation techniques.

5.3. Computational Structures to Support Analysis and Evaluation

Van Smaalen (2003) argues that any given geographic phenomenon can be classified under three headings: network features, areal features forming exhaustive tessellations of map space, and discrete, relatively small rigid structures which are typically anthropogenic in form. Methods of analysis and evaluation criteria are required both within and between each of these classes. Often computational structures are required to model and make explicit the various characteristics that define each of these three phenomena types. For example, Figure 5.2 illustrates how a range of techniques can be used to make explicit properties inherent among a set of discrete points: neighbourhoods, densities, shapes formed by a set of points, relative angles, distances, and the shortest path connecting a set of points. Examples of research that utilises these structures include work by Regnauld (1998) who developed techniques for extracting properties inherent among groups of buildings based on the minimum spanning tree (Figure 5.3). Ruas (1999) used Delaunay triangulation to model proximity between blocks of buildings (Figure 5.4 left), and in combination with other techniques, to create measures of congestion (Figure 5.4 right). Such information can be used to control optimal directions of movement in displacement algorithms or to select the best building to remove in cases of high building density. Peng et al. (1995) and Regnauld (1998) demonstrate how size and shape of the Delaunay triangles can be used to reveal patterns of distribution and the grouping of features.

A similar set of structures exist for modelling the topology of networks. Graph theory can be used to characterise many qualities in a network (Heinzle and Anders, this volume

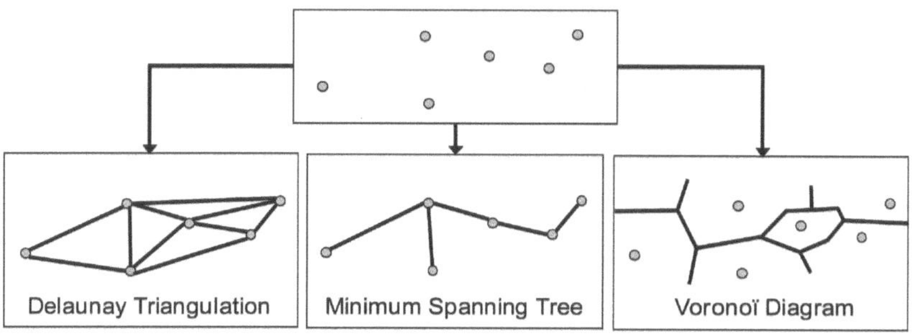

Fig. 5.2. Using a variety of techniques to make explicit the characteristic structures among a set of points.

Fig. 5.3. Groups of buildings "connected" together based on the concept of the minimum spanning tree.

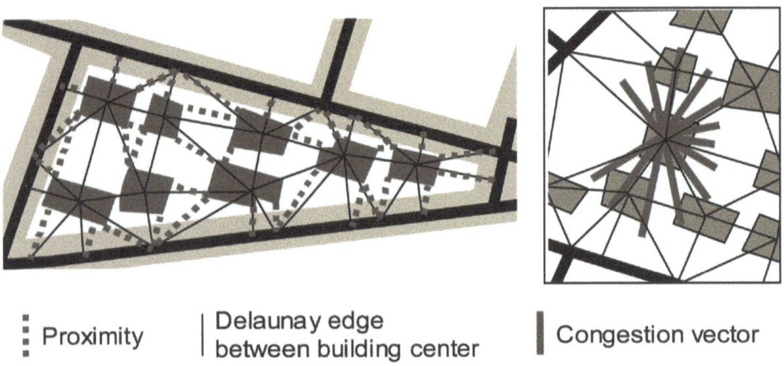

Fig. 5.4. Modelling building proximity and congestion used Delaunay triangulation techniques (Ruas, 1999).

(Chapter 12); Thomson and Brooks, this volume (Chapter 13)). Mackaness and Mackechnie (1999) also demonstrated the value of graph theory in generalising road junction structures of various size and scale, and in characterising and modelling the changes in topology of a network as it is simplified.

Instead of connecting objects or places by means of a graph, sometimes it is necessary to partition the space into cells. Voronoï is a tried and tested structure used to create a partition of cells between points or even lines. As an example, Hangouet (1998) used a Voronoï diagram created from streets and building boundaries to analyse built up areas and to typify buildings.

Though the computational overhead of these underlying structures is high, it is argued that they are essential to the analysis and evaluation of generalisation solutions. The reason being that if you want to conserve or give emphasis to any given characteristic trait of a phenomenon, it is first necessary to discern what those characteristics are. This is equally true for the characteristic relationships that exist *between* phenomena. In the absence of such structures, one is forced to rely on the human eye for appraisal – something that can be tedious and inefficient.

5.4. The Calibration of Evaluation Criteria

Assuming the existence of a large library of evaluation measures, a number of outstanding issues remain. Any given solution is considered to be a compromise between clarity, providing sufficient context, and conveying the salient message among selected phenomena. Therefore a method is required by which we can calibrate and prioritise evaluation criteria according to a notional scale and intended theme. Where it is hard for human cartographers to articulate preference, researchers have invariably resorted to comparison of output against paper maps as a basis for measuring the success of their algorithms and as a basis for "tuning" their solutions. In many instances the desired outcome has been identified through the evaluation of hand drawn paper based products, and the algorithm designed and parameterised to produce a result that mirrors that solution. The output is then presented to human cartographers who, in qualitative terms, somewhat instinctively and subjectively, comment on the quality of the solution. This was the approach adopted by Downs and Mackaness (2002) in the development of techniques for categorical generalisation of geological maps. The argument for evaluating output against paper maps being that a paper map reflects many years of perceived wisdom and iterative advances in production techniques. In trying to emulate the paper map however, we must be sensitive to the suggestion that we are using advanced digital technologies to create nothing more than cartographic kitsch – maps designed to appeal to popular sentimentality that fail to acknowledge the impact of technology in the way we visualise and interact with geographic information.

This begs the question: what alternatives exist to comparing output against paper products? How might we measure the transformations that arise during the generalisation process? At the detailed level we might measure the amount by which a feature has moved or changed shape. At a more general (or collective) level we might ask if we have successfully transformed the data such that we can now see different patterns, and relationships that characterise the features when viewed at a more synoptic level. Here lies a conundrum at the very heart of evaluation – we view maps at different scales precisely because we seen different information – some sort of transformation has taken place. It is a requirement – indeed an expectation, rather than an unwanted byproduct of the generalisation process, that content changes. For example, by removing or reclassifying features, we change the topology of the map. The map may contain different classes of features (represented by different data type) that have different topological relationships within and among other classes. What form of evaluation criteria do we use in these contexts? We observe that "detailed" topology is lost, but that the more "general" topology is indeed preserved. For example, buildings within a city may be amalgamated, parts of the road network pruned, but at a more general level, the juxtaposition of cities and their relationship with inter city motorways remains unchanged. Therefore it would seem that evaluation in topological modelling needs to be applied differentially; that we need hierarchical topological modelling techniques capable of identifying and differentiating broad topological patterns from the topology discernible at the fine scale. In this context generalisation is about compromise – losing fine topological detail, and by doing so, giving emphasis to broad scale topological information and patterns. From this we conclude that to model compromise we need techniques both for modelling fine scale changes and broad "gestalt" level changes. Furthermore in defining evaluation criteria that assess both the impact of a generalisation technique at the fine scale/individual level, and at the broad synoptic level, we need to incorporate the idea that maps fundamentally change over scale (an idea encapsulated in Muller's "conceptual cusps" (1991)). It is clear that much remains to be done in this area. In the following section we present a formalisation framework (§5.5) and current research work (§5.6) that enables us to describe more precisely the changes at different levels of analysis and according to different criteria.

5.5. A Formal Framework of Evaluation

Visual assessment of solutions generated by generalisation systems is subjective and qualitative. One approach to evaluating the quality of solutions is to ask the user to grade the quality of the solution, from say "excellent" to "very poor" (Figure 5.5). On a sliding scale this can be readily translated into a numerical value, from say 5 to 0, which is of a form more readily modelled within evaluation systems. From a system design perspective, the task is then to build evaluation functions that compute equivalent numbers, from 5 to 0, at both the global and the local scale and to relay these values back to the user. Of course, grading a generalised map with a single value is very complex if not simplistic; the aggregation of very different concepts into a single value is problematic, and to communicate to the user the meaning of this term raises all sorts of problems. But a *divide and conquer* approach to evaluation helps to evaluate different aspects that can at least be computed.

The challenge is to break down the evaluation process into a set of values that are automatically computed and that remain meaningful to the user. This process forces us to formalise an evaluation methodology and to define just what is meant by good generalisation. Beyond

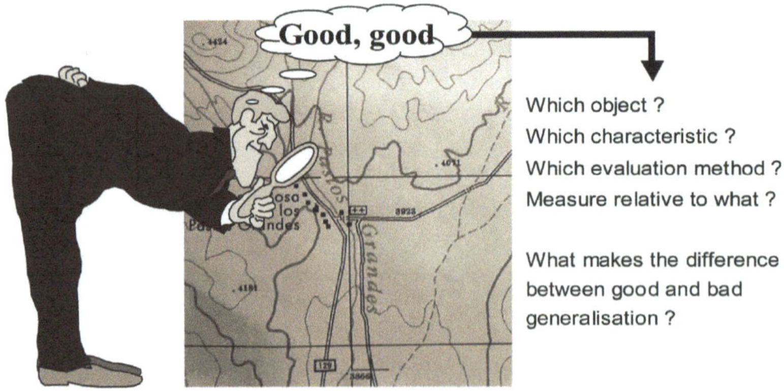

Fig. 5.5. From human qualitative statements to computed values.

the idea of "being fit for purpose", there are general expectations that are assumed by the user. For example the user assumes that the solution will be clear, make effective use of symbology and take account of the resolution of the screen display when providing a solution. Additional to these implicit assumptions, there are explicit needs that relate to how the user will use the data. The challenge is in building up a detailed set of requirements from a simple user request. Is it sufficient for the user to simply specify the task of the map, or does the user need to specify the theme, together with the level of detail (scale) required?

As Borges or Magritte might argue, a representation is not itself reality, but seeks to reflect that reality – seeking to simplify the message (Foucault, 1983). In the process of creating that representation, we must avoid the accidental introduction of errors, since the generalised data are a representation by which the user constructs their understanding of the real world (Figure 5.6). If the representation is false – is badly generalised – the user's mental representation of the real world will be wrong. Any automated solution has to deal with the complexity inherent in the phenomenon being represented and seek a compromise among a multitude of objectives. That compromise had to avoid what MacEachren and Ganter (1990) refer to as Type I and Type II errors. In this context a Type I error would be where information such as a pattern or association is conveyed but that does not, in reality, exist. A Type II error is where the solution fails to convey the essential characteristics of a feature. There are many points throughout the generalisation process where Type I and Type II errors may arise. The idea of cartographic licence (where some distortion is allowed in the interests of clarity) is all about solutions that avoid Type I and Type II errors. There is something of a fuzzy boundary between what is acceptable and what is not – a boundary that "shifts" according to both the task (scale and theme) and the interpretive skills of the user. It is this issue that raises so much debate among cartographers when asked to review automated solutions!

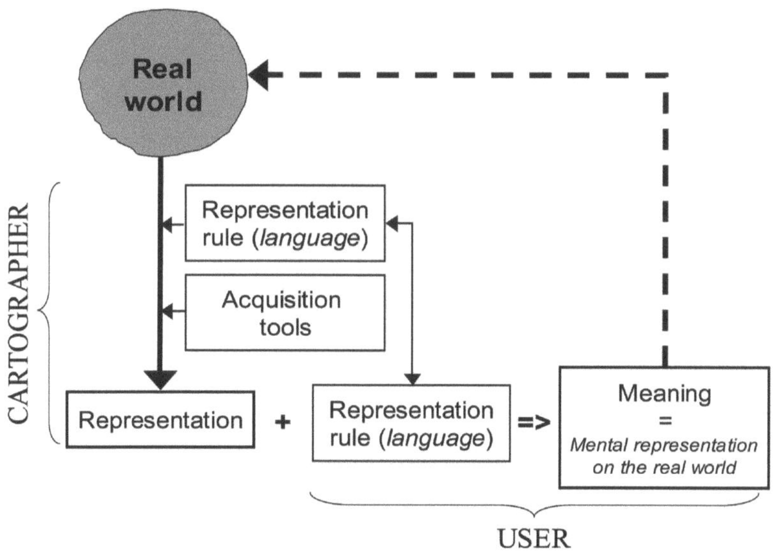

Fig. 5.6. Reality, representation and meaning.

The evaluation process must include assessment of both the implicit and explicit needs of the user. As a consequence, the evaluation functions should incorporate:

- a representation of the "real world" to ensure that the generalised data still "looks like" reality;
- a representation of the user's needs in terms of the level of detail and information content required;
- a schema of the representation rules used including those coming from cartography (to ensure that the map is legible) and others coming from the data base (to ensure that the data base is coherent).

These ideas are summarised in Figure 5.6: The representation (in analog or digital form) is a generalised view of the reality created by means of acquisition tools and specific representation rules. The representation can be understood only if it follows graphic representation rules known by the reader. This representation should be such that it allows the reader to build a faithful mental representation of that part of the world.

The easiest way to represent the real world is to take non-generalised data as the reference and to check that the generalised data still resembles that reference. The assumption is made that the non-generalised data have an acceptable quality and that it reflects reality at a specific level of detail. Moreover we know that it is not possible to simply evaluate the generalised form of each individual object but that the generalisation process requires us to consider the context in which the entire process takes place. As a consequence any formalisation of the evaluation process requires a reference frame that links generalised output to the real world and enables quality assessment in the context of the overall objective of the map. In the following section, we present and extend a model first proposed by Bard (Bard, 2004a, 2004b; Bard and Ruas, 2004). It is based on the qualitative description of the links between the generalised and the non-generalised objects taken as reference objects.

5.5.1. *Formalising the evaluation process*

As a precursor to later discussion, we first formalise the process of generalisation. Let us name Og_i as representing one or several generalised object (where $i \in [1, n]$). O_j is one non-generalised object and (O_j) (where $j \in [1, p]$) represents several non-generalised objects. The group of objects (Og_i) is the result of the generalisation of (O_j), and conversely (O_j) can be viewed as being the *source* of the generalised objects (Og_i). The following notation is used:

- $(Og_i) = Generalisation(O_j)$,
- $(O_j) = Generalisation^{-1}(Og_i) = Source(Og_i)$.

The process of aggregation and elimination, means that n and p, in most cases are not equal $(n < p)$ (Figure 5.7).

In generic form, an evaluation function **f** could be defined as follows:

- **f** is applied to generalised objects: to one object, a part of one object or several objects (Og_i);
- **f** may integrate non-generalised related objects (O_j) in order to compare the generalised objects with the original;

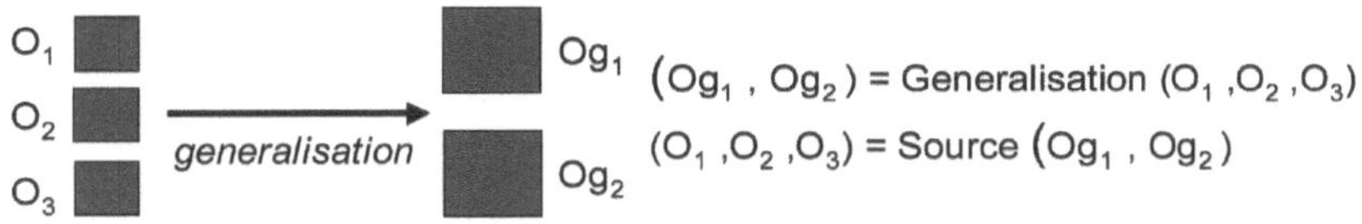

Fig. 5.7. Relationships between non-generalised and generalised objects.

- **f** applies to one specific characteristic (for example the shape, position, or density of several objects) measured by a specific algorithm which we call measure-k. measure-k computes the value of character-k, on one or several objects (e.g. the size of a building, the density of a group of buildings);
- **f** incorporates thresholds that represent the required amount of generalisation or the Level of Detail (LOD);
- **f** returns a quality value reflecting the quality of the solution that is both understandable to the system and the user;
- the returned value lies between a minimum and a maximum value (the function is bounded). It may be normalised or qualitative (for example 6 integers from 0 to 5 or ordered words from *very poor* to *excellent*).

$$\begin{aligned} Evaluation &= Quality_{\text{character-}k}(Og_i) = \mathbf{f}_{\text{measure-}k}\big((Og_i),(O_j),LOD\big) \\ &= \mathbf{f}_{\text{measure-}k}\big((Og_i),Source(Og_i),LOD\big). \end{aligned}$$

Consider the following example. Let us evaluate the size and the shape of a generalised building Og_8 which is the generalised representation of the building O_4. The minimum size is assumed to be 400 m^2 measured by the function $\mathbf{f}_{\text{area}}$. The shape is measured by the function $\mathbf{f}_{\text{elongation-1}}$ and the theme or other user requirements require that elongation should not change more than 10% from the original (i.e. 0.1). It can be formalised in the following way:

- $Quality_{\text{size}}(Og_8) = \mathbf{f}_{\text{area}}(Og_8, 400)$,
- $Quality_{\text{shape}}(Og_8) = \mathbf{f}_{\text{elongation-1}}(Og_8, O_4, 0.1)$.

Research and empirical analysis are necessary to identify the appropriate forms of measurement (measure-k) as well as the thresholds that ensure translation to the required level of detail. Section 5.6 illustrates some recent research work in this domain.

5.5.2. *Evaluation at varying levels of detail*

Usually, a process of evaluation consists of measuring a distance between the data being evaluated and a reference. The term nominal ground ("NG" in Figure 5.8) is the reference, and performing an evaluation consists of measuring the errors and imperfections (Δb in Figure 5.8) – i.e. the differences between the generalised data and the nominal ground. The Nominal Ground is defined as a "A theoretical view of the reality defined by the specification of the geographic dataset and forming the ideal geographic dataset to which the actual geographic dataset will be compared for evaluating its quality. The nominal ground is considered to represent the true value for the values contained in the geographic dataset" (ISO, 1996). When

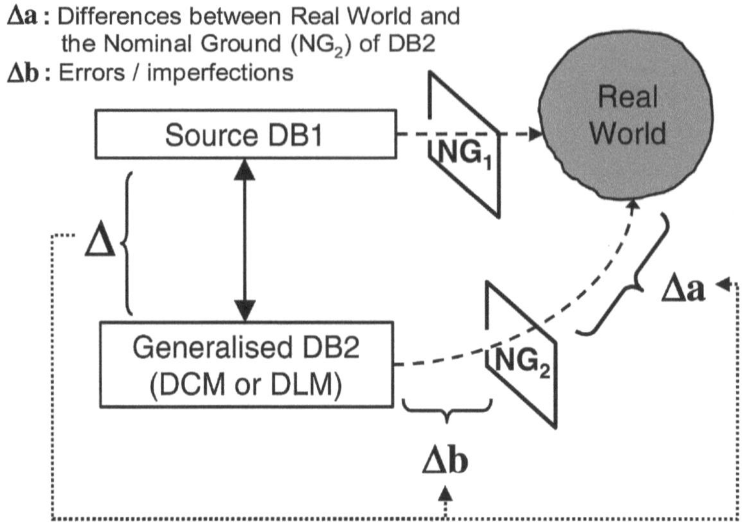

Fig. 5.8. The generalised data (DB2) are compared with their source data (DB1).

we evaluate generalised data, we have no such reference: unfortunately the nominal ground is unknown. This makes evaluation of any solution particularly problematic. The only references we have are the non-generalised data ("DB1" in Figure 5.8).

As a consequence, instead of measuring the errors ("Δb" in Figure 5.8), we measure Δ the differences between DB1 and DB2 and we check if these differences are coherent with respect to the change in the level of detail.

- We do not have the "classic evaluation problem" (being able to ground truth a subset of the data). This would require selection of the appropriate choice of samples, data matching and the choice of the extrapolation function. Instead we have all the data before and after generalisation and we measure and interpret the differences between both data sets.
- The drawback is that the reference is highly detailed:
  - The functions of evaluation `f` are more complex than the identity ($x_{\text{gen}} = x_{\text{ini}} \pm \varepsilon$). These functions should integrate the appropriate and required level of detail (LOD). For example, during generalisation the number of objects should be reduced (i.e. reduction function) and the size of the polygons should be greater than a certain value (i.e. threshold function) (Figure 5.10);
  - In cases where some objects are removed or aggregated, the evaluation may need to occur at a more global level.

In summary, if we compare the evaluation of generalised data with the classical evaluation of geographical data, the former does not require ground truth because we have a good and exhaustive set of reference objects. Nevertheless it remains complex because the functions `f` are unknown. More specifically the difficulty comes from the required quantity of information lost that is reflected in the cardinality of the relationships between objects before and after generalisation. As a consequence, and in addition to different levels of analysis required to automate the generalisation process (Harrie and Weibel, this volume (Chapter 4); Ruas and Duchene, this volume (Chapter 14)), it is necessary to use different levels of analysis for evaluation. The evaluation could occur at the following object, group or population levels:

- We call *micro-evaluations*, the evaluations that occur at the object level. We compare one non-generalised object with one generalised object (*1-to-1* relationship);
- We call *meso-evaluations* the evaluations that occur at the group level: we compare a group of non-generalised objects with a group of generalised objects (*n-to-m* relationships, n greater than m; n and m having small values). An example of a meso-evaluation would be to compare two groups of buildings, or the street network of a town before and after generalisation;
- We call *macro-evaluations* the evaluations that occur at the data set level. We compare all the data of a particular type before and after generalisation in order to control object removal and maintenance (*n-to-m* relationships, n greater than m; n and m having large values). Macro-evaluation is used primarily to control changes in the quantity of objects (e.g., in terms of the number, length, and area of objects). It is also used to check for maintenance of relative order between micro- or meso-objects. For example that the relative density order and size order between two cities is preserved during generalisation.

Thought should be given to two specific and very common cases:

- the *1-to-0* relationship is, in effect, object removal. It is studied at the micro-level to determine if an object should be removed according to its *own* properties and at the meso-level to check if this object should be *contextually* removed (due to spatial considerations);
- the *n-to-1* relationship corresponds to the process of aggregation: several objects are merged into a single one. This particular case can be viewed as a micro- or meso-evaluation. Whatever the case, the generalised object should represent the n non-generalised objects. It should inherit from their geometry, their type and if possible, their attributes.

5.5.3. *Classification of measurements*

As seen above in §5.5.1, evaluation measures a criteria or character (e.g. size) by means of a measurement performed on one or several objects ($Evaluation = Quality_{\text{character-}k}(Og_i) = \mathtt{f}_{\text{measure-}k}((Og_i), (O_j), \text{LOD})$). Consequently, evaluation is dependent on a set of spatial analytical tools.

As a first step, it is important to distinguish the criteria of a specific algorithm used to measure a value. For example, there are different measurements used to compute the shape of a building (such as compactness, elongation, or medial axis) and different ways of implementing those specific measures – algorithms. Each algorithm calculates a criterion with a certain degree of complexity, sensitivity and in a specific unit of measurement. Based on the studies presented in §5.2.2, we propose the following synthetic classification of measurements:

- A measurement can be internal or external:
  - Internal: measure performed at one Level of Detail;
  - External: measure performed between two Levels of Detail.
- A measurement can be Micro, Meso or Macro:
  - Micro: on one object or a part of it;

  - Meso: on a group of objects;
  - Macro: on all objects of a specific type (all roads, all buildings).
- At the micro-level, the measurement used can be classified in the following way:
  - Position; e.g. deviation between one position and a reference;
  - Orientation; e.g. deviation of the orientation of a building;
  - Size; e.g. the area of a building, the length of a road, the width of a bend;
  - Granularity: size of the smallest shape;
  - Shape; e.g. the elongation, the convexity of a building, the sinuosity of a road.
- One object with respect to its neighbourhood:
  - Congestion (Figure 5.4 illustrates how congestion can be measured).
- Between two objects:
  - Distance, relative position (e.g. parallelism);
  - Topological relationships: intersection/touching/inclusion;
  - Comparison with the micro criterion (e.g. size comparison).
- Meso (a set of close objects):
  - Topological relationships: intersection/touching/inclusion;
  - Spatial arrangement: existence of clusters, of free space, patterns such as building alignment;
  - Density; e.g. density of buildings within a building block, density of streets in a urban area;
  - Network specific measurements (see Thomson and Brooks, this volume (Chapter 13)).
- Macro:
  - Number of objects;
  - Total area, total length.

5.6. Types of Evaluation

In this section we differentiate between the different types of evaluation and briefly present recent research work in this area. Evaluation can take place prior to generalisation and as a basis for setting parameters and "tuning" of generalisation systems (§5.6.1). It can act to control the process during generalisation (§5.6.2), and it can take place after generalisation – in determining the overall quality of the solution (see §5.6.3), and in choosing between different generalised data in cases where there is a choice of output. Evaluation can also be used to compare between different systems by using the same input data, and comparing outputs, or to improve the operation of a system.

During the evaluation of generalised data, we wish to accomplish the following three objectives:

- To detect inconsistencies, – objects or groups of objects that have not been correctly generalised. In such cases, the aim of the evaluation is to automatically detect such inconsistencies in order to allow a user to correct the generalised data as a final step in the generalisation process. We call this "evaluation *for editing*". The errors may highlight the need for additional algorithms, or refinement to existing algorithms. For example, a building simplification algorithm may require modification if it is found to badly generalise "round shaped" buildings.

- To improve the description of the final product (presented in the form of metadata). The aim of such evaluation is not to describe what is wrong but what the characteristics of the generalised data are. We call this evaluation "*descriptive evaluation*". This information describes the "distance" between the reference and the generalised data (Δ in Figure 5.8). This type of evaluation is also fundamental to tuning or improving the system and its algorithms. More precisely, it describes:
  - What has been removed from the reference: the type and the quantity, the reasons why, and the conditions by which some objects have been removed or aggregated (for example, why the smallest buildings have been removed, irrespective of their type);
  - The type and quantity of the changes that have been made to those objects that have been retained (e.g. "the average displacement from their initial position is 10 m").
- To mark the evaluation with a unique value that defines the quality of the solution. This is useful when comparing different solutions of generalisations obtained with different sets of parameters or with different generalisation software. This is termed "*evaluation for grading*".

To summarise, we differentiate between the three types of evaluation and three subtypes:

- *Evaluation for controlling*: Evaluation that occurs during generalisation;
- *Evaluation for tuning*: Evaluation that occurs prior to generalisation;
- *Evaluation for assessing* the quality of generalised data after processing. This can be further subdivided into:
  - *Evaluation for editing* which aims to identify errors and mistakes. It can be an automatic process that occurs in the final stage of the generalisation process;
  - *Descriptive evaluation* which provides summary information on what has been removed, emphasised or altered;
  - *Evaluation for grading*, which seeks to derive an aggregated value reflecting the quality of the solution overall.

5.6.1. *Evaluation to support the tuning of a generalisation system*

The aim of research by Hubert (2002, 2003) was to propose an interface that helps a user to choose the appropriate parameter values of a generalisation system. The underlying system was based on a constraint based methodology, requiring the user to set constraint values of the system (rather than the parameter value of specific algorithms). Instead of requiring a user to set the parameter values of various generalisation algorithms, the aim of the research was to develop a method to understand the required quality for each object type and each criterion. To give an example, for a constraint such as Size(Generalised-Building) $> \lambda$ m^2 (i.e., the building after generalisation should be bigger than λ m^2) the aim is to find the value λ that corresponds to the user's needs. These values define the required Level of Detail of the final generalised data. This research was done at the IGN-France COGIT laboratory and the aim was to propose a system that automatically tuned the AGENT package according to each user need (Ruas and Duchene, this volume (Chapter 14)). A prototype was developed by Hubert (2003) focusing on building generalisation.

The system works by presenting the user with a set of pre-generalised samples and asks the user to choose those that best fit their requirements. When the user chooses an example, the constraint values associated with the creation of that example are stored and analysed. In the prototype system, building samples were generalised by the AGENT prototype. Each building was generalised in approximately thirty different ways corresponding to different parameterisation of a set of constraints. In essence the chosen sample reflects a particular set of parameter values.

Tuning (i.e. finding the optimal set of parameters) is made difficult by the fact that (1) the geometric characteristic of each sample influences its generalisation and (2) that different parameterisations could equally be acceptable for any given building. In order to overcome this difficulty, the user is asked to choose a set of samples of the same type (i.e. a set of generalised buildings). From this, the system is able to select the most appropriate parameterisation. After each set of selections (i.e. a step), the system analyses the user's choice – comparing the values of the constraints for different building samples. If the chosen values are nearly the same, we can assume that the parametrisation is good. If not, the system would need to propose more samples to try and find a better cluster. In other words the system has to find the best values in a parameters' space. The best value is supposed to be a concentrated cluster of user choices. This process of selecting the core parameter values (i.e. the concentrated cluster) is akin to a negotiation process: The system proposes samples and each time it finds a cluster of values, it asks the user to validate the choice. As a consequence the system reduces the solution space step by step. This mechanism of convergence by negotiation was one of the most innovative parts of the research (Hubert, 2003).

5.6.2. *Evaluation for controlling and converging upon solutions*

Evaluation is a fundamental part of the automation process. In the model proposed by Weibel and Dutton (1998), a verification step is used to estimate parameter values and to detect conflicts.

- The continuous optimisation model (Harrie and Weibel, this volume (Chapter 4)) summarised in Figure 4.6 shows that convergence can be driven by measures that are computed iteratively in order to count the number of conflicts and to reduce them progressively. As an example, in the algorithm proposed by Harrie and Sarjakoski (2002) based on least square adjustments (Harrie and Weibel, this volume (Chapter 4)), the aim of the process is to reduce iteratively a residual vector.
- In the combinatorial optimisation process proposed by Ware et al. (2003a) and presented by Harrie and Weibel, this volume (Chapter 4), each new potential solution is evaluated by means of a cost function that calculates the number of remaining conflicts. This simulated annealing technique allows temporal degradation, the idea being to gradually solve as many conflicts as possible. Here again the evaluation of the conflict is central to the process of convergence.
- In the Agent process (Lamy et al., 1999; Ruas and Duchêne, this volume (Chapter 14)), the evaluation plays an essential role since:
- Each agent evaluates itself in order to select the next possible action,
- Each action is validated according to the evolution of the conflicts: if the situation improves, the solution is validated – at least temporally, until another solution is tried.

These functions of evaluation, which are based on spatial analytical tools, always guide the process. As a consequence they govern and limit the rate of convergence. If the evaluation function is not good the system can hardly converge on good solutions, unless it does so by chance.

5.6.3. *Evaluation after generalisation: evaluation for assessing*

The first experiments and research work explicitly focused on evaluation of generalised data was undertaken by João (1998) using Ordnance Survey Data and by Ehrliholzer (1995) who undertook tests to evaluate the quality of some generalisation algorithms. These experiments as well as the ones developed during the OEEPE work (Ruas, 2001) and the AGENT project (AgentDC1, Agent DC4, 1999) helped the community to develop evaluation methodologies. More recent work has also been done by Brazile (1999, 2000) and Bard (2004a). Despite these significant developments, there is a continued need for work in this area. In the following section we present a theoretical framework to build quality functions as described in §5.5.1 as proposed by Bard (2004a). The aim of this research was to find generic functions for evaluation. These generic functions are patterns that are used to build the specific evaluation functions for each feature type and each character, such as the function for the evaluation of the size of buildings. The functions are tuned according to the scale change. In the following discussion we only present these functions and give some results for the evaluation of the generalisation of buildings at 1 : 50 000 scale.

The evaluation of the generalised data is decomposed into a set of evaluation criteria for each object (at a micro-, meso- or macro-level) according to various characteristics. The quality function is decomposed into two functions F1 and F2 ($Quality_{\mathbf{x,obj}} = \mathbf{f2}(final\_value, \mathbf{f1}(initial\_value))$) where:

- *Initial value* refers to the value of a character of an object before generalisation (for example the initial size of a building),
- *final value* refers to the value of a character of an object after generalisation,
- F1, the function of reference, computes the *ideal final* value of a character **x** according to its initial value (this is the value the character should have according to the final scale constraints),
- F2, the function of interpretation, interprets the difference between the final value and the ideal one.

In Figure 5.9, we give an example of a simple case in the context of building size evaluation.

- F1 computes the ideal value according to the following rule: "The generalised size should either be equal to the initial size if bigger than the threshold, or equal to the threshold."
- F2 says that if the generalised value is nearly equal to the ideal value computed by F1 (i.e. the difference is less than ε), the generalisation result is Good, otherwise Bad.

The aim of Bard's research (2004a) was to identify a set of reference functions and the association between object type, object character, the evaluation function F1 and F2, and the threshold values of these functions according to the level of detail. Figure 5.10 shows a variety of possible forms that reference function F1 might take. These functions are patterns that describe the computation of ideal value (Y) according to the initial value (X-abscissa) of a character. For example the function of reference for building size is the "threshold function" (5.10d: the size should be at least bigger than a threshold value). The reference function for

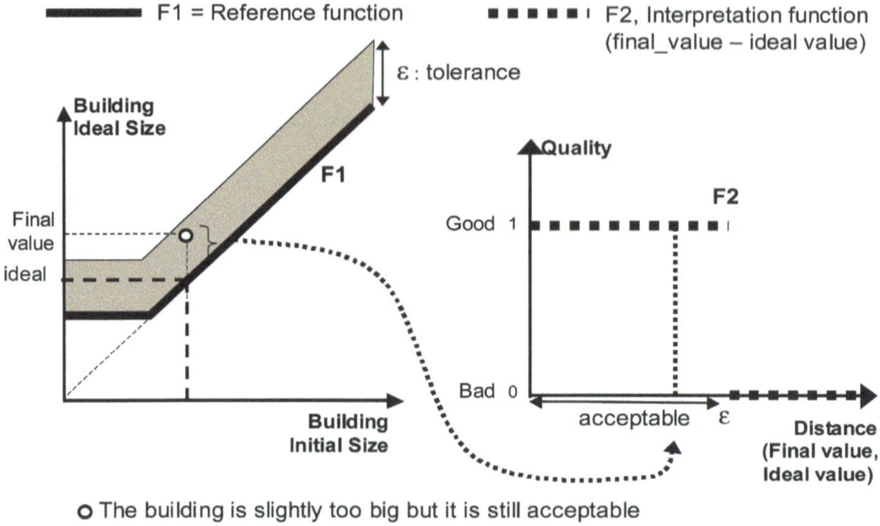

Fig. 5.9. Reference and interpretation functions for building size (Bard, 2004a).

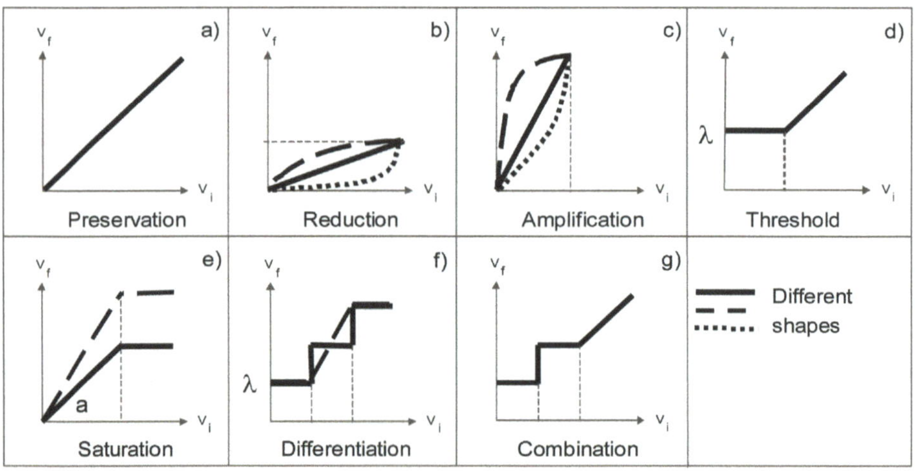

Fig. 5.10. Identification of the shape of the reference functions F1 (Bard, 2004a).

the shape might be its preservation (5.10a: the shape should not change very much) and for the building density it might be saturation (5.10e: the density of buildings within a urban block should not be higher than some saturation value).

Bard (2004a) proposes a set of functions and thresholds for building generalisation at the micro- and meso-level. Figure 5.11 summarises his proposals for building micro-generalisation. For example in Figure 5.11 the function F1 of granularity is a threshold func-

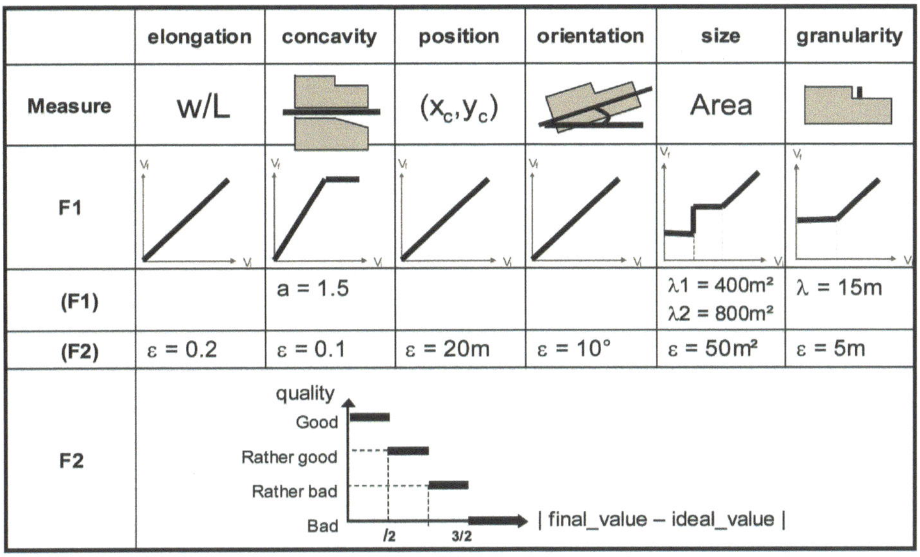

Fig. 5.11. Synthesis of the measure and the functions F1, F2 for building generalisation (Bard, 2004a).

tion with $\lambda = 15$ m. It means that the minimum size of a stair shape should be greater than 15 m. The function of interpretation F2 in Figure 5.11 is the same for all criteria. It defines four levels of quality from *very bad* to *good*, but the value ε depends on the criteria. For granularity $\varepsilon = 5$ m. As an example let us take a building that has an initial granularity of 2 m, and 12 m after generalisation. By means of F1 we can compute that the ideal value should be 15 m. With the interpretation value F2, we can compute the quality of this generalisation. 3 m $(15 - 12)$ is between $\varepsilon/2$ and ε; this means that the result is *rather good*. This qualitative description of the quality is necessary to aggregate the evaluation of all objects.

This research work has been implemented on a research prototype (presented in more detail by Ruas and Duchene, this volume (Chapter 14)) and is a first attempt to integrate evaluation as a final step of the generalisation process. Specific user interfaces have been defined to allow the user to specify the final scale as well as the evaluation functions (F1 and F2) and their parameters (λ and ε). For more details on this prototype see Bard (2004a). Though the focus has been on evaluation of building generalisation, it is anticipated that this system can be extended to include evaluation of other classes of features at the micro-, meso- and macro-levels of evaluation.

One of Bard's objectives was to develop a framework in which the relative quality of generalisation solutions could be graded (*Evaluation for Grading*). It became clear that *Descriptive evaluation* and *Evaluation for editing* were more meaningful and useful in the context of this research. The theoretical framework proposed by Bard allows for the correcting and description of generalised data. This work seeks to link Descriptive Evaluation with the conception of metadata. Such links will hopefully help foster much greater work in this field.

5.7. Discussion and Conclusions

Generally speaking the process of generalisation can be said to de-emphasise qualities of individual objects, whilst emphasising the inter-class and intra-class qualities of the phenomenon being represented. For example the fine scale allows us to represent detailed metric qualities of a section of river but at the coarse scale, the emphasis is on the river as a network feature, and how it relates to, say, land morphology. Therefore the process of evaluation requires us to examine representational issues from the fine (micro), to the broad (macro) scale. This in turn, requires us to utilise a full spectrum of cartometric analysis techniques, extending from the metric and the topological through to pattern and distribution analysis techniques that can measure the broader, gestaltic qualities of the map. This includes evaluation techniques that can measure qualities such as contiguity and association, trend, connectedness and impediment, isolation, orientation, density and distribution. Techniques in pattern recognition are therefore pertinent to continuing developments in evaluation methodologies (Heinzle and Anders, this volume (Chapter 12)). It is not so much that there is a shortage of these evaluation techniques, more the issue is of selecting and prioritising evaluation criteria according to theme and the intended level of detail. The problem is that the representational form of the phenomenon changes with scale, making some criteria less relevant to the process of evaluation, whilst others take on an added importance.

The order and degree of application of map generalisation techniques can have both synergistic and counter intuitive effects. Content can be varied using a mix of model and cartographic generalisation. The symbology can be varied between iconic and pictorial (mimetic) form and the solution often manifests itself as a compromise. Thus an adequate solution might be reached by a variety of routes. Therefore, an evaluation framework must also be able to handle the notion that design is one of compromise among a set of sometimes competing design objectives. Given these complexities, it is often hard to state what is required as an end product except by example or in rather general terms, such as requirements that it must be "correct, and easy to understand". The whole process of evaluation is made still more challenging because of the need to interact with the user. The challenge being to elicit from the user – in an easy, intuitive way – their requirements and then convert those requirements into a set of evaluation parameters. More work is required both to develop an interface to support this process, and more broadly a framework by which evaluation criteria can be calibrated and linked to specific map tasks.

For a variety of reasons it appears to be the case that systems capable of complete autonomous evaluation are neither achievable (due to complexity) nor desirable (evaluation being intrinsically linked to a user's needs which then change in response to the information presented). When incorporating the human, we must find an optimal division of labour between the human and the machine – a balance that needs to acknowledge the limits of cartographic knowledge on the part of the human. In such circumstances, evaluation techniques can play a critical role in directing the user to those parts of the solution that are deemed unsatisfactory.

This chapter has made no mention of evaluation techniques with respect to multiple representation. The creation of multiple representations does not obviate the need for evaluation methodologies. Indeed every idea explored in this chapter would apply in equal measure to the creation of multi-representational databases. We also note that this chapter has focused predominantly on cartographic evaluation in support of map generalisation, but it is important to acknowledge the existence of other evaluation methodologies relating to algorithmics (such as speed and scalability of map generalisation algorithms) and informatics including interface design and the optimal use of humans in the evaluation process. It is important to stress that output from evaluation processes can support a range of tasks that extend beyond just visual assessment. For example, output from evaluation can be stored as metadata and used in reasoning, data quality management and interoperability activities. In summary, it is clear that evaluation remains critical to the development of autonomous systems. The absence of more complete evaluation frameworks in which visual characterisation and evaluation can take place remains as a major stumbling block to more fully automated map generalisation solutions. Contrary to evaluation being an afterthought in the design of automated solutions, there is a real need for evaluation to be considered as an integral component in the design of any generalisation solution.

Generalisation of Geographic Information: Cartographic Modelling and Applications
W.A. Mackaness, A. Ruas and L.T. Sarjakoski (Editors)

Chapter 6

Database Requirements for Generalisation and Multiple Representations

Sébastien Mustière[a], John van Smaalen[b]

[a]*Institut Géographique National, COGIT Laboratory, 2-4 av. Pasteur, 94165 Saint-Mandé, France*
e-mail: sebastien.mustiere@ign.fr
[b]*Universiteit van Amsterdam, Institute for Biodiversity and Ecosystems Dynamics (IBED), Computational Bio- and Physical Geography, Kruislaan 318, 1098SM Amsterdam, The Netherlands*
e-mail: smaalen@science.uva.nl

Abstract

In this chapter, database requirements for generalisation and multiple representations are presented. In particular the paper focuses on modelling requirements. Some general concepts in the field of databases are reviewed before elaboration on the modelling of geographical databases and discussion of the modelling of multiple representations for geographical databases. The chapter stresses the requirements for efficiently modelling data before and during the generalisation process. The reader will first encounter a description of the basic notions of model, schema and objects. Then geographic modelling and multiple representation will be described through the example of the MADS model. A discussion is presented on the different requirements for modelling multiple representations (1) during the generalisation process, or (2) to store the generalisation result, or (3) to perform the integration of independent databases. Then the meaning and role of topology, composite objects and hierarchies will be emphasised in the requirements for generalisation, with aggregation taken as an important example to illustrate this process.

Keywords: modelling, schema, conceptual model, multiple representation, database, entity-relationship, object-oriented, integration, aggregation, taxonomy, partonomy, object hierarchy, composite object, class generalisation

6.1. Introduction

In the field of computer science, a database is a set of organised data that represents a part of the world in order to support digital applications. Those data must be accessible through their content, which means that one should be able to know the structure of the data, and to find all the data respecting certain constraints expressed in this structure. In order to manipulate databases, a database management system (DBMS) is required for saving, querying, presenting and transforming the data. These are the primary requirements of a DBMS. Additional complex requirements include securing the data, optimising the speed performances and allowing users to share data (Gardarin, 1999).

In addition to conventional databases requirements, managing geographic data requires us to manage geometries representing the shape and location of objects (e.g. the shape of a single house), continuous geographic fields of information (e.g. altitude) and spatial relationships between data (e.g. topological relationships). This specific information requires specific tools: display and mapping tools, tools to handle the geometry, tools to manipulate the common topological relationships, specific models to express spatial phenomena (Parent et al., 1998; Bédard, 1999b), and spatial indexes to optimise speed performance (van Oosterom, 1999). Geographic Information Systems contain these specific tools. In earlier developments, GIS concentrated on provision of geographic tools and the data were saved in classical file structures. Today, GIS has evolved from computer assisted design programs to full information systems that either contain their own DBMS, or rely on an external DBMS, in order to ensure comprehensive data handling within the system.

Geographic databases have other very important qualities. First, the geographic world is complex: geographic phenomena are numerous and highly interrelated. That makes the issue of *modelling* geographical phenomena particularly important (Worboys, 1999; Bédard, 1999a). Second, when one looks at a set of displayed geographic data, one can immediately identify a lot of spatial relationships between objects, groups of objects, or parts of objects. For example one can identify that two lakes are close, that some buildings are aligned, or that two rivers from two different databases represent the same entity. Those relations can be identified visually, but they are usually not explicitly represented in geographical databases: they are implicit and need to be detected through the analysis of the geometric and thematic description of the objects. This makes the issue of *manipulating* objects of geographical databases particularly complex.

The complexity of modelling and manipulating geographic data becomes very apparent when we consider issues of generalisation and integration of databases. First, generalisation is concerned with the transformation of a representation of a part of the world. Before generalisation, the implicit relations between objects must be explicitly manipulated, in order to be kept or enhanced. One of the reasons for the complexity of generalisation is the need for data enrichment in order to identify and represent the implicit geographic phenomena before manipulating them. For example, one must identify aligned buildings in order to keep the alignment during generalisation. Second, the integration of different spatial databases into an integrated one (referred to integration, fusion, unification or conflation) is a complex process. It also requires identification and manipulation of the implicit geographic phenomena represented in both databases, in order to compare them. For example two different representations of an alignment of buildings must be identified before determining that those representations correspond to the same real-world entity.

Furthermore, both processes require manipulation of different representations of the world simultaneously. We are faced with the issue of multiple representations, where different database objects represent the same real-world phenomenon (Vangenot et al., 2002). First, during the generalisation process, both initial and transformed representations are manipulated. After the generalisation process has been completed, links between initial and transformed representations should be kept. Second, the integration process compares two representations of the world. The result of integration may be a single database, or a multi-representation database if the diverse points of view on the world have to be preserved (Devogele et al., 1998; Bernier and Bedard, this volume (Chapter 9)). Thus, generalisation and multiple representations are faced with several challenges: they must handle the complexity of the real world, the manipulation of implicit geographical phenomena, and the manipulation of different representations of the world. In this context, the database issues, and particularly the issue of data modelling, are particularly important.

In this chapter, we address the database requirements for generalisation and multiple representations. We particularly concentrate on the modelling requirements. §6.2 reviews some important concepts in the field of databases: the notions of model, schema and objects are addressed, as well as the differences between conceptual and storage models, and the particular case of geographical modelling. §6.3 focuses on the modelling of multiple representations for geographic data. Different requirements are highlighted if multiple representation models have to be used either during the generalisation process, or to store the generalisation result, or to perform the integration of independent databases. Section 6.4 presents some essential modelling concepts for generalisation. It reminds us of the importance of a proper and consistent spatial description of the objects. Composite objects are introduced and the meaning and role of three different types of hierarchies (object hierarchy, partonomy and taxonomy) are explained. Then, object aggregation is presented as an example of an operation that is based on these concepts. The section concludes with an overview of methods for storing the composite objects that are the result of the aggregation. Finally, §6.5 concludes the chapter with a summary of database requirements for generalisation and multiple representations.

6.2. Modelling and Geographical Modelling: Database Principles

6.2.1. *Model, schema and instances*

The terminology adopted in this chapter is the one commonly used in the database community. In the GIS community, another terminology is sometimes used: model in the database terminology may be called meta-model in GIS, and schema in the database terminology may be called model in GIS terminology (Balley et al., 2004). As database management becomes more and more of an issue in GIS, we believe that using the database terminology could fruitfully avoid misunderstandings between these communities.

The main goal of modelling is to represent the real-world entities as objects in a database. Work in the field of knowledge representation have proposed some formalisms to develop these representations. These formalisms, called *models*, define the basic building blocks that can be used to describe data: a set of representation concepts and a set of rules defining how to combine those concepts. Numerous models exist; they must be chosen according to their intended use, in order to reach a good compromise between expressiveness and simplicity.

One approach, Object-Oriented (OO) (Booch, 1994), relates to models based, amongst others, on the concepts of object, class of objects, attribute of objects, link between objects and inheritance between classes. Another approach, Entity/Relationship (E/R) (Chen, 1976), relates to models based on the concepts of entity and entity type, relationship and relationship type, and attributes of entities or relationships (sometimes extended with the notion of inheritance between types). The Relational model is another model with the basic concepts of relation, attribute, tuple and key.

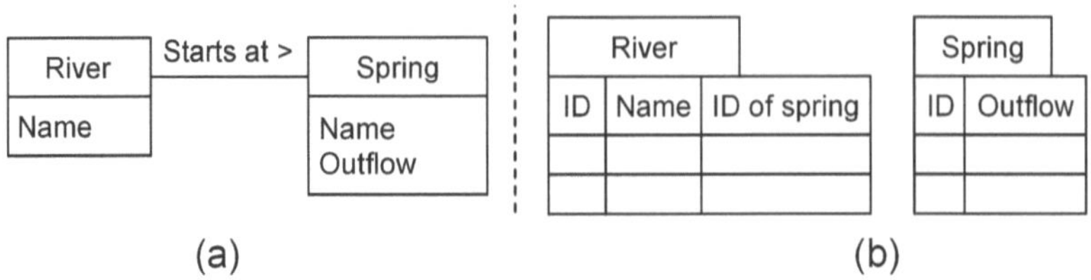

Fig. 6.1. Example of an OO schema with the UML notation (a) and a relational schema (b). The considered types are Spring and River; springs are described by a name and an outflow, rivers are described by their name; a river starts as a Spring.

In order to concretely manipulate a model, it usually has an associated *modelling language*. For example, UML (Unified Modelling Language for object-oriented development) is a modelling language used to draw a type of OO model, representing classes with a rectangle, relationships with lines, and so on (Booch et al., 1999). UML defines simultaneously graphical notation and a standardisation of concepts for OO approaches (i.e. one particular OO model). In the same way, the relational model is usually associated with a language based on the notion of a table (a relation = a table, an attribute = a column of the table).

The description of a particular set of data, expressed in the modelling language, is called a schema. The schema defines the specific types of objects necessary for the application; such as classes in OO models, entity types and relationship types in E/R models and relations in the relational model. One particular object of a type is called an instance of the type. Figure 6.1 shows an OO schema expressed in UML, and one of the possible ways to express equivalent information by means of the relational model.

6.2.2. *Conceptual and internal models*

It is essential to differentiate between two different schemas when manipulating a database; the conceptual schema and the internal schema (Worboys, 1999; Bédard, 1999a). The conceptual schema is used to reason about the world. It is platform independent, whereas the internal schema describes how the data are stored. It is platform dependent. For the data user, reasoning and ideally manipulating data should only be done at the conceptual level. Consequences on the stored data should be transparently propagated to the internal level.

In reality, many schemas can be defined and there is a global tendency to mulitply them. At the internal level, several sublevels can be added to differentiate the schema between the file storage on the hard disk from an intermediate level close to the conceptual level but including additional constraints governed by the storage capabilities of the DBMS. At the conceptual level, one can differentiate a global schema describing all the data, from some user-dependant schemas, named views, describing data in a way more adapted to some specific application. Other approaches introduce an ontology on top of the conceptual schema to better express the semantics of the elements of the schemas. Whatever the number of considered levels, consensus now exists on the necessity of defining different schemas at the conceptual and internal levels.

These different schemas do not necessary rely on the same model. A current tendency for modelling is to express the internal schema in the relational model, and the conceptual schema in an object-oriented or entity-relationship model. Conceptual schemas can be defined by means of the relational model, but OO or E/R models are much more expressive as data become more complex, as is the case with geographic data. Conversely, internal schemas can also be defined using the OO model since OO DBMS do exist, but relational DBMS are known to be more efficient at optimising the speed of response when handling large datasets. The so-called Object/Relational DBMS has now become more widely used; these are relational DBMS extended using some concepts of OO; see Gardarin (1999) for more details on object and relational DBMS.

6.2.3. *Geographical models*

There has been a trend in geographical modelling whereby the geometry is separated from other information. Classical attributes of objects are modelled with classical models, geometries are modelled with a dedicated model, and typically there is a special link between classical objects and geometries. In early developments, managing geometry and topology was a new challenge leading to the concentration on the optimisation of the storage and manipulation of these concepts (for studying the geometric consistency of planar graphs for example). Now the global trend in geographic modelling is to consider spatial and non-spatial information as much more closely related. Geometry is but one part of the model. For example, geometry can be thought of as an attribute just like any other. This provides more convenient models, as we do not intuitively consider the spatial part of geographic entities separated from the rest of their description. This also provides richer models, allowing for example an object to have several geometric attributes. In this latter case, it also raises new challenges, such as the management of topology among multiple geometries.

In order to define a geographical data schema, one can use a classic data model though this requires effort since every basic geographic concept must be defined, such as the definition of the geometry of a line. A better approach is to use models specially developed for dealing with geographic data. These models define how to represent and manipulate the different types of geometries used to describe geographic objects. Some of them also support definition of rules to represent and manipulate the temporal aspects of the objects, or topological relationships. Two approaches exist to extend classic models for the purpose of modelling geographic data: defining patterns (the idea of reusable schemas) or extending the model.

The first approach is to use a conventional model and to reuse pre-defined patterns usually encountered in the schemas of geographic databases. These patterns are recurring combinations of modelling elements, and can be thought of as a set of parts of schemas that can be copied and reused for different purposes (Gamma et al., 1995; Lisboa Filho et al., 2002). Patterns and specific parts of the schema may be linked by means of conventional associations or inheritance relations. Figure 6.2 shows a pattern used to represent lines, points and surfaces, organised in the manner of a topological graph. In this example, the schema is used to represent rivers as lines, springs as points and island as surfaces, by means of links between the classes defined by the user and classes of a classical pattern of a graph. It can be noticed that the schema of Figure 6.2 results in a separation of the geometry from other information because the geometry is only represented in the pattern.

The second approach to modelling geographic databases is to use a new dedicated model. Most of the time this model is an extension of a classic model, but new concepts have been

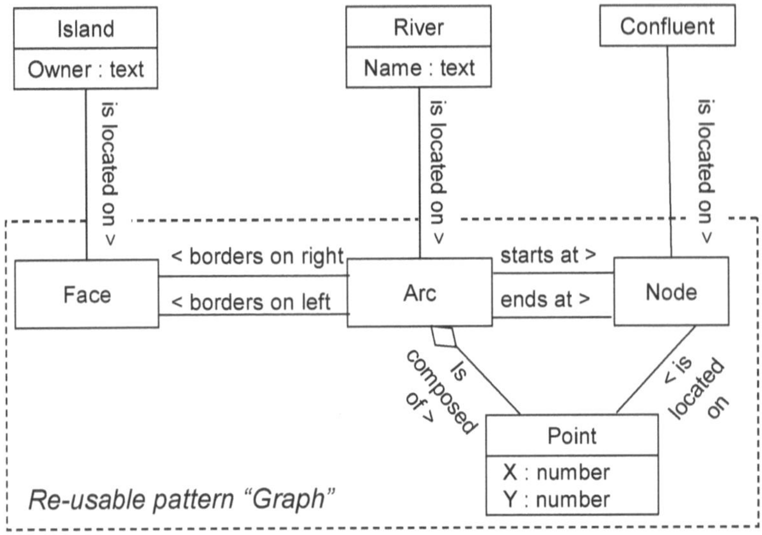

Fig. 6.2. A re-usable pattern to define 2D nodes, arcs and faces, and their topological relationships (lower half), and a specific data modelling of rivers connected to this pattern (top half).

added to the model. For example, MADS based on the Entity-Relationship approach (Parent et al., 1998) and Spatial-PVL based on UML (Bédard, 1999b) are extensions of modelling languages dedicated to the creation of conceptual schemas for geographic data. They define, among others, types of geometries and provide a graphical modelling language. For example, Figure 6.3a shows a set of icons defined in the MADS modelling language to represent different types of pre-defined geometries (other icons are defined to represent temporal attributes in MADS and Spatial-PVL, as well topological relationships in MADS). Figure 6.3b illustrates the use of those icons to define the types of *river* (geometry of rivers is modelled with a line) and *spring* (geometry of springs is modelled with a point). Other solutions can be used in MADS: for example icons could also be assigned to attributes instead of (or in addition to) being defined at the object type level. This can be used to model multiple geometries for a single object.

These conceptual models support the definition of abstract types and the definition of an associated graphical modelling language. Other models exist for different purposes. Rather than being dedicated to conceptual modelling, they may be dedicated to the manipulation of data at the internal level. This is for example the case of SFS (OpenGIS Simple Feature Specification for SQL) that defines basic geometric types, how to store them in different formats, and basic operations to manipulate them with a query language (such as the definition of topological predicates) (OGC, 1999a).

Alongside geographical modelling, has been the development of CASE tools (Computer Assisted Software Engineering) used to manipulate those models. Examples include Perceptory for using the Spatial-PVL model (Bédard et al., 2004) and tools used to manipulate the MADS model developed during the MurMur and Amber projects (MurMur Consortium,

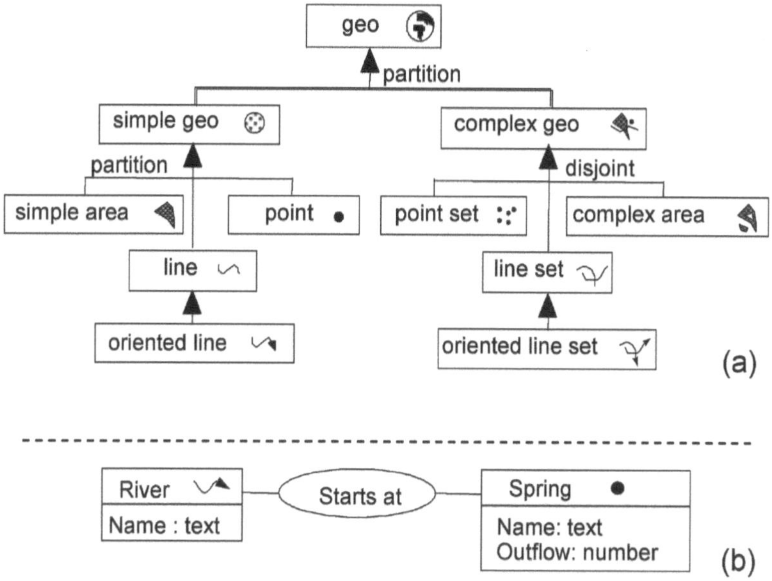

Fig. 6.3. (a) MADS basic hierarchy of spatial abstract data types (Parent et al., 1998), and (b) their use in defining rivers and springs.

2002; Stonykova, 2003). However there is a need for greater integration of CASE tools within GIS in order to support the full development of spatial databases, from the conceptual and internal schema definition, to the query, manipulation and eventual visualisation of data.

6.3. Multiple Representation Databases

6.3.1. *Why multiple representations?*

The many geographical databases that exist worldwide reflect the diversity in how we perceive and model the world around us. These databases have different levels of detail, a concept for geographic databases related to the notion of scale for maps. For example houses may be represented by their boundary or by a single centre-point in a less detailed representation. These databases may contain different objects (for example houses may be represented in one database and not in another) and they may represent the same entities but in different ways, without any hierarchical ordering between the levels of detail. For example houses may be represented by their ground surface for the purposes of calculating property tax, or by their roof surface for topographic modelling.

These databases may be created independently of one another or one database may originate from the generalisation of another one. Data producers may have to simultaneously manage these databases, and data users often need to simultaneously analyse them. In this context, there is a growing need for the integration of different databases into a single one making explicit the relationships between them, possibly with the use of multiple representations. These multiple representations make explicit the links between homologous objects, i.e. objects corresponding to the same real world entity. Various authors have highlighted the benefits of the explicit modelling and storage of multiple representations (Kilpeläinen, 2001; Hampe and Sester, 2002; Sheeren et al., 2004). These include:

- Ease of database maintenance and propagation of updates:

  The cost of updating can be minimised by integrating changes only once in the database and propagated, at least semi-automatically, to different representations of the same geographic phenomenon, either between different levels of detail or between geographic databases and cartographic representations (Kilpeläinen, 2001; Badard and Lemarié, 2000).

- Assessment of quality:

  Integration of databases may be useful in the control of quality. If one representation is known to be of better quality than the other one, the former can be used to control the latter. At the very least, integration allows us to detect inconsistencies between representations and thus to flag possible errors (Egenhofer et al., 1994; El-Geresy and Abdelmoty, 1998; Paiva, 1998; Sheeren et al., 2004).

- Increase in the efficiency of applications using these databases:

  The use and analysis of data coming from different databases is facilitated by a close linkage between them. This allows navigation between levels of details according to the type of information required. The analysis of objects through the analysis of their corresponding objects at a higher level of detail becomes possible. Even when applications only require visualisation of the data, the existence of links between the objects of different representations can help in development of visualisation tools.

6.3.2. *Modelling multiple representations*

In geographical databases, it is common to manipulate different representations of the same object. This is the case where it is necessary to manipulate data at different scales. For example, an entity may be manipulated as a whole compound object or as its individual parts. Thus, some research has focused on definition of principles in order to better model geographical multi-representation databases. Some models rely on the notion of view, which is a virtual subpart of the database schema (Claramunt and Mainguenaud, 1995; Bernier and Bedard, 2002). Some are dedicated to temporal databases (Peuquet, 1999) – a kind of multiple representation where an entity of the world has different representations at different times – while others are focused towards multi-scale databases (Timpf, 1998; Stell and Worboys, 1998; Zhou and Jones, 2001). Some other approaches are dedicated to the federation and maintenance of distributed databases (Friis-Christensen, 2003).

In order to illustrate the modelling of multiple representations, we describe hereafter the approach based on the extension of the Entity/Relationship and MADS models (Vangenot et al., 2002), that could also be adapted to other models. This approach is quite generic and proposes different methods by which to model multiple representations (Figure 6.4, modified from (Balley et al., 2004)). The first extension of the model is the addition of "stamps" associated with every piece of the schema (entity types, relationships types, and attributes). Those

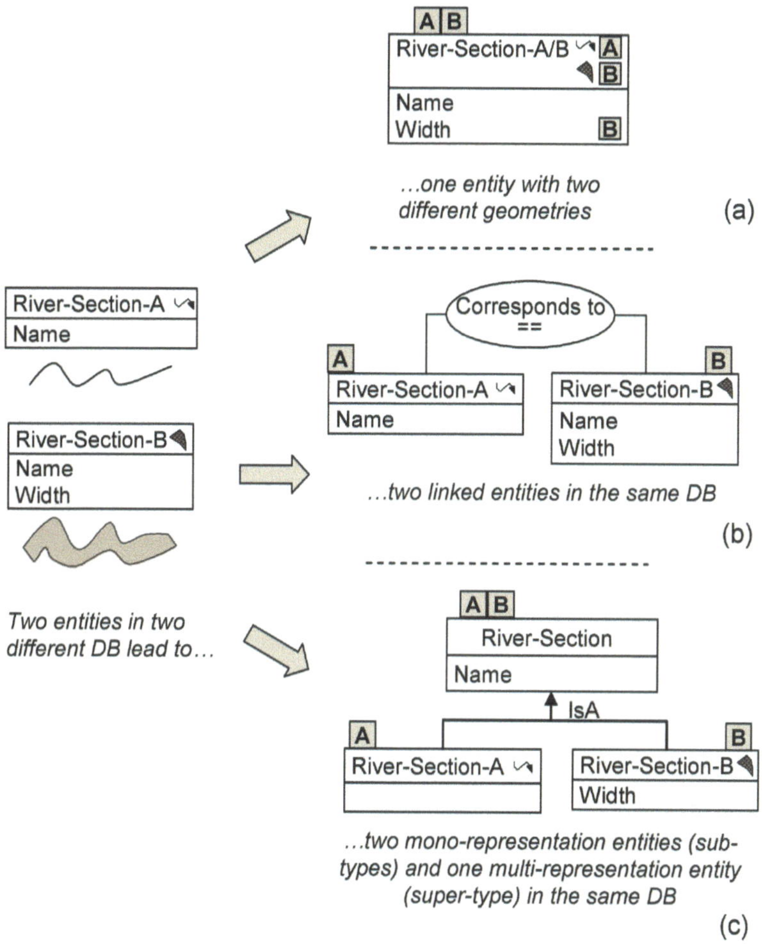

Fig. 6.4. Different representations of multiple representation (adapted from Balley et al., 2004, with permission from the authors).

stamps determine to which representation the piece of the schema belongs. In Figure 6.4, those stamps are labelled A and B. In this example, three modelling solutions are proposed to explicitly represent the links between two originally independent entity types River-Section-A and River-Section-B:

- The first solution (Figure 6.4a) is to make only one entity type with different attributes (geometrical or not) corresponding to the different representations. In this case, only one entity type exists, but it holds the two different representations.
- The second solution (Figure 6.4b) is to add a relationship type between the two entity types. This relationship type (*corresponds to*) has been defined for the particular purpose of multiple representations and has the particular semantic of "the database entities linked

by this relationship correspond to the same real world entity". In this case, there are two entities in the database that represent the same real-world entity.

- The third solution (Figure 6.4c) is to add a third federated entity type, holding the common characteristics of the different representations. Then both individual entity types are linked to this common type. The link is made by means of a *IsA* relationship – a notion related to inheritance in OO models. In this case, three database entities exist to represent the same real world entity.

Relationships between homologous entities (the "corresponds to" in Figure 6.4) may have different cardinalities. This is expressed in the MADS model through the definition of different types of relationships: *Equivalence*, *Aggregation* and *Set-to-Set:*

- *Equivalence* is used when one entity of one database may correspond to only one entity of the other database (1–1 relationship). This may be a usual case in two databases where the same basic real-world entities are represented, but with different points of view or different selection criteria. For example, each house represented with a punctual surface in DB1 may correspond to a house with a surfacic geometry in DB2.
- *Aggregation* relationship is used when one entity of one database may correspond to several entities of the other database (1–*n* relationship). This may be a usual case with two DB with different levels of details. For examples, each urban area of DB1 may correspond to a set of buildings in DB2.
- Finally, *Set-to-Set* is used when several entities of one database may correspond to several entities of the other database, without being able to separate this relationship into several simpler relationships (*m*–*n* relationship). This is a usual case between digital landscape models (DLM) and digital cartographic models (DCM). For example a set of five buildings in the DLM may be represented in the DCM by another simplified set of three buildings. *Set-to-set* is also a usual case when segmentation principles are different in two databases. For example, three road sections in DB1 (where one road section is made as soon as the number of lanes changes) may correspond to two road sections in DB2 (where one road section is made as soon as the speed limit changes).

6.3.3. *Requirements of multiple representations*

The models previously described show that a good modelling of multiple representations may require much more than just the ability to define different geometries for one single object. The choice between the different models, or between the different modelling patterns, must be made in order to create rich but simple schemas, adapted to the data and their intended use. Multiple representations may be useful either during the generalisation process, or to model the result of generalisation, or to model the result of the integration of originally independent databases. Those different cases raise different requirements for the choice of model. This section details these particular requirements. Of course, requirements for the different cases are not always clearly separated, and requirements for one case may apply to another case. In any event, we try to emphasize the most important requirements for each case.

6.3.4. *Multiple representations for the results of generalisation*

As previously stated, it is of prime importance to keep the links between the original data and generalised data. General principles for modelling multiple representations have been explained in the previous section. In the particular context of generalisation, multiple representation models would benefit from adhering to four main requirements:

– Store data with *multiple levels of details.*

  Generalised data are usually less detailed than the original data. Therefore multiple representation models should emphasise multi-level modelling: generalised objects (less detailed) must be linked to the original objects from which they have been derived (more detailed). This particular case of multiple representations is mostly hierarchical: one generalised object often corresponds to one or several original objects, and one original object has been used to derive only one generalised object. Thus, models dedicated to the representation of the results of generalisation should emphasize this hierarchical aspect.

– Store *additional objects*, not necessarily present in the original or generalised database, but deemed useful during the process.

  Generalisation requires the making of some objects and relationships explicit, such as some groups of objects or the compound/components relationships (Ruas, 2000b). These objects are useful to the process, but do not initially exist in the original database, and may not even exist in the generalised database. But these objects are necessary to model the links between the original and the generalised database. Let us take the example where an alignment of five building has been transformed into an alignment of three buildings, for some cartographic reason. In this instance, in order to propagate future updates concerning the original buildings, it is more useful to store the links between the object's "alignment" and the links between the alignments and their components, rather than storing a direct link between the houses (Ruas and Holzapfel, 2003). Thus, models of multiple representations need to have the ability to represent new objects concerning the original database, while keeping them in some way separate from the original data (since the original data managers may not want to change their modelling).

– Support a strong *interconnectivity* between representations.

  Bidirectional links between levels of detail may be of prime importance in order to propagate updates between representations, and to optimise multi-level analysis (Kilpeläinen, 2001). For example, when detailed objects are updated, one needs to identify the associated generalised objects that also require updating: links from detailed to generalised data are therefore necessary. But data transformation at the generalised level may impact the surrounding objects. In such a case, in order to update those surrounding objects, it may be necessary to re-generalise them, or at least to consider again their detailed counterpart: links from generalised to detailed data are therefore required.

– Manage *on-the-fly and stored generalisations.*

  Some generalisation processes are long and may require interactive work. Some other processes are fast and it thus may be useful to only generalise data when they are explicitly queried. This may be the case if the original data are being frequently modified such as road traffic information. In this case, efficient models should allow for the simultaneous management of pre-computed and stored data, and on-the-fly generalisation operations.

6.3.5. *Multiple representations during the generalisation process*

Multiple representations do not only relate to the results of generalisation: it is also very useful to define a proper modelling during the generalisation process. There are several important issues in modelling the links between the original objects and the corresponding generalised ones during the generalisation process. The first requirement is that we keep track of the links between original and generalised objects in order to easily build the resulting database with multiple representations. It is much easier to keep track during the process than to rebuild the corresponding links from scratch once the data have been generalised. The second issue is to control the process. Indeed, changes created via generalisation must be validated in some way (see Mackaness and Ruas, this volume (Chapter 5)). Those changes can be verified at the end of the process, but it may be necessary to assess the quality of the solution during the generalisation process. This concerns generalisation approaches capable of backtracking and searching among a number of alternate solutions (something discussed in greater detail by Ruas and Duchêne, this volume (Chapter 14)). These approaches require us to efficiently manipulate different representations of the same object. This places some special requirements on how to model the multiple representations during the generalisation process, in addition to requirements for the result of generalisation:

- Allow multiple representations between *many representations* (more than two).

  During some generalisation processes (see for example Ruas and Duchêne, this volume (Chapter 14), it may be necessary to select among a number of different representations of the same object (possibly generated using different algorithms)). Thus the model must be able to handle multiple representations between more than two objects during the generalisation process.

- Manage simultaneously some objects *effectively stored* in the database, and some *intermediate ephemeral objects* only manipulated during the process.

  Some of the intermediate results of generalisation, such as the intermediate states mentioned above, are only useful at specific times. They do not need to be stored indefinitely. Thus a good model in this context should allow the user to decide which objects need to be stored in a persistent manner.

- Allow control over the activation of *integrity constraints.*

  Some constraints may be defined within a data schema (see for example Object Constraint Language (OCL), a formal language for defining constraints in UML schemas). For example, one can decide that a certain attribute must have its value within a range of predefined values. In such a case, the DBMS will refuse to store (or flag) an inconsistent object. In the case of geographical data, one may decide for example to define a topological constraint that prevents roads from crossing lakes without a bridge. These constraints are useful in controlling the process and preventing generation of inconsistent results. But such constraints need not to be strictly taken into account during the process. For example a road may be displaced, temporarily crossing a lake, but the lake will only be displaced at a later stage. Thus constraints defined on the model should not stop the process during the generalisation, as temporarily violating constraints are sometimes useful or even necessary. In some generalisation models, the analysis of the violation of constraints can even be the central concept guiding the process (see Ruas and Duchêne, this volume (Chapter 14)).

- Deal with *missing and evolving objects*.

  During the generalisation process, some objects will be transformed, some will be deleted, and others will be created. This presents two difficulties. The first is to decide how to model evolving objects: are objects before and after generalisation the same object that moved, or are they two different objects? We are thus faced with the same problem as for spatio-temporal modelling (Peuquet, 1999). The second difficulty is to handle missing objects, as generalised objects are progressively created. Logical models manage incomplete information quite well, but database models usually do not. This is particularly challenging when dealing with the relationships between objects: how do we update and manipulate topological relationships when objects are progressively created or displaced? How do we query properties and relationships of objects that are only created during the generalisation process?

These constraints are quite difficult to handle, and the creation of a suitable model may therefore be one of the most challenging issues for generalisation.

6.3.6. *Multiple representations for integration*

Multiple representations from the integration of independent databases is also very useful, but raises some issues other than the ones encountered for generalisation. First of all the integration process is complex, as the matching between the original data must be discovered from the analysis of data, schemas and specifications of the original data. Some work addresses the issue of schema matching: which schema element of one DB corresponds to the schema element of another (Devogele et al., 1998; Rodriguez et al., 1999; Uitermark, 2001)? Once the schema matching is completed, the challenge is one of instance matching: which objects of one DB corresponds to the objects of another DB (Devogele et al., 1998; Walter and Fritsch, 1999; Mantel and Lipeck, 2004)? An extensive description of the database integration process is beyond the scope of this chapter, but in summary we argue that multiple representations for integration would benefit from the following special requirements:

- Support the *complex modelling* of patterns.

  The original databases may be very different in their modelling, and very heterogeneous. Thus, models used to support multiple representations after integration must be able to model complex cases. One may require richer models than the ones required to model multiple representations after generalisation, where schemas and data can be very different but still remain closer than when dealing with independent databases. These complex cases include: modelling 1–1, 1–*n*, and *n*–*m* links; modelling the absence of links for certain objects; modelling links between groups of objects; modelling links between only parts of objects; modelling links between attributes to express constraints between values of attributes; and modelling fuzzy links. These fuzzy links may either be vague (this object is *somehow* related to this one) or uncertain (this object *may be* related to this one).

– Support *interoperable* access.

  In order to integrate two independent databases, one must access them simultaneously. One solution is to first integrate the two databases in a unique DBMS. Another solution is to access these independent databases only when needed. In this case, one must be able to query the independent databases (each one with its own DBMS) from the system that supports integration. This need for remote and transparent access is an issue of interoperability. One way of achieving interoperability is to standardise the description of data and ways to exchange it. Web Feature Services (WFS) developed by the Open GIS Consortium is a standard increasingly used to remotely access geographic data within a spatial data infrastructures (OGC, 2002a).

– Keep some levels of *independency* between databases.

  Even if the resulting multiple representation database makes the links between the two databases explicit, it may be useful for some applications to keep a certain level of independency, at least at the internal level of storage. The original databases may be produced and managed by different organisations, and these organisations may want to retain physical control of their databases (Friis-Christensen, 2003). In this instance, the concept of federated databases may be useful (Stonebracker and Hellerstein, 1998). Federated databases are a set of databases, each one managed by its own DBMS (whether or not it is the same physical location, the same software or the same model), but that can be transparently accessed through a unique system as if it was only one database.

– Support a strong *interconnectivity* between representations.

  Firstly, like in the case of modelling the results of generalisation, navigating between the representations in both directions is important. For example in order to make a combined analysis of representations, or propagating updates. Second, if more than two databases are integrated, one must be able to manage multiple links (one object of one database may be linked to several objects in several databases). The greater the number of representations involved, the more it may be useful to link each representation to a pivot representation, rather than connecting representations two by two. For example, Friis-Christensen (2003) proposes a model to federate independent databases and to describe how schemas of independent databases are linked to the schema of a centralised view of the data.

6.4. Database Principles for Modelling Data for Generalisation

Multiple representation databases can be created by integrating existing databases at different levels of detail or by generalising a database. One of the central issues in the first approach is instance matching (Sarjakoski, this volume (Chapter 2)). This section focuses on the second approach, more specifically on the database requirements for *model* or *geo-database generalisation* as a fundamental part of the map generalisation process. We focus on one of the essential operations in model generalisation: *aggregation*. Whereas integration of databases may lead to 1–1 or m–n, aggregation always leads to 1–n relationships (see §6.4.2), or in this instance an m–1, since the process is such that several component objects create one composite object. The first subsection reviews the role of modelling concepts needed for the generalisation of geo-databases. The second subsection explains the relation between the concepts in the first section and the aggregation process. The third subsection focuses on one of the concepts, the object hierarchy, and how to model this in the database.

6.4.1. *Modelling concepts for the generalisation of geo-databases*

Objects

Objects have properties. Properties can be static or dynamic. Static properties are recorded in the form of attributes. For example attributes such as population – in the case of cities – or maximum allowed speed – in the case of roads. Dynamic properties or *behaviour* of the object are stored as *methods*, computer programs that are linked to the objects. Besides thematic or non-spatial properties that all objects have, geographic objects also have geometry. This particular attribute consists of one or more *geometric primitives*. In the traditional 2D vector GIS environment the geometric primitives are points, lines, and polygons. In the case of 3D GIS, volumes are added. Object geometries can consist of more than one primitive and the primitives can be of different geometric types. An obvious example of the latter is a river of which the narrow parts are represented as lines and the wider parts as polygons.

Topology

Topology describes the connectivity relationships between objects. Topological models have been used extensively in the past. Examples of these is the Arc/Info vector format (similar to the graph of Figure 6.2), and Formal Data Structure (FDS) (Molenaar, 1998a). Figure 6.5 shows the latter with an extension to include composite objects. For a more elaborate discussion on topology see Theobald (2001). Topology serves an important role in the map generalisation process (van Smaalen, 2003); it is therefore vital that these relationships can be queried efficiently. Most current GIS however are still hampered by the fact that different types of geometric primitives are stored separately and each type of geometric primitive requires different operations. This makes it difficult to manipulate objects comprised of different data types (for example the river represented by lines and polygons). Instead, it should be possible to retrieve topological relationships irrespective of the types of geometric primitives involved. In the case of the river example, a single function should allow manipulation of the river's lines and polygons simultaneously. This geometry independence can be achieved by means of the concept of polymorphism in OO, i.e. a method defined for a superclass – geometry in general – can be re-defined for specific classes such as points, and lines.

The current trend in GIS is towards calculating topological relationships when needed, not storing them permanently. Thus, data schemas may be different for storage and processing, the topological schema only being used during the generalisation process. However, having to calculate topological relationships on the fly may be disadvantageous for generalisation since generalisation operations use topology repeatedly.

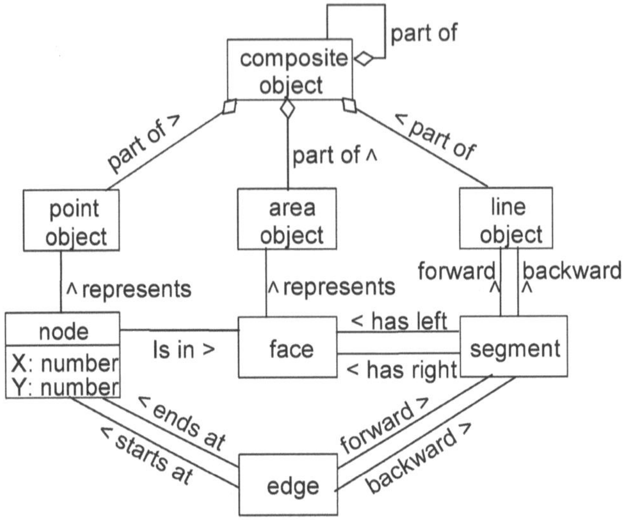

Fig. 6.5. A model describing topological relationships between the geometric primitives and their relationship to geographic objects and composite objects, using UML notation, adapted from (Molenaar, 1998a).

Topology is not the only kind of spatial relationship that can play a role in the generalisation process. Proximity or distance and orientation, such as parallelism, may also be used to determine inter-objects relationships. This requires other data models that support the querying of such relationships. The Extended Formal Data Structure (EFDS) (Peng, 1997) and the Integrated and Extended Formal Data Structure (IEFDS) (Liu, 2002) are examples of such models that include adjacency relationships between unconnected objects. Displacement operations might need a data model that relates unconnected neighbour objects such as buildings and roads through flexible and/or rigid beams that allow deformation within defined tolerances (Højholt, 1998). Simplification of (road) networks, on the other hand, may be based on a data model of nodes connected by segments with additional information about the orientation of the segments at the nodes where they meet. This is required in order to assure "good continuation" of the route (Thomson and Richardson, 1999; Thomson and Brookes, this volume (Chapter 13)).

Composite objects and object hierarchies

Objects can comprise several geometric primitives but also other – lower level – objects. These composite objects consist of components that are in some way, thematically or spatially, related and form a meaningful unit at a higher level of abstraction. A composite object or aggregate is a relationship between two or more objects seen as a new object (Smith and Smith, 1977). Composite objects are considered important for the generalisation process (van Smaalen, 1996; Ruas, 2000b; Boffet, 2000). The relationships between component and composite objects are of the "part-of" type and form an object hierarchy (Molenaar, 1998a).

The process of grouping multiple individual objects to form a new composite object is called "aggregation" (Frank and Egenhofer, 1988). In Figure 6.6 we see a collection of com-

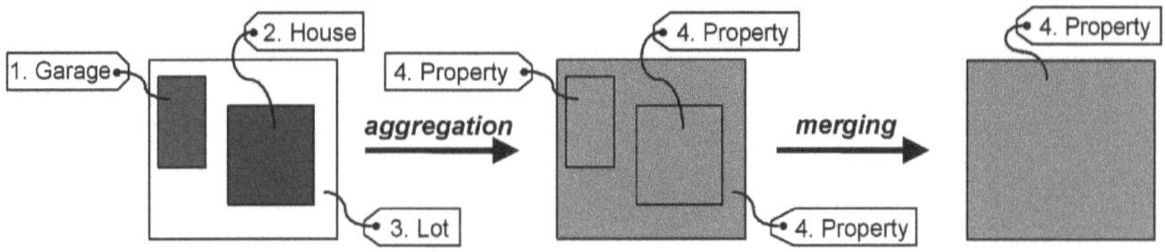

Fig. 6.6. Aggregation followed by merging of the components' geometry.

ponent objects from which a single composite object is formed at the next higher aggregation level in the object hierarchy. If the component objects are adjacent, their spatial descriptions are usually merged, but it is also possible to assign a common composite object identifier to the components, indicated by the number 4 in Figure 6.6. Merging and aggregation are therefore two distinct operations. Aggregation mainly involves combining the thematic information attached to the objects, whereas merging only relates to the spatial component. Composite objects are mostly, but not necessarily contiguous. They can consist of several spatially disconnected parts.

Taxonomies

In order to comprehend large amounts of data, people tend to categorise. We speak of houses, roads, trees because we deal with them in different ways. These general notions refer to a number of actual objects that can be represented and operated on in the same way even though not a single one is completely identical to another. This concept of categorising things into groups, or classes, based on similarity is called classification. We call objects that are member of a class, instances of that class. The relationship between objects and classes is called instantiation. Some classifications are more specific than others. If an object is said to be a building, one roughly knows what it will look like, but it can still be many diverse kinds of building. If it is referred to as being a house, one can make a far more accurate estimation of its properties. The class of buildings is called a superclass of the class "house", the class "house" is a subclass of the class "building". The subclass presents a more detailed description of the object. Subclasses and superclass are related in the generalisation plane (Smith and Smith, 1989), and the resulting structure is called a classification hierarchy or taxonomy.

Moving up in the taxonomy is called class generalisation, and moving down is called class specialisation. As we move up and down in the taxonomy not only the class label changes, but more importantly the object is referenced in different ways, revealing different levels of detail (Figure 6.7). Class generalisation should not be confused with the broader use of the word generalisation in map generalisation, which encompasses many other operations. The similarities between objects expressed in the taxonomy may be used for aggregation and for other generalisation operations. An example is the aggregation of adjacent objects with a common superclass (Figure 6.8). To support this, the database structure should allow classes to have their own specific attribute structure. This enables determination of object similarities by individual attributes as well as class similarity.

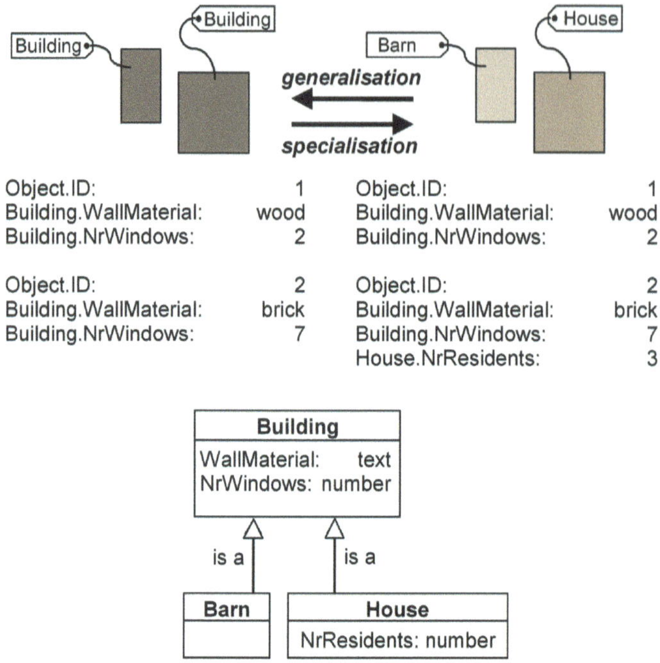

Fig. 6.7. Subclasses and superclasses reveal different levels of detail.

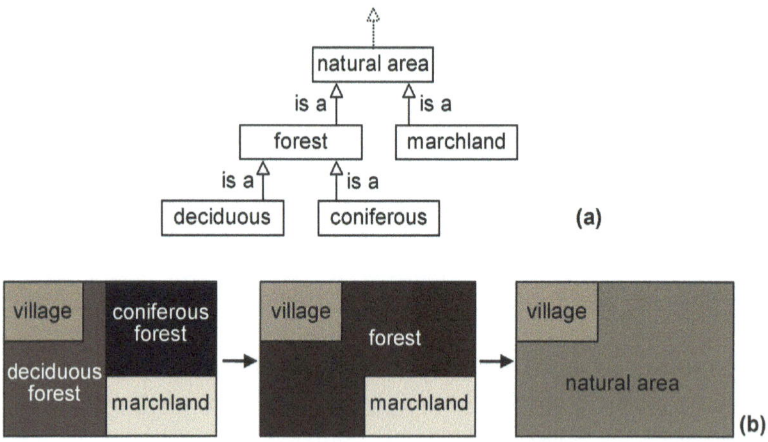

Fig. 6.8. (a) Part of a taxonomy or classification hierarchy, and (b) an example of class-driven generalisation based on this taxonomy.

Partonomies

Classes are only valid within a limited range of spatial resolutions. The class "building" is a good example since there are generally no buildings larger than a few hectares. If we want to describe areas at higher levels of abstraction, we will have to switch to classes such as "built-up area", composed of objects that bear little similarity, such as buildings and roads. Small steps towards higher levels of abstraction can usually be accomplished by merging similar objects. The larger steps are often based on other relationships, such as functional ones. Functional relationships may play an important role in a spatial abstraction process, but they are only occasionally mentioned in the map generalisation literature. One example related to topographic data was described by Robinson (1995). He described an aggregation hierarchy or partonomy of buildings, aggregated buildings, building groups and blocks. Note that aggregation hierarchies or partonomies are different to object hierarchies. Partonomies deal with classes and describe relationships that generally occur between objects of those classes. Object hierarchies refer to actual objects. Ruas and Lagrange (1995) observed that hospitals are generally composed of sets of buildings and areas. Ruas (2000b) later calls these constructs meso-objects. It appears however that some generalisation processes described in the literature comprise relatively small generalisation steps, which can still be realised by elimination and reclassification based on similarity.

Partonomies are based on the functional coexistence of the objects that belong to the classes involved. These functional relationships are generally much more interesting than relationships based on similarity. Take the example of a leopard. The leopard depends on a varied habitat with both forested areas and more open vegetation. Assume that we have a satellite image with a spatial resolution of 1 m. In the image we will be able to recognise trees, shrubs, and grassland. We can aggregate adjoining trees and shrubs into larger areas of higher vegetation, keeping the larger areas that emerge and eliminate the smaller ones. However, in this process we will lose the specific domain of the leopard that is, in fact, characterised by the variation in higher and lower, denser and more open vegetation. To find the leopard we will have to look at spatial units which contain both types of vegetation, high and low, and we have to look at the right level of abstraction; the open spaces must be large enough but not too large. To the leopard, the shrubs and open fields are functionally related, the shrubs providing cover for the animal to stalk its prey grazing in the open.

Figure 6.9 shows an example of partonomy-driven map generalisation. Note that the relationship type in the partonomy is "part of" as opposed to "is a" in a taxonomy. Also note that the part-of relationships in a partonomy connect classes whereas the aggregation process establishes part-of relationships between the component objects and the composite object in an object hierarchy.

The functional relationships that create a partonomy can play an important role in map generalisation and accommodate adjustment of the generalisation process for specific purposes. Just as it should support taxonomies, the database structure should support the storage of partonomies too.

6.4.2. *Creating composite objects*

Model generalisation has two aspects: thematic and spatial abstraction; they are often closely related. We speak of "thematic abstraction" if the number of distinct attribute values is reduced, i.e. the domain is limited. For attributes with a quantitative domain this can be achieved by grouping values into classes. If the attribute values refer to classes already, the domain can be limited by referring to superclasses rather than classes, since there are less superclasses

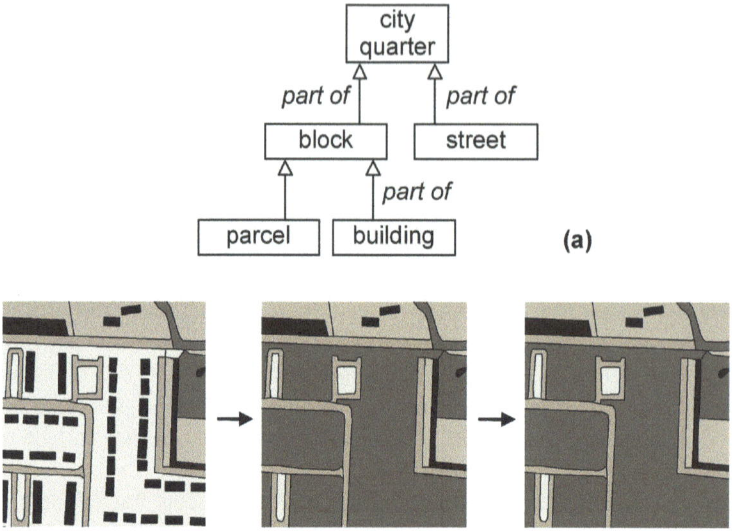

Fig. 6.9. Example of a partonomy (a) that drives aggregation and merging of dissimilar objects (b).

than there are classes. Thematic abstraction is the conceptual GIS equivalent of combining legend items in traditional cartography. Spatial abstraction is achieved by reducing the number of objects by means of aggregation or elimination. Spatial abstraction can be compared to simplifying the map image in map generalisation. Thematic abstraction can trigger spatial abstraction when objects with an identical class value after classification or after class generalisation are subsequently aggregated. For that reason, spatial abstraction frequently depends on a preceding thematic abstraction.

Thus, in a spatial abstraction process some objects are eliminated, others are combined. Aggregation is an important operation given its ability to reduce spatial complexity while at the same time retaining most thematic information of the component objects (van Smaalen, 2003). With elimination on the other hand, all information of the eliminated object is lost in the process. Elimination can be thought of as an aggregation in which a contaminated object is created, i.e. the thematic properties of an "unimportant" constituent object are not reflected in the properties of the composite object it has become part of. Figure 6.10 illustrates this; the lot with number 4 is still called "lot" even though it is different from the lot with number 3 since it encompasses the garage. When elimination is treated as a special case of aggregation you can first evaluate the influence of the "eliminated" object on the composite object before deciding to erase or retain its thematic properties.

Aggregation can be based on similarity of the objects involved or on functional relationships between them. Similarity can be defined in the following four ways:

- Direct comparison of attribute values with a ratio scale. Attributes with a ratio scale cause few problems since the "distance" between two attribute values can be measured objectively. For example, a river with a width of 12 meters is clearly more similar to a 10 m wide river than it is to a 90 m wide river.

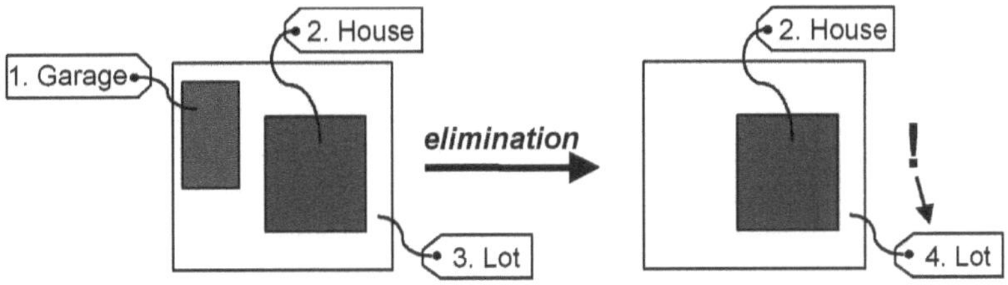

Fig. 6.10. Elimination (nominal attribute).

- More challenging are attributes with a nominal scale. These nominal data raise questions such as: what is similar and which combinations are more similar than others? This requires the definition of similarity measures. Bregt and Bulens (1996) encountered this problem and introduced the idea of a similarity matrix, a matrix in which a similarity measure is assigned to every possible combination of class values. One drawback of this approach is the labour-intensiveness of creating a similarity matrix. Moreover, similarity matrices are application-specific and can therefore not often be reused. More recently Rodriguez et al. (1999) presented a method for defining similarity measures between classes based on set intersections.
- Nominal attributes often refer to classes, the values being class identifiers. Class-driven aggregation (Molenaar, 1998a) is a special case of similarity-driven aggregation. Class-driven aggregation is based on the use of an existing taxonomy to establish whether the classes share a common superclass. In Figure 6.8a, for example, we can see that "coniferous forest" and "deciduous forest" are both linked to the superclass "forest". These relationships can be used as a basis for aggregation. Since "deciduous forest" and "coniferous forest" are both types of forests, adjacent lots of both types can be aggregated, resulting in composite objects of the type "forest" (Figure 6.8b). In an object-oriented or object-relational database the different classes will be part of a taxonomy of object classes, each class having its own attribute structure. Such a database will also support inheritance so that generic attributes can be defined at the superclass level.
- Class-driven generalisation can follow the strict rules of object modelling, with the superclasses having a less intricate attribute structure. But often the classes involved are mere data classes and the taxonomy takes the form of hierarchical attribute values or a related table. Richardson's method for the generalisation of land cover data is based on this principle (Richardson, 1993). Another example is Bregt and Bulens' attribute class method (Bregt and Bulens, 1996). Many datasets contain hierarchical attribute values that will support this practise. The Dutch topographic "TOP10vector" dataset, for example, contains a hierarchical classification for roads.

But as stated in §6.1.1, similarity can only be used within a limited spatial range. Therefore we also need other relationships such as functional ones. These functional relationships are stored in partonomies. Partonomies can either be user defined, or derived from statistical properties in the dataset itself:

- User defined. Disadvantages are that defining such a partonomy is labour intensive and it requires significant knowledge of the content of the database – just as in the case when defin-

ing a similarity matrix. The partonomy, in combination with the topology of the dataset, defines potential composite objects.
- Derived from statistical properties of the dataset itself. van Smaalen (2003) used a spatial data mining approach by using global adjacency measures between the objects of two classes to detect functional correspondences. He assumed that the spatial correlation between the instances of classes can indicate an associative relationship, in conformity with Tobler's "first law of geography": "everything is related to everything else, but near things are more related than distant things" (Tobler, 1970). These associative relationships indicate functional dependencies. High global adjacency measures of the objects of two classes thus indicate that adjacent objects of those classes may form a meaningful unit at a higher level of abstraction and may therefore be aggregated (Figure 6.9). Topology serves a double purpose in this approach, both for defining the global adjacency measures and for merging the geometry of adjacent components of a composite object.

Whichever approach is used, they have to be made available in the database. The database structure should provide for the storing these hierarchies. This is easily done by defining a database table in which classes are linked to composite classes. The structure should allow repetition in order to support hierarchies with more than two levels.

6.4.3. *Storing composite objects during and after the generalisation process*

In preserving the links between composite objects and the component objects that they are composed of, the aggregation process builds an object hierarchy; this effectively creates a multi-resolution dataset that allows navigation through the levels of abstraction. The composite objects may have their own geometry, or not. It is possible to store the aggregation results solely in the form of a database table (Table 6.1), containing the links between component and composite objects, and to create the geometry of the composite objects only when needed. The two columns in Table 6.1 both contain unique objects identifiers. Objects 34 and 234 are both components of the composite object 822. Composite object 822, in its turn, is a component of composite object 824. Object 25 is another component of 824. For reasons of performance however the results of each aggregation step may also be stored in the form of full GIS datasets that include the merged geometries of the composite objects. The composite objects may thus be stored as:

- Objects without their own geometry. The composite objects are just references in a table (Table 6.1). Object 822 (Table 6.1) illustrates that composite objects can be part of other, higher-level composites.
- Objects with their own geometry. The geometry being the merged geometry of its components. The component objects contain references to the composite.
- A new dataset for each aggregation level. This might be disadvantageous for the reason of redundant storage in the sense that objects that have not changed between the levels are stored more than once. Moreover, it may not always be possible to identify distinct aggregation levels.

Intermediate objects, for example to compare different solutions, may be stored in similar ways. They may either be stored with the permanent objects carrying a special annotation or in separate classes/tables, again depending on what is most effective for the operations that have

Table 6.1. Table containing object-to-composite relationships.

object.ID	composite object.ID
25	824
34	822
140	823
234	822
333	823
623	822
822	824
823	
824	

to be performed on them. If composite objects are derived through aggregation they initially do not have their own geometry nor do they have other attributes of themselves. Similar to their geometry, thematic attributes of the composites may be derived from the components' attributes. These too may be stored permanently or derived when needed.

6.5. Conclusions

The efficiency of GIS would greatly improve if they were able to support complex processes such as generalisation and management of multiple representations. This requires a rich modelling of data and processes, and so database management becomes a key issue in GIS. In this chapter we described some requirements to deal with this issue. To summarise, an ideal database system for generalisation should:

- Support the use of the Object-Oriented model or another rich conceptual model such as Entity/Relationship, in order to support complex data modelling and processing. Data schemas expressed within these models should discern a sufficient number of classes and preferably also super-classes, in order to be suitable for generalisation purposes.
- Support the use of conceptual models especially dedicated to geographical data. In particular, such models should allow rich modelling and manipulation of geometry and topology. According to these models, data schema may concentrate more on the natural organisation of geographic data, rather than on the management of geometries that becomes much more simplified in terms of modelling.
- Rely on a database management system, in order to take advantage of the data management capabilities (data modelling, data sharing, versioning, query capabilities, speed optimisation, and security management).
- Support the use of classification hierarchies and inheritance, in order to better organise data and to efficiently guide some of the generalisation processes such as aggregation.

- Allow the building and retention of links between component and composite objects, in order to reorganise data to make explicit the information implicit within geographic phenomena, and thus guide generalisation processes such as aggregation through composite objects.
- Support topology that allows the relationships between (large) sets of objects to be queried efficiently (not just the topological relationship between two objects). Most likely this means that the schema that is used during the generalisation process must support stored topology.
- Deal with multiple representations for a single real-world entity. This is helpful in guiding the generalisation process and to efficiently store results. Furthermore generalisation systems should have an interface for displaying and manipulating multiple representations.

W.A. Mackaness, A. Ruas and L.T. Sarjakoski (Editors)

Chapter 7

A Real-Time Generalisation and Map Adaptation Approach for Location-Based Services

Tapani Sarjakoski, L. Tiina Sarjakoski

Department of Geoinformatics and Cartography, Finnish Geodetic Institute, P.O. 15, FIN-02431 Masala, Finland
e-mail: {Tapani.Sarjakoski,Tiina.Sarjakoski}@fgi.fi

Abstract

Utilisation of the geographic locations of a user and services, or points of interest, lies at the heart of location-based services. In most of the current location-based services, a map on a mobile device is used to communicate information to the user. This chapter studies related map generalisation issues. Real-time generalisation and map adaptation utilising context parameters for controlling the data content and presentation of mobile maps are discussed. The chapter reflects on experience gained in a research project called GiMoDig (2001–2004), tasked with developing a prototype service using mobile technologies and generalisation methodologies. The prototype served as a comprehensive demonstrator for setting up a web-based layered system for providing vector-based maps as part of a location-based service. The role of the real-time generalisation layer in the overall system architecture is presented, together with the service interface (itself based on existing web-specifications). The map adaptation functionality utilised a knowledge-based approach for controlling the creation of the map view in real-time. The results from the GiMoDig project showed the relevance of real-time generalisation to mobile applications requiring timely and focused information, customised according to context parameters, user preferences and location. The project and the developed prototype system also highlighted the importance of linking the architecture and implementation to the conceptual process model. The proposed model seems flexible enough to support a wide variety of implementation approaches.

Keywords: mobile map service, location-based services (LBS), real-time generalisation, adaptive generalisation, adaptation

7.1. Introduction

Geoservices are defined as web services that provide, manipulate, analyse, communicate and visualise any kind of space-related information (Reichenbacher, 2005). Being a special type of geoservice, a location-based service (LBS) includes an additional parameter: location. An LBS may be described as a type of geoservice that utilises the geographical location of a mobile user, and operating via a mobile device delivering a range of services based on positional information. Typically LBSs deliver geographic information between mobile and/or static users via the Internet and/or wireless networks.

7.1.1. *User scenario*

The following user scenario (Figure 7.1) is drawn from the project "Geospatial info-mobility service by real-time data-integration and generalisation" (GiMoDig, 2001). It serves as an application example of mobile use throughout this chapter and shows how a variety of information might be accessed in order to undertake a task.

Hiker in Nuuksio National Park

A hiker, 22-year-old geology student Annu, goes on a camping trip to Nuuksio National Park in southern Finland. She uses topographic maps displayed in real-time, on a mobile device – a Personal Digital Assistant (PDA), provided by the GiMoDig map service. Using maps of different scales she is able to determine her location, navigate to her destination and obtain information on specific areas and features on the map. She starts her trip from Helsinki city centre and first requests route guidance for Nuuksio National Park. The service shows her an overview roadmap, upon which is highlighted the route from her current location (a bus stop in Helsinki city centre) to Nuuksio. She asks the service: "Where is the nearest bus and at what time does it leave?" The GiMoDig service communicates with a public transportation service and shows a map, which informs her that the next bus leaves at 10:10 a.m. from Platform 8 in Terminal B.

When she arrives in Nuuksio Park, Annu wants to know the rules governing the park. Meteorological information is delivered to her mobile device informing her that currently hikers are not allowed to build open fires since this may cause a forest fire. By selecting the Nuuksio area on the screen she obtains more detailed information on the area: people are not allowed to pick flowers, cut trees or

Fig. 7.1. A hiker zooms from an overview mobile map into a detailed map when looking for a camping place.

disturb animals. First she wishes to see a map of the whole of the Nuuksio area and then zoom in on some interesting places (Figure 7.1). All the maps she uses are delivered to her mobile device by the GiMoDig service in real-time, including the most up-to-date map information, such as changes to paths as compared with last year. Depending on the scale there are different symbols shown on the map, which she can click on to get more detailed information (such as detail sketches of campsites, photos and links to related information sources). One of the links shows a 3D model of the hill area where she is going to hike. If she wishes, the service can adapt the symbols on the map related to her special interests. For example it can overlay places of interesting geological formations on top of the topographic map. She can also scroll the map to view areas outside the National Park. Annu can request information on nearby campsites. She can also look in detail at the services of a campsite and see whether it is unoccupied and has a water supply. The PDA offers shortest route information to the chosen campsite provided via a detailed topographic map which also shows waterways, hills and features to avoid such as swamps. The map is oriented along the direction she is walking and gives instructions if she strays from the route. Finally, she arrives at the campsite by Haukkalampi lake and looks at the beautiful view over the lake, just as the device has predicted.

7.1.2. *Background*

Great interest in LBSs has emerged in tandem with an increased use of cell phones and PDAs. It is worth remembering that the need for such services existed prior to digital technology and their development has gone through several step changes. Before the era of digital communication, printed catalogues displaying driving information and maps typically showing motels, restaurants and stores, were typical. There then came car navigation systems using GPS, digital maps and route information typically stored on a CD-ROM. Over the last ten years, the same information has become available over the Internet. Currently, the arrival of high bandwidth mobile networks offer the potential to disseminate a huge variety of information via LBSs.

Much of this information is spatial in nature, and naturally lends itself to communication in map form though it could be conveyed in the form of coordinates, street addresses or named locations. The basic design principles for a map on a mobile device are the same as for any map: some underlying topographic information, with thematic information (such as Points of Interest, PoIs) overlaid on top of the map. Additional design constraints include taking into account the ambient conditions and the fact that the user is mobile.

The process of map design requires us to consider various interdependent processes of map generalisation and visualisation. This chapter reports on an EU funded project, GiMoDig, that seeks to illustrate the challenges of designing an LBS capable of disseminating real-time mobile maps in vector format at various scales, for different user groups and tasks (Sarjakoski and Sarjakoski, 2005). The chapter starts with a brief review of the research and development of maps in location-based services, followed by a discussion of the generalisation requirements for real-time delivery of this LBS. The GiMoDig service architecture and the generalisation interface of the system are described, concluding with a proposed conceptual process model for generalisation in LBSs.

7.1.3. *Previous research*

The rapid development of mobile map applications began in the mid 1990s, though with little discussion as to the relevance of generalisation methodologies. Abowd et al. (1996) presented one of the first research works on mobile guides, the CyberGuide system, in which a GPS was used to locate outdoor locations. The CyberGuide project focused on how portable computers could assist in exploring physical spaces and cyberspaces. It provided simple static maps in black and white, which were locally stored in the device. Baus et al. (2001) presented a mobile guidance system providing static maps, and Deep Map (Zipf, 1998) was also developed in the LBS context presenting the user location on a map displayed on a mobile device. Additional information, such as PoIs, photos and text on landmarks were displayed on top of the map (Kolbe, 2003).

In its infancy much of the LBS development relied on the enthusiasm of computer scientists, and had an emphasis on technical development. Cartographic issues were often neglected in the development of LBSs (Reichenbacher, 2001), but have since received more attention. Early versions of LBSs contained static images, in raster form and often suffered from poor legibility. They were adopted from existing digital maps, which were originally intended for larger displays. Other lines of research have included the incorporation of multimedia technology in mobile devices. For example, Gartner et al. (2003) reported on the Navio project, in which multimedia cartography presentations were used to support wayfinding and navigation for pedestrians. Pedestrian guidance was improved by embedding active and/or passive landmarks. Landmarks are prominent objects, which act as marks for places and can be used as reference points. As explained by Gartner and Uhlirz (2005), active landmarks are places equipped with technology to communicate with the pedestrian's mobile device when within the range of the active landmark. For more detailed discussion on landmarks see Sester and Elias, this volume (Chapter 10). Luley et al. (2003) describe a multimedia based system for tourist guidance that includes 2D, 3D maps and satellite images. The TellMaris research prototype reported by Schilling et al. (2005) provides 3D mobile map views for boat tourists. In this system the user is able to choose among three different viewing modes: a pedestrian perspective, looking directly down from above as in a 2D map and finally slightly downwards which is here called a bird-perspective or flying mode. In user tests the flying mode was found to be much easier than the walking mode for navigational purposes. The GUI allows switching between the 3D view and the 2D vector map providing a smaller scale overview.

There are several commercial applications for maps on mobile devices, in which the maps are displayed on the screen of a PDA, or a cell phone (Sarjakoski and Nivala, 2005). Most of the applications are for car navigation purposes, but there are also navigational guides for cyclists and pedestrians. Examples include Navman GPS 3300 Terrain (2005), Outdoor Navigator (2005), TomTom CityMaps (2005), Falk City Guide (2004) and MapWay (2005). Developers of LBSs, being clearly service oriented, have given great importance to the end users when creating user-friendly mobile applications. The usability issues are emphasised in the European Union's (EU) Information Society Technologies (IST) programme (IST, 2001). Examples of projects of the IST programme are WebPark (Edwardes et al., 2003a); LOL@ (Pospischil et al., 2002); Crumpet (Poslad et al., 2001); LoVEUS (2005); and GiMoDig (Sarjakoski et al., 2002).

Adaptiveness and context sensitivity are regarded as essential ingredients for LBSs (Meng et al., 2005). Cheverst et al. (2000) studied how context awareness can be utilised in a tourist information system, though maps played a minor role in their study. In their GUIDE project, based on the approximate location of the device, guidance and information services were provided through a browser-based interface. An adaptive system should be capable of changing its own characteristics automatically according to the user's needs (Oppermann, 1994). Baus et al. (2002) describe a system that determines the location of the user and adapts the presentation of route directions according to the characteristics of the user's mobile device as well as to the cognitive resources of the user. Reichenbacher (2004) studied the process of adaptive and dynamic generation of map visualisations for mobile users.

The need to improve the quality of maps displayed on mobile devices is widely recognised. Experiments have been carried out in several research projects on maps based on the Scalable Vector Graphics (SVG) technique (SVG, 2004). SVG maps were produced in the TellMaris project (Schilling et al., 2005), as well as in the RIMapper test-bed application for online risk indicator maps as reported by Köbben (2003). While vector data offers good flexibility for application development, the time to transfer the data over a narrow bandwidth connection is often an issue. Developments have been pursued for solving the problem, especially by using compression methods. Persson (2004) describes a method (implemented in RaveGeo software), which makes geographic vector data available at wide scale ranges over wireless networks and the Internet, and supports efficient streaming of compressed vector data at different resolution levels. The work described by Persson is based on earlier studies by Bertolotto and Egenhofer (2001) for disseminating geographic vector information. The task of transmitting vector data progressively has also been studied by Buttenfield (2002).

The rapid development of LBSs has brought additional challenges for cartographers. The field of TeleCartography has emerged to respond to the requirements of mobile map applications (Urquhart et al., 2003; Gartner and Uhlirz, 2005). Although many research papers address the relevance of map generalisation in LBSs (for example Reichenbacher, 2004), only a few results on generalisation functionalities have been reported (but see Burghardt et al., 2003). The WebPark project is concerned with delivery of thematic information over mobile devices. The thematic foreground data on PoIs and wildlife observations are supplied by an information service, generalised and overlaid on a topographic background map (Edwardes et al., 2005). In the project, the generalisation operators applied on the point data include selection, simplification, aggregation and displacement (see Edwardes et al., this volume (Chapter 8)). In a study reported by Dillemuth (2005) generalised maps were compared with aerial photographs in a route-following task, with the maps/photographs on a handheld computer used as a navigation aid. The use of generalised maps resulted in overall faster route completion, less map browsing, and fewer stops compared with the use of aerial photographs. The results illustrated the need for appropriately designed map representations for mobile devices.

7.2. Generalisation of Maps in Location-Based Services

In this section we list the characteristics of LBSs and discuss the role of generalisation in map-based services, with reference to the GiMoDig project.

7.2.1. *Characteristics of using maps in Location-Based Services*

Gartner and Uhlirz (2005) and Reichenbacher (2004), among others, have studied and identified characteristics of map-oriented LBSs. The following characteristics previously discussed in the literature are expanded upon:

- Positioning and its utilisation;
- Mobility and activity of the user;
- Personalisation and adaptation of the services for various user groups and contexts;
- Information processing and presentation.

Positioning and its utilisation

Using information about the mobile user's position is central to most location-based services. A range of technologies provide positional information, with varying positional accuracy, varying technological complexity and varying availability. While we mainly discuss here LBSs using maps in outdoors situations, it is worth studying the most central requirements for positioning. The positional accuracy of 5–10 m, offered by standard GPS single point positioning using simple handheld receivers in unobstructed outdoor situations, is sufficient in most cases. Based on this, users can locate and reference themselves within their surroundings. Certain applications, such as lane-level vehicle navigation, would require even higher precision, about 1 m accuracy offered by differential GPS. Higher accuracy requirements seem to be applicable only for technically oriented and professional tasks. Besides GPS positioning, simple network-based positioning techniques are also widely used for LBSs. These offer usually much lower accuracy, thus limiting their applicability for example for navigation purposes but are sufficient for locating near-by services. More detailed surveys on the issue can be found for example in Gartner (2005).

Mobility and activity of the user

User mobility is the essence of LBSs. Research on modelling movement patterns of users has been conducted by Mountain and Raper (2002). While moving around by walking, cycling or driving a car, the users do not often know beforehand the exact place from where they will need map information, nor the type of information required. Thus users need access to information regardless of place and time and in a range of forms (Kopomaa, 2000). User activities have been discussed by Dransch (2005, p. 36). Having the basis in activity theory (Leontjew, 1978), "mobile geoservices are regarded as artefacts that mediate activities which are related to or executed during movement. They can either be used continuously throughout a mobile activity... or discretely during a mobile activity at a special place or/and time to get information about nearby spatial objects. The leading idea of mobile geoservices is presenting information with relevance to a user in a specific mobile activity and context. This goes far beyond the possibilities of a traditional analogue or digital map." The main issues in designing mobile geoservices are: What activities, goals and sub-goals have to be supported by the service? Such questions have important ramifications for generalisation.

Personalisation and adaptation

Oppermann (1994) discusses adaptive systems, which are capable of changing their own characteristics automatically according to the user's needs. Reichenbacher (2004) notes that personalisation can be regarded as a special case of adaptation and refers to Searby (2003) who explains that personalisation occurs when a system is modified in its configuration or behaviour by information about the user. An example of an adaptation of a map is given by Harrie et al. (2002a). Their approach applies adaptive generalisation, based on the current location

Fig. 7.2. A variable scale map is an example of an adaptive generalised map for LBSs (modified from Harrie et al., 2002a, 2002b). Printed by permission.

of the user. Figure 7.2 shows a variable scale map for small mobile devices, where the level of map generalisation increases towards the edges of the map.

Map adaptation may be controlled by context parameters. Information supplied to the LBS often relates to the user's immediate vicinity, therefore the user location is the most prominent context parameter for mobile usage situations. Dey (2001) defines the context as any information that can be used to characterise the situation of an entity, where entity means a person, place, or object, which is relevant to the interaction between a user and an application. A mobile system needs to be aware of the user's context and characteristics to assist the user in a mobile environment (Reichenbacher, 2004). Besides the location, the other contexts relevant to mobile map usage (as proposed by Nivala and Sarjakoski, 2003) include: orientation, time, navigation history, purpose of use, social and cultural situation, physical surroundings and system properties. Sarjakoski and Nivala (2004, 2005) discuss how these context parameters may influence the visualisation of mobile maps.

Information processing and presentation

Since LBS displays are small, the content of the maps and their design must be simple enough for legibility. Only the most essential information can be displayed. A typical resolution for a PDA display is 240 × 320 pixels. Klippel et al. (2005) compare the purpose of generalisation with schematisation from a cognitive perspective of knowledge representation and reasoning. The expanding mass of mobile applications, of which maps are a substantial part, introduce new user groups who do not necessarily have much previous knowledge of using and interpreting maps. The map information must often be strongly generalised and adapted to the specific usage situation, so that information overload can be avoided.

Although mobile maps have similarities with web maps (Kraak and Brown, 2001), creating them according to the same design principles does not work properly. Mobile devices are typically used in outdoor situations, and it is often difficult for the user to interpret map colours and patterns due to light reflection. Since there is no room for a map legend, alternatives such as "mouse-over" functions must be used. Screen dynamics and interactive maps can be used, as in web maps, so that not everything has to be shown in a single image. Clickable map features can be used to access additional information accessible via the network the mobile device is connected to. Different levels of map views or scales are needed in order to interpret the relational distances between real world objects while moving at varying speeds.

The mobile user is by default interested only in timely map information. Therefore, one of the most essential characteristics of the mobile maps is the need for information that is current.

7.2.2. *Need for generalisation in real-time*

Solutions based solely on downloadable pre-processed maps may not be ideal for typical mobile users. The user often needs tailored maps, adapted on-the-fly to the current usage situation. Generalisation is integral to this (§7.3.2). The most useful generalisation operator is selection, which underpins creation of thematic maps such as for cycling, hiking, or skiing (Figure 7.5). The other operators serve to improve the legibility of the map and are especially needed due to the limited display sizes. If we compare the process of on-the-fly generalisation with traditional cartographic generalisation, several differences can be identified (Lehto and Kilpeläinen, 2000; Lehto and Sarjakoski, 2005a). These characteristics define a category of map generalisation, which in the GiMoDig project was called real-time generalisation.

- Computational processing of generalisation is carried out in real-time, during the request–response dialogue from the client to the server in the network;
- The current location of the user is taken into account in the generalisation computations;
- Arbitrary, not predefined scales are provided by the system;
- The resulting maps are personalised;
- The display scale of the mobile device is taken into account during the generalisation process.

The design principles applied to mobile maps in comparison to paper and web maps are summarised in Table 7.1.

In the following section we describe the GiMoDig project, in which these principles were further elaborated and tested using a prototype system.

7.3. The GiMoDig Project

7.3.1. *Background to GiMoDig*

In the EU project GiMoDig (Geospatial info-mobility service by real-time data-integration and generalisation, GiMoDig, 2001; Sarjakoski et al., 2002), the overall objective was to develop methods for improving the accessibility and interoperability for using national topographic databases in mobile contexts. From a methodological point of view, real-time generalisation and data-integration were the focus of the work. An extensive prototype system was implemented and serves as an example of how real-time generalisation can be used to improve the cartographic quality of the maps delivered to different types of mobile devices and users.

The overall assumption for the project was that national topographic maps could play a key role in mobile map services. The traditional supplier-centric view should gradually be replaced by a more flexible and operational workflow, in which the individual user's needs are given a high priority. For this vision to be realised, the National Mapping Agencies (NMAs) must provide the network not only with pre-designed maps but also spatial data services that

Table 7.1. Distinguishing characteristics of paper maps, web maps and mobile maps.

	Paper map	Web map	Mobile map
Generalisation			
• Technology	Pre-processing offline	Pre-processing offline On-the-fly/real-time	Pre-processing offline On-the-fly/real-time
• Level of generalisation	Map specification dependent	On-demand specific	On-demand specific, high level of generalisation typical
• Operators	Spatial and attribute transformations	Spatial and attribute transformations for pre-processed maps Processing time controls the usage of on-the-fly solutions	Spatial and attribute transformations for pre-processed maps Processing time controls the usage of on-the-fly solutions Location parameter as an additional parameter in the array of variables for real-time solutions
User guidance	Map legend	Map legend Mouse-over functions Pop-up menus Help menus	Mouse-over functions Pop-up menus
Interaction	No interaction	Clickable objects to additional network information sources	Clickable objects to additional network information sources
Map adaptation	Thematic pre-processing	Thematic pre-processing Adaptation interactively to preferences given by the user	Thematic pre-processing Adaptation interactively to preferences given by the user Adaptation to mobile contexts in real-time

can deliver data in real-time. On top of such data services, various end-user applications can be built by value-added service developers. In the GiMoDig project a prototype system was designed and implemented for verifying the validity of the approach. The prototype system provided basic geospatial data and a number of applications were implemented on top, based on use-cases identified at the beginning of the project. A special focus of the research was on adaptability according to the needs of different mobile users.

7.3.2. *From field testing to adaptive generalised maps*

The idea of adaptive maps was developed following an analysis of early field-testing in the project. Here we limit ourselves to a few results relevant to map generalisation (for complete discussion see Nivala et al., 2003 and Sarjakoski and Nivala, 2005). The purpose of the evaluation was to identify design principles for maps on small displays as well as the main benefits and obstacles to using topographic maps on mobile devices. Typical questions were: What are the deficiencies of the maps? Can the user recognise the map symbols? What map scales are mostly used?

Among the results from field testing, users believed that one of the main advantages of mobile maps was the possibility of zooming between different map scales. The users wanted the step between the scales to be smooth enough so that no one would lose the sense of place in the map overall. The design solution should not show the user a totally different-looking view, and visual representation should be consistent between scales. It was also noted, quite naturally, that small-scale maps were used for planning a route whilst the larger-scale maps were used when walking along the route. The overview map should contain general information on the terrain, routes and services in the area. When using large-scale maps, people were interested in seeing more specific information of nearby areas, landmarks and services.

The test revealed the need for maps adapted to specific usage situations. It was observed that the actual environment surrounding the user essentially affects the use of maps. The map is strongly related to situations in which the users try to find their way in an unfamiliar environment, and the adaptation of map visualisation and contents within the usage context greatly improves the usability of mobile topographic maps. The results of the user tests were taken into account in the implementation of the GiMoDig Graphical User Interfaces (GUIs) and adaptive maps. A PDA, a laptop computer and a cell phone were used as client platforms (Figure 7.3). The PDA implementation was based on Scalable Vector Graphics (SVG) (Sarjakoski et al., 2003) using a mobile SVG viewer called Embedded Scalable Vector Graphics (eSVG, 2004). The client on a laptop used an SVG plug-in within a Web browser. Finally, the client on the cell phone was based on Java MIDP. The cell phone and PDA maps were designed for outdoor usage and therefore the colours should be visible in different light conditions. The level of detail (contents and level of generalisation) on the maps varied depending on the size of the display.

Based on the defined user requirements and the results from the field-testing, four different use cases were identified for implementation during the project: (1) A Hiker in a National

Fig. 7.3. The GiMoDig maps were adapted for different devices and GUIs needing generalisation for different display sizes.

Park, (2) An Emergency Case, (3) A Cyclist, and (4) Expert Use. The test area for Use case 1 covered a national park in southern Finland. The Use case 2 test data covered Hanover city, the cycling routes crossed the Danish–German border area in Use case 3 and finally expert use in Use case 4 was tested for all our four test areas, including a cross-border area between Sweden and Finland. Throughout this chapter we focus on the example of "A hiker". For further applications, see Sarjakoski and Sarjakoski (2005).

The GiMoDig prototype system accommodates personalisation based on: user's age group and language, activities, time, test area and device (Figure 7.4). The right column of Figure 7.4 demonstrates how personalisation of the service is achieved. Based on the personalisation, the maps (contents, generalisation, visualisation) are adapted on the server in real-time according to different users in different usage situations (in the right column of Figure 7.4, to teenagers hiking in winter time). The users may define the season and time of day (day/night), according to which map information they are interested in displaying. The choice of the user's age group governs the displayed icon style (right, bottom of Figure 7.4: 11–17 years; left, bottom of Figure 7.4: 45+ years). The topographic map data are displayed together with the PoI symbols overlaid. The level of generalisation varies while zooming. For the overview map (left, middle of Figure 7.4) generalisation of roads, trails and lakes, is done in real-time. The icon and feature class selections are also done adaptively in real-time. For example on a summer base map (left, bottom of Figure 7.4) swimming places are displayed but no skiing tracks are shown. In the GiMoDig service, adaptive generalisation is controlled by context parameters such as the display size of the device (Figure 7.4), location of the user and the level of detail chosen by the user.

7.3.3. *Architecture of the GiMoDig prototype*

The prototype system in the GiMoDig project is based on a layered web-architecture (Figure 7.5). The overall service architecture and the functionality of the GiMoDig map service have been described in detail by Lehto and Sarjakoski (2004, 2005b). A Mobile Client, running on a mobile device, stands at the top in the hierarchy of service layers. It communicates with a Value Added Service layer through any wireless or wired channel serving IP-protocol. The Value-Added Service returns an SVG-formatted map, based on the request from the Mobile Client. The map content is based on the desired Point of Interest (PoI) information and desired elements from the topographic data. PoI data are provided by a PoI database while the topographic data are provided primarily by the NMAs' topographic databases (TOPO DBs). In between we have three additional layers: Integration Service, Generalisation Service and Portal Service. The role of the Integration Service layer is to transform the topographic data to be compliant with the GiMoDig Global Schema. This database schema has been designed in the project and serves as a means for supplying a unified view to the topographic databases varying from country to country. Generalisation Service is responsible for the generalisation of the topographic data jointly with the PoI data while Portal Service makes the rendering by converting the GML-formatted spatial data to an SVG-formatted map. The specifications for these maps are given in parameter files, accessed by the Value-Added Service layer and passed down to the Portal Service layer, and further down when appropriate. The specifications include those to control the Generalisation Service.

Fig. 7.4. The GiMoDig service adapts and delivers maps in real-time to mobile users.

Communication between the layers is based on international standards and specifications by ISO, W3C and Open Geospatial Consortium (OGC, 2005) wherever possible. As Figure 7.5 shows, Web Feature Service (WFS, 2002) requests, with spatial data represented in Geogra-

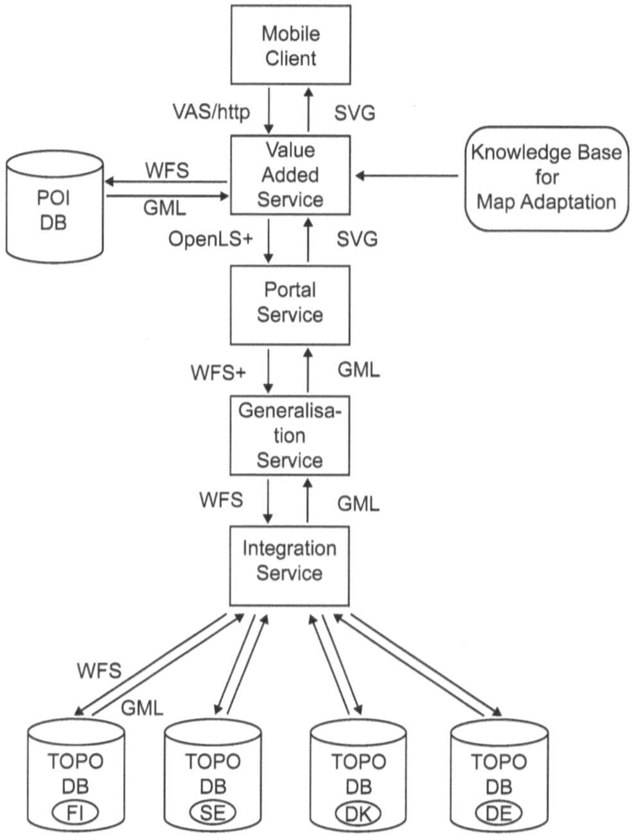

Fig. 7.5. The GiMoDig service architecture. The format of the requests and replies are indicated next to the arrows.

phy Markup Language (GML), are used extensively. For the Generalisation Service the WFS interface is extended in order to be able to transfer the parameters and data guiding the generalisation process (Sarjakoski et al., 2005b). The Open LS Presentation Service (2005) is also extended in order to be able to carry the generalisation parameters through this layer from the Value-Added Service (VAS) layer. Interestingly, the Open LS Presentation Service specification already makes it possible to transfer the PoI data along with the request such that the PoI data can be integrated into the returned map.

The prototype system is largely based on XML-technology. For example, the schema transformation in the Integration Service and the rendering in the Portal Service are processed as XSL transformations. In some cases these transformations are augmented by Java functions, for example where coordinate system transformations are required. The implementation of the Generalisation Service layer is based on the usage of JTS Topology Suite (2005) and Unified Mapping Platform (JUMP, 2005), open source packages for topological and geometric processing of spatial data. Although not covered in this chapter, it should be noted that the GiMoDig project also included a research line on how multiple representation could be used as an alternative approach for mobile applications (Hampe et al., 2004), instead of real-time generalisation.

7.3.4. *From context parameters to generalised maps*

As described in §7.3.2, the implementation accommodates personalisation based on: user's age group and language, activities, time, test area and device. Based on the personalisation, the maps (contents, generalisation, visualisation) are adapted in real-time to different users in different usage situations (Figure 7.4). Personalisation as such is a part of the Mobile Client's functionality, and is reflected in a query from the Mobile Client to the VAS layer. This query comprises context parameters, described in Table 7.2.

A query to the VAS layer represents a high level description of the map requested by the Mobile Client. The role of the VAS layer is to translate this into a map specification such that the underlying service layers can produce the requested map. The translation mechanism is described in §7.3.6. In the case of GiMoDig, these specifications dealt with the following issues:

1. Selection of the topographic feature classes to be shown on the map;
2. Selection of the PoI classes to be shown on top of the topographic data;
3. Specification of the generalisation functions to be executed for the selected features;

Table 7.2. The parameters for a query from a Mobile Client to the Value-Added Service layer.

Parameter name	Domain of the parameter	Explanation
Usecase	ennumerated: *outdoors, cycling, emergency, expert*	A use case implemented in the prototype
Time	ennumerated: *spring, summer, autumn, winter*	Season
Lod	ennumerated: *basic, intermediate, overview*	Level of detail
Scale	integer value	Numeric scale of the map
Center	two floating point values	Map coordinates of the center point of the requested map
Device	ennumerated: *iPAQ, PC*, etc...	Device, the properties of the supported devices are known by VASL
Age	integer value	User's age
Position	two floating point values	User's position (optional)
Example query: http://geoinfo2.fgi.fi:8080/VASL/get2? &usecase=outdoors &center=362600.0,6688300.0 &time=autumn &age=31 &lod=basic &scale=20000 &device=iPAQ		

```
<?xml version='1.0'encoding='iso-8859-1'?>
<gen:GetGeneralised xmlns="http://www.opengis.net/wfs"
              xmlns:gen="http://gimodig.fgi.fi/gen"
              xmlns:xls="http://www.opengis.net/xls">
  <gen:WFSQuery>
    <GetFeature>... </GetFeature>
  </gen:WFSQuery>
  <gen:GeneralisationParameters>
    <gen:Parameter name="PARAM" value="VALUE"/>
    .
    .
    <gen:Operator name="OPERATOR_NAME" order="EXEC_ORDER">
      <gen:Parameter name="OPER_PARAM" value="OPER_VALUE"/>
      .
      .
      <gen:Target filter="FILTER_TYPE">
        <gen:FeatureType name="FEATURE_TYPE" />
        .
        .
      </gen:Target>
    </gen:Operator>
    .
    .
  </gen:GeneralisationParameters>
  <gen:Overlays>
    <xls:Overlay>... </xls:Overlay>
    .
    .
  </gen:Overlays>
</gen:GetGeneralised>
```

Fig. 7.6. The structure of the query for the Generalisation Service layer.

4. Specification of other functions to be executed (e.g. icon placement);
5. Specification of map visualisation (e.g. colours, line widths).

The selection of topographic feature classes (item 1) is dependent on the database schema that is supported by the Integration Service. In the GiModig prototype the schema is fixed and hard-coded in the VAS layer. The VAS layer requests the appropriate PoI data from the PoI database server (item 2). Regarding other items, a request will be sent to the Portal Service layer along with the other specifications. This request is expressed in a form of an extended OpenLS Presentation Service request. The specifications for visualisation (item 5) are given as SVG styles that are used by the Portal Service layer once the data has been returned from the Generalisation Service layer.

In an automatic cartographic generalisation process, it is usually necessary to assign the responsibility of the overall control of the generalisation process to a certain system module. In the GiMoDig prototype this responsibility is given to the VAS layer. The Generalisation Service as such has a generic interface used to control which generalisation operators should

```
<?xml version='1.0'encoding='iso-8859-1'?>
<gen:GetGeneralised>
  <gen:WFSQuery>
    <GetFeature outputFormat="GML2" handle="GiMoDigQuery">
      <Query handle="query0" typeName="Road" version="1.0">
        <PropertyName>centerLineOf</PropertyName>
      </Query>
    </GetFeature>
  </gen:WFSQuery>
  <gen:GeneralisationParameters>
    <Operator name="Douglas-Peucker" order="0">
      <Parameter name="threshold" value="5.0"/>
    </Operator>
    <Operator name="Gauss-filtering" order="1">
      <Parameter name="distance" value="10.0"/>
      <Parameter name="kernelShape" value="0.00001"/>
      <Parameter name="kernelSize" value="3"/>
      <Parameter name="numPoints" value="2"/>
      <Parameter name="extrapolation" value="1.0"/>
    </Operator>
  </gen:GeneralisationParameters>
  <gen:Overlays>
    <Overlay>
      <POI ID="POI.Finland.1762059" POIName="Campsite">
        <gml:Point>
          <gml:pos>362236.937 6688534.245</gml:pos>
        </gml:Point>
      </POI>
    </Overlay>
    <Overlay>
      <POI ID="POI.Finland.1762060" POIName="Fireplace">
        <gml:Point>
          <gml:pos>362240.384 6688496.551</gml:pos>
        </gml:Point>
    </Overlay>
  </Overalays>
</gen:GetGeneralised>
```

Fig. 7.7. An example of a query to the Generalisation Service layer.

be applied and which parameter values should be used. The query contains a WFS query for querying the data and parameters for data processing. It also enables the transfer of PoI-data and the location of the user into the data processing layer. The structure of a data processing query is given in Figure 7.6

A WFS query is inside the gen:WFSQuery element and it is passed directly to the WFS Integration Service. PoI-data and user position coordinates are inside the gen:Overlays element.

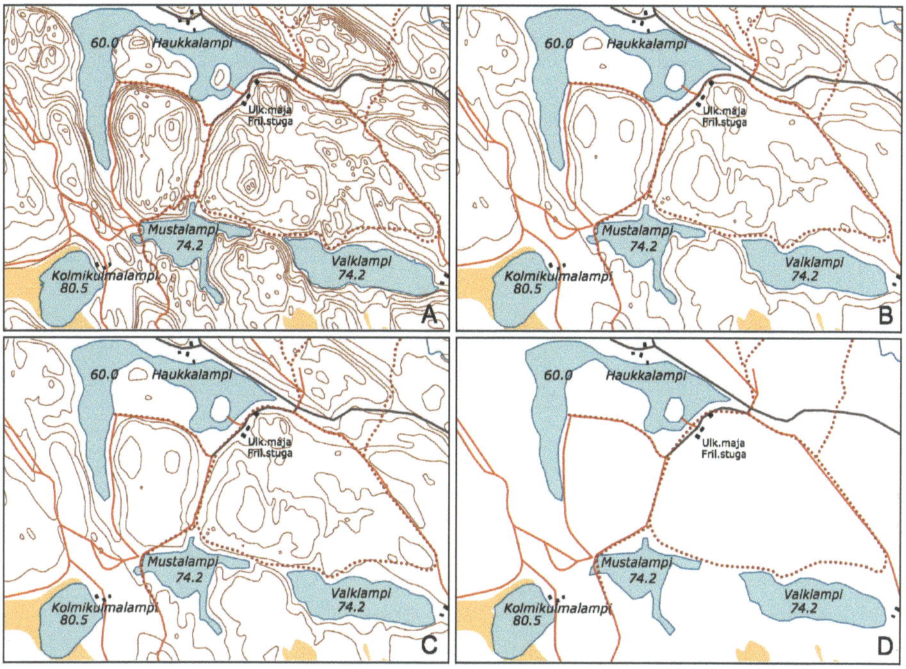

Fig. 7.8. Screen shots showing real-time generalisation functionality applied in the GiMoDig service for maps in a scale range of 1 : 10 000–1 : 50 000.

The structure of xls:Overlay elements complies with OpenLS Presentation Service specification.

Actual data processing is described with the Operator elements. There can be many operators – even with the same OPERATOR_NAME. The EXEC_ORDER attribute describes the execution order of operators. The Operator that has the smallest execution order value will be executed first. The parameters of each operator are described by the Parameter elements inside the Operator element.

The target features of an operator are selected by the element Target. The FILTER_TYPE attribute can be either Include or Exclude. In an inclusion target, the group will be the feature types listed in FeatureType elements. In an exclusion target, the group will be a group of all feature types except the listed ones.

Figure 7.7 gives an example of the query. The query requests road features and generalises them using line simplification using the Douglas–Peucker algorithm and line smoothing with Gauss-filtering. In the overlay part, "Campsite" and "Fireplace" have been included as PoIs to be merged into the generalisation process.

7.3.5. *Real-time generalisation operators in GiMoDig*

The real-time generalisation functionality implemented in the prototype service includes the following operators:

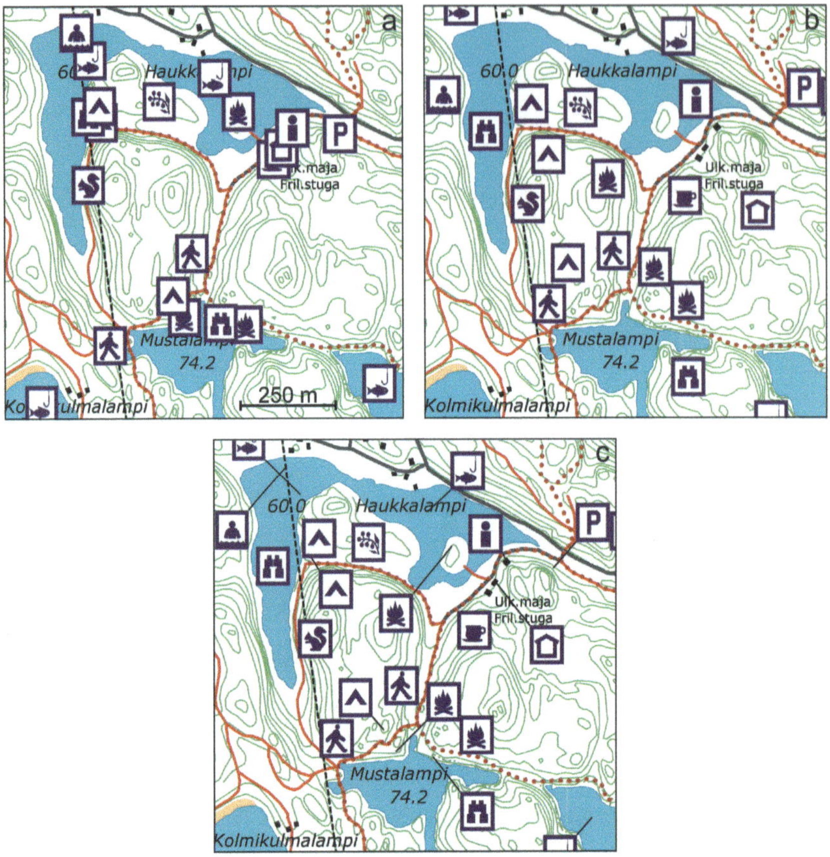

Fig. 7.9. Icon placement in the GiMoDig service is based on real-time processing.

- Selection by feature class;
- Area selection by min/max value;
- Line selection by min/max length;
- Contour line selection by interval;
- Line simplification with Douglas–Peucker algorithm;
- Line simplification with Lang algorithm;
- Line smoothing with Gauss-filtering;
- Building outline simplification.

These functions define the basic functionality of the system. The example in Figure 7.8 shows how real-time functionality is applied: the contour lines are generalised, as well as the lakes, roads, tracks and paths. The figures are screen shots from an application for a laptop device, where the map data is delivered from the National Land Survey of Finland to the GiMoDig service for processing and finally displayed as an SVG image on the client's display. The original data are shown in Figure 7.8a at a scale of 1 : 10 000. Figure 7.8b shows the data after generalisation at a scale of 1 : 20 000 (for comparison purposes the scale of presentation is equal in all Figures 7.8a–d). The contour lines have been omitted by real-time selection, based on their interval. The tracks and roads have been simplified using the Douglas–Peucker algorithm and lakes using the Lang algorithm. Finally, Gaussian filtering is used for the smoothing of trails and roads. In Figure 7.8c, the generalisation has been done for a scale of 1 : 30 000 using another set of parameters for simplification and smoothing. In Figure 7.8d, generalisation is undertaken for a scale of 1 : 50 000. All contour lines are omitted by real-time selection, and more extreme parameter values are applied in the generalisation of the line features. Here, the maps are not stored in advance, but are processed during the request from the client in real-time, thus making it possible to deliver maps at a variety of scales.

In addition to the generalisation operators described above, more advanced functions can also be incorporated, such as least squares adjustment based generalisation (Harrie, 2004). The generalisation layer also includes functions for optimal placement of icons representing PoI data (Harrie et al., 2004) (Figure 7.9). In Figure 7.9a, icon placement is not applied, compared with Figure 7.9b, where the icon placement is applied. In Figure 7.9c, the icon placement is done in real-time and reference lines to the original positions of the PoIs are automatically drawn from the displaced icons.

The icon placement functions can be called via the generalisation interface. Although icon and text placement have traditionally not been considered to be part of generalisation, in our case this integration is useful. The same JTS/JUMP environment is used for implementing generalisation and icon placement functions. Functions in both these categories can also be called through a single interface to the Generalisation Service layer. This gives the benefit of being able to call generalisation and icon placement functions in any order to achieve the desired result. The task of text placement is closely related to icon placement, and important for the legibility of the resulting map, but was not in the scope of the GiMoDig project.

7.3.6. *Knowledge base for map adaptation*

Map adaptation is controlled by the Value-Added Service layer in the GiMoDig system architecture. This reflects the idea that we must be able to tailor the maps in detail for each use case. This kind of organisational and system architecture approach allows a single map service to be used by several value-added services. Context parameters control map adaptation (§7.3.4). In principle, the task comprises functional mapping from the context parameters through to the map specifications for each use case. Enumerating all the combinations of the context parameters is not practically feasible. A knowledge-based approach resembling prototype-based programming is used to simplify this task. The knowledge base consists of so-called prototype maps (or prototype-map specifications), which are used by an inference engine to ascertain the specifications for a map with given context parameters. The inference engine uses matching techniques, priority rules, and an inference mechanism in this task (Sarjakoski et al., 2005a). When the knowledge base is created, the prototype maps are defined, starting from the most general ones, then progressing to more and more specialised ones, until the inference engine produces desired specifications for any given combination of the context parameters. This idea is summarised in Figure 7.10.

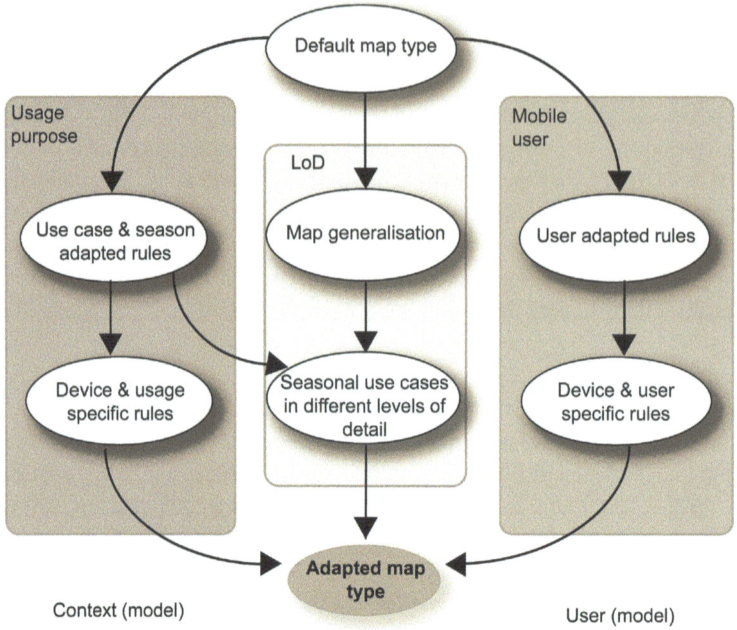

Fig. 7.10. The principle of defining the prototype map specifications. A default prototype map is specified, which can be used as a basis for all other prototypes.

The knowledge base is created and updated by a Map Specification Editor (Figure 7.11). Prototype map specifications can be tailored and entered into the knowledge base of the server. The service administrator selects a prototype map via the context parameters (left side of Figure 7.11) and then gives the specifications (right side of Figure 7.11). The more context parameters are left unassigned (n/a), the less specialised any given prototype map is. In an operational environment the editor is intended to be used by the designer and the system administrator of a value-added service provider. The innovative aspect of this knowledge-based approach is that the service can deliver various types of maps, which match in real-time with the current context parameters and user preferences.

7.4. Conceptual Process Model for Generalisation in Location-Based Services

In the sections above, we have discussed the role of map generalisation for LBSs. Map generalisation can be understood as a subclass of context-sensitive mobile services. The location of the user is the most fundamental context parameter for the process but many others can be utilised as well, such as orientation, time, navigation history, purpose of use, social and cultural situation, physical surroundings and system properties. The overall process for generalisation in mobile services can be described conceptually (Figure 7.12). The process for creating a mobile map is initiated by a request from a mobile client. Selection of the data content for the base map and for the overlay data are the first steps, followed by generalisation,

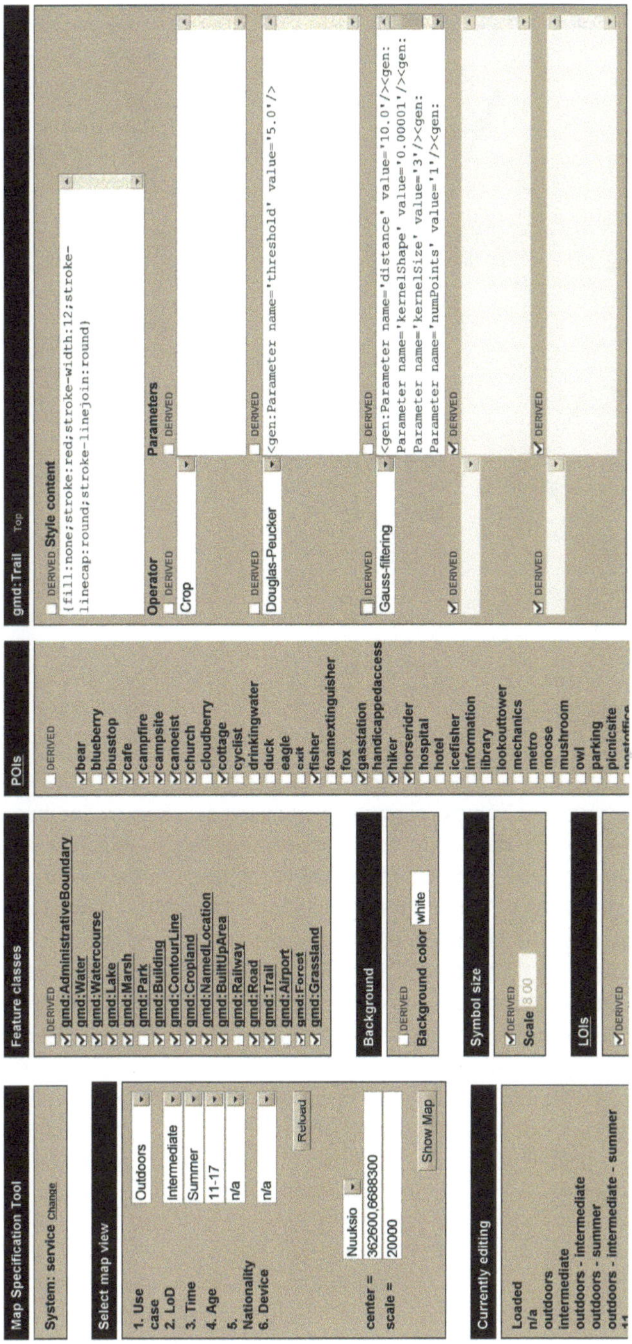

Fig. 7.11. Screen shots of the graphical user interface of the Map Specification Editor.

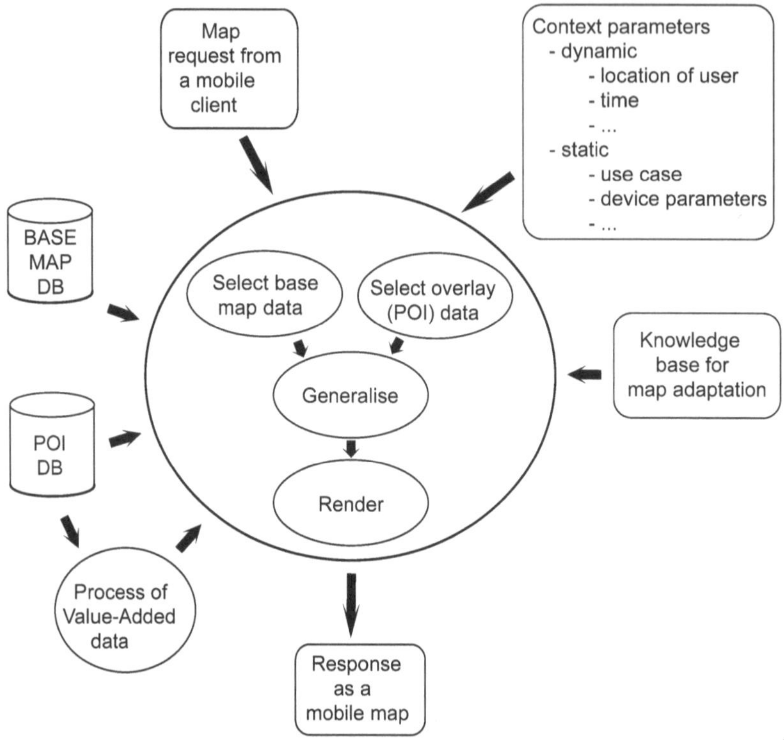

Fig. 7.12. A conceptual process model for map generalisation in LBSs.

and finally map rendering. The context parameters are used by a knowledge-based mechanism to control the selection of the data content and processing steps. Selection of map data and overlay data, generalisation and graphical rendering of the map are the essential sub-processes involved. The overlay data may come directly from an external PoI database or an external process for the value-added data. Independently of the source, the overlay data is dynamic with respect to the map data. Therefore, generalisation using both overlay data and map data can only be carried out by real-time processing. Dynamic context parameters have a similar consequence, whereas in the case of static context parameters, generalisation could be carried out in advance using pre-processing. As already pointed out in the introduction to this chapter, the process of creating a mobile map for LBSs is a holistic task. Map generalisation has to be tightly coupled with other tasks. This is especially true because of the integration of the PoI data. In the GiMoDig prototype system, icon placement, for example, is integrated into the generalisation environment.

7.5. Conclusions

Location-based services build upon such recent technological developments as positioning, mobile networks, and mobile handheld devices. LBSs are dependent on good-quality graphics, including maps. It is clear that maps with different levels of detail are needed, making map generalisation a necessity. Personalisation of services is a driving trend in making widely accepted and usable services, and is applicable for LBSs using maps. Maps in LBSs dependent on value-added data, presented on top of the base map, for example as Point of Interest. It is especially important that the presented value-added data is current, and adapted to the user's needs. As discussed in this chapter, use of context-based, adaptive generalisation is a natural reflection of the underlying requirements. It also appears that location-based services call for real-time processing. Real-time processing is needed not only for generalisation but also for all the other processing phases included in the service chain.

This chapter is largely based on a comprehensive description of the prototype system implemented in the GiMoDig project. The prototype demonstrated the feasibility of the approach and revealed technical solutions that can be used in the implementation. At the end of the chapter, we have also presented a conceptual process model for map generalisation in LBSs. The architecture of the prototype system in the GiMoDig project follows this conceptual model, which is flexible enough to support a wide variety of implementation approaches.

Acknowledgements

This research is part of the GiMoDig project, IST-2000-30090, which was funded by the European Commission through the Information Society Technologies (IST) programme. The authors want to thank colleagues in the GiMoDig Consortium for their co-operation.

Generalisation of Geographic Information: Cartographic Modelling and Applications
W.A. Mackaness, A. Ruas and L.T. Sarjakoski (Editors)

Chapter 8

Experiments in Building an Open Generalisation System

Alistair Edwardes, Dirk Burghardt, Moritz Neun

Department of Geography, University of Zurich, Winterthurerstr. 190, CH-8057 Zurich, Switzerland
e-mail: {aje,burg,neun}@geo.unizh.ch

Abstract

The increasing complexity of methods used in generalisation research together with growing demands for generalisation processing to support myriad of new geospatial services and technologies has led researchers to investigate how open architectures might better support such changing needs. The chapter is motivated by two such requirements: to share generalisation techniques amongst researchers within the field and to present generalisation functionality externally to geographic services. The chapter first discusses the issue of openness in relation to the requirements of the generalisation community and then in the broader context of open standards, open architectures and open source for geographic information. Two open architectures for providing access to generalisation processing are then discussed based on these considerations. The first takes the approach of a middleware component that presents generalisation functionality through a coarse, web mapping interface. The second adopts the model of a web-service registry that provides access to generalisation operations in a distributed and platform independent way. Implementations of the two systems are then described. The middleware approach reuses existing protocols for requesting maps and representing map specifications and geographic phenomena. It is based on open-source technologies for its core infrastructure. The registry model implements a set of web services standards that allow generalisation researchers to register and expose their operations. These can then be accessed independent of language bindings or of the location where the code is being executed. Interaction with the registry is possible with plug-ins implemented for different desktop mapping software or via dynamically generated web information. In its conclusion the chapter considers the successes of each approach and the main limitations. Foremost of these is the lack of a formal conceptualisation for the generalisation domain.

Keywords: open architectures, open systems, standards, protocols, open source, on-demand mapping, on-the-fly generalisation, web mapping, location-based services, web services, publish-find-bind, Open Geospatial Consortium, middleware, registry, common research platform, portrayal model, point-of-interest maps, interoperability

8.1. Introduction

There is a growing consensus within the map generalisation research community that open research platforms will allow closer integration in terms of collaborative research, data abstractions, interoperability of functional components and the augmentation of geo-spatial applications with generalisation capabilities. This desire has been evidenced through discussions at the various meetings of the ICA Commission on Generalisation and Multiple Representation (Beijing 2001, Ottawa 2002, Paris 2003, Leicester 2004, ICA 2005a), described in Edwardes et al. (2003b).

This chapter describes the authors' experiences in developing systems to support collaborative research in map generalisation. Two models using open architectures to facilitate cooperation amongst researchers are described. In the first openness is explored in relation to the interfaces and concepts of mapping services. In the second, opening access to generalisation techniques using web services is investigated.

The chapter provides insights into models for designing open systems and the concepts used by these. It also describes how standards for geographic information concepts and services can be used to help share generalisation research. It then presents practical experiences gained from implementing such systems and suggests useful technologies to assist the development of these. Finally, it discusses interoperability issues associated with improving collaboration amongst generalisation researchers and makes suggestions as to the best way for future progress.

8.1.1. *Motivation*

The growth of new technologies for the retrieval, handling and visualisation of geo-spatial data has brought with it increasing demands on automated cartography to provide theory and methods that will allow their coherent operation (Meng, 2003). As well as classical needs for cartographic generalisation, new challenges are raised by the necessity for more usable (Nivala et al., 2003), flexible portrayals of geographic data (Barkowsky et al., 2000; Avelar and Müller, 2000; Elias, 2002), matching different levels of interaction (Crampton, 2002), device capabilities (Chalmers et al., 2001; Arleth, 1999), modes of use (Fuhrmann and Kuhn, 1999) and needs of users for geographic information (Raper et al., 2002). These include: The provision of cartographic products on devices with differing display capabilities such as personal computers and small screen devices (Sarjakoski and Sarjakoski, this volume (Chapter 7), the "ego-centric" adaptation of information portrayal according to an individual's context and preferences (Zipf and Richter, 2002; Reichenbacher, 2003, 2004) and activities (Sester and Elias, this volume (Chapter 10), dynamic geo-spatial information retrieval and data conflation for "Point of Interest" mapping (Arikawa et al., 1994; Edwardes et al., 2005).

At the same time, the techniques used in more conventional generalisation research have become increasingly more complex. Algorithm development has become more focused on satisfying constraints and modelling knowledge (Beard, 1991a; Weibel, 1997; Weibel and Dutton, 1998; Ruas, 1999). This requires both the intelligence to optimise amongst numerous design constraints (Burghardt and Meier, 1997; Ware and Jones, 1998; Sester, 2000; Harrie and Sarjakoski, 2002) and better representations of geographic phenomena, space and spatial relations (Ruas and Plazanet, 1997; Regnauld, 1996; Mustière, 2001). Considerable effort is needed to meet these requirements. Researchers must spend significant amounts of valuable time gathering together tools, designing a platform, integrating tools etc. just to reach the research frontier. Moreover, platforms developed through this process tend to be institute specific and differ greatly, leaving little opportunity for sharing research.

8.1.2. *Requirements for an open generalisation system*

These motivations highlight two areas where openness in research systems could enhance collaboration and cooperation in research. On the one hand, there is the need to support cooperation by sharing of techniques *within* the research community, for example new algorithms, data structures, measures and control architectures. On the other, there is the need to support external collaboration through the *application* of generalisation in other GIS research areas, for example in on-demand mapping, geo-visualisation and location-based services. Within the community openness needs to be at very detailed and technical levels. Outside openness should be to support the inclusion of generalisation concepts and techniques within the broader body of GI research. Internal to generalisation research, requirements include the need:

1) to support sharing and comparing of techniques and results;
2) to allow researchers to share and access complex spatial data-modelling, analysis and restructuring functionality so that these may be used as the basis for new techniques;
3) to enable access to libraries of algorithms that can be used to study the procedural knowledge required for sequencing and orchestrating generalisation operations;
4) to allow researchers to test and demonstrate new or improved functionality within a holistic generalisation framework made from shared common components.

Externally they include the need:

1) to allow generalisation functionality to be presented and accessed in such a way that it can be integrated transparently with geographic applications from other research areas;
2) to express generalisation concepts, such as map constraints, multi-scale representations of information, generalisation operations and alternative portrayal types, in a formal machine-readable manner;
3) to perform generalisation in real-time on transient data (e.g. search results or the result of a route calculation) respecting both dynamic restrictions, such as the map user's current location or the time of day, and static associations, such as inherent spatial relationships between the retrieved features and the base map features;
4) to link between sets of multi-scale representations of the same data.

In addition, the two areas also have many requirements in common, for example:

1) the ability to support researchers independent of their choice of platform or development environment;
2) the ability to access, encode and transfer data without loss of information;
3) the ability to describe and encode map specifications and styling rules.

8.2. Context and Evolution in Computer Science

8.2.1. *Open architectures*

Open Architectures are collections of components and their interfaces whose specification, at some level of granularity, has been made public. Interfaces can be thought of as contracts setting out the obligations between a component's user and provider (Vckovski, 1998). To share generalisation techniques amongst researchers, interfaces need detailed definitions, presenting many generalisation operations and their parameters. For generalisation applications, interfaces may be coarser, perhaps with only a single "Generalise Map" operation. Open architectures based on lightweight distributed platforms have attracted significant interest in the generalisation community (Lehto and Kilpeläinen, 2000, 2001; Harrie and Johansson, 2003; Badard and Braun, 2003; Burghardt et al., 2005).

8.2.2. *Open protocols*

Open protocols provide the language syntax and concept formalisation for communicating within an open architecture. Protocols are defined to encode data modelling concepts relevant to the domain and the operations of an interface. They may define formats for describing an operation, for encoding information to be exchanged across the interface or for coordinating communication. Protocols should be based on a single domain abstraction common to all relevant components, though their logical encoding (e.g. Java classes, XML, SQL structures) can differ throughout the architecture. For generalisation research, protocols must be able to represent models of geographic phenomena (Sondheim et al., 1999) that encapsulate geometric, topological and semantic concepts and which may be encoded as parameters of operations. To present a generalisation service to other applications, protocols must also represent cartographic concepts such as: map specifications (e.g. map extent and scale), map content and styling, map theme and spatial region-of-interest, and different map types.

8.2.3. *Open standards*

Open standards formalise the definitions of open architectures. They are defined through open, international, participatory processes, are publicly documented and available to use without licensing or royalty restrictions. Standardisation eases interoperability between components from different researchers (Hjelm, 2002) and allows platforms to be designed without reference to particular software. Most importantly, they formalise definitions for geo-spatial data abstractions, easing communication amongst researchers and minimising duplication in design decisions. In this regard, the work of standardisation bodies such as the Open Geospatial Consortium (OGC, 2005) and the World-Wide Web Consortium (W3C, 2004) has revolutionised how GI Science is now practised. There are many standards that relate to open generalisation systems. Table 8.1 illustrates some of the more salient ones.

The Web Mapping Service (WMS) and Web Feature Service (WFS) specifications aid the deployment of generalisation functionality through geographic application services. The Styled Layer Descriptors (SLD) and Filter Encoding Specification (FES) protocols allow requests to these services to be characterised and provide a mechanism for triggering generalisation. GML is a protocol for encoding and exchanging spatial data. It is important for both

Table 8.1. Current relevant standardisation efforts.

	Standard	Description
Mapping	Web Map Service (WMS, 2004)	The WMS specification defines an interface to allow mapping services to be made accessible over the web.
	Styled Layer Descriptors (SLD, 2003)	The SLD allows map specifications to be defined. It encodes concepts for specifying the content, the map "layers", and presentation, the layer "styles" of a map.
	Scalable Vector Graphics (SVG, 2004)	SVG is an XML encoding to describe graphics using vector primitives (c.f. Lehto and Kilpeläinen, 2000, 2001; Cecconi and Galanda, 2002; Reichenbacher, 2002; Takagi et al., 2003).
Data Handling	Web Feature Service (WFS, 2004)	The WFS specification defines an interface for accessing spatial data, as geographic features, over the web.
	Geography Markup Language (GML, 2004)	GML is a protocol for encoding geographic feature descriptions in XML. GML uses standard conceptual abstractions (ISO 19107; ISO 19109) as a data schema for classifying features, their attributes and geometries.
	Filter Encoding (FES, 2004)	The filter encoding specification provides a neutral protocol for constraining spatial and semantic queries on spatial data resources (e.g. a WFS).
Web Services	Web Services Architecture (WSA, 2004) – Simple Object Access Protocol (SOAP, 2003) – Web Services Definition Language (WSDL, 2001)	The Web Services Architecture (WSA) is a collection of protocols and standards to allow software applications to interoperate over the Internet in a platform independent manner. Two of these are: – SOAP is a light-weight implementation of a WSA protocol. It allows access to objects, operations or data over a network using structured messages in XML. – WSDL a WSA protocol to describe the interface of a Web Service. It allows the service to expose the operations and message formats it supports.

coarse and fine-grained systems. For mapping and data services it encodes query responses, describes transient data to be portrayed dynamically and describes regions of interest related to the map theme and user. For fine grained operations it describes parameter data types. The web services standards are generic mechanisms for accessing computational objects and operations over the web. In a platform for sharing research the main aim is to provide access techniques for comparison rather than to provide a framework of operations that can be integrated into a complete system. Hence, these standards can be used to express generalisation operations at fine-grained atomic levels. These can be accessed in a neutral XML format, independent of particular development environments.

8.2.4. *Open source*

There are many definitions of open source software (c.f. OSI, 2004), mainly differing in their description of licensing conditions and how the software can be modified and redistributed. Typically the software is provided together with its source code. Such software might therefore be considered as an open architecture that has been specified at a very detailed level. Because open source software can be modified, it is highly adaptable and can be closely integrated with other custom code. It is also often provided free of charge. Often research efforts can be hastened because open source tools can supply core functionality necessary to build

Table 8.2. Relevant open-source projects.

Software	Description
GeoAPI (2004)	GeoAPI implements OGC protocols for geographic information in Java. This simplifies the low level integration amongst the different frameworks described below. The work is being made in concert with the OGC Geographic Objects initiative (GO-1 2004).
Deegree (2004), GeoTools (2004), GeoServer (2004)	These are different Java frameworks implementing OGC Web Services specifications (including WMS and WFS).
JTS Topology Suite (JTS), JTS Conflation Suite (JCS) and JUMP (Vivid Solutions, 2004)	JTS and JCS are tools for spatial analysis and spatial data structuring. They use standard based geometry abstractions (Harrie and Johansson, 2003; Sarjakoski et al., 2005b). JUMP is a desktop mapping platform that uses JTS and JCS.

a mapping platform, which is otherwise not related to research aims. A number of GIS open source projects exist that are of interest to the field of generalisation. Some examples of these are listed in Table 8.2.

Amongst the most useful types of open source software are *frameworks*. Frameworks define architectures that modularise functionality through the definition of responsibilities and collaboration. Their aim is to encourage the reuse of the architectural design in order to create similar types of software rather than simply encourage the reuse of code (Gamma et al., 1995). Frameworks are often implemented in response to open standards, with their modular design enabling compliance at very fine levels of granularity. GeoApi is an example of this (Table 8.2). The Frameworks implementing the WMS standards are also particularly useful for an open generalisation system since much of their design can be reused.

8.3. Architectural Models for Open Generalisation Systems

The two types of research requirements suggest two different architectural approaches. Generalisation provided as part of a wider portrayal strategy is best deployed through extensions to existing models for geographic services. This is because these use interfaces and protocols that are standardized and well understood, and open-source implementations mean that developers can focus on augmenting existing systems rather than starting from scratch. This sort of deployment is termed *middleware*, since the services sit in the middle of the architecture, between client applications and data servers. Research requiring atomic access to generalisation techniques is better supported by services that can expose operations and their parameters in very detailed ways. The web services model (a set of operations that can be accessed over the Internet using XML, possibly via an intermediary *registry*), provides a good method to meet these needs. This is because: (1) operations, offered by the generalization services, can be described and used independently of language bindings, (2) data types for parameters can be described using existing standards for geographic information exchange, and (3) functionality can be shared without users needing physical access to either the code implementing the operation or the server performing the computation.

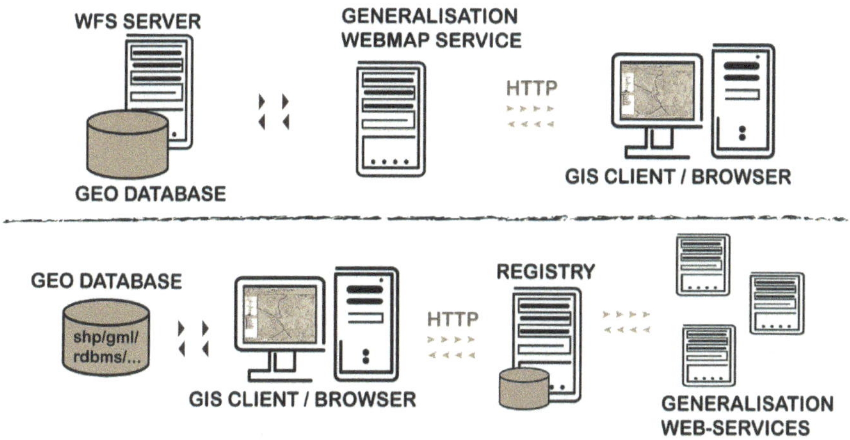

Fig. 8.1. Two types of Generalisation Service: Middleware (top half) and Registry (lower half).

Figure 8.1 illustrates the two types of generalisation services. In the first scenario access to the service is by specifying a map, with the service handling its own data requirements (Lehto and Sarjakoski, 2004; Illert and Afflerbach, 2004). In the second case the client must handle its own data and data validation issues. Additionally, the mode of interaction in each scenario differs. The middleware model runs completely automatically whereas in the registry model the user has more interactive control.

8.3.1. *Middleware*

A categorisation of basic services for architectures portraying spatial information is described by the OGC's "Portrayal Model" (OGC, 2003a, p. 23) (also known as the "Cuthbert Model" (Cuthbert, 1998)). It describes a pipeline of four sequential processing steps:

- FILTER: Accessing geographic features from a database through spatial and semantic filters;
- DEG (Display Element Generator): Combining geometric and semantic feature information with styling rules to generate styled graphical vector primitives (e.g., postscript instructions, SVG elements or, Java graphic objects);
- RENDER: Drawing the display elements on an image canvas. In essence, projection, clipping, rasterisation and anti-aliasing of graphic vectors (Foley et al., 1996, Ch. 3 and 19);
- DISPLAY: Making rendered images visible on the output device (Foley et al., 1996, Ch. 4).

Adopting the concepts of this decomposition is useful because it allows generalisation to be discussed within the broader context of portrayal. Several open source implementations for web mapping also follow this model as a *de facto* standard. Generalisation functionality can be added to this model at various points. Figure 8.2 illustrates this, presenting the portrayal model on the left with possible modifications in boxes on the right.

At the data source level the concern is with model generalisation with respect to data resolution and data volume rather than graphical generalisation, which would require associ-

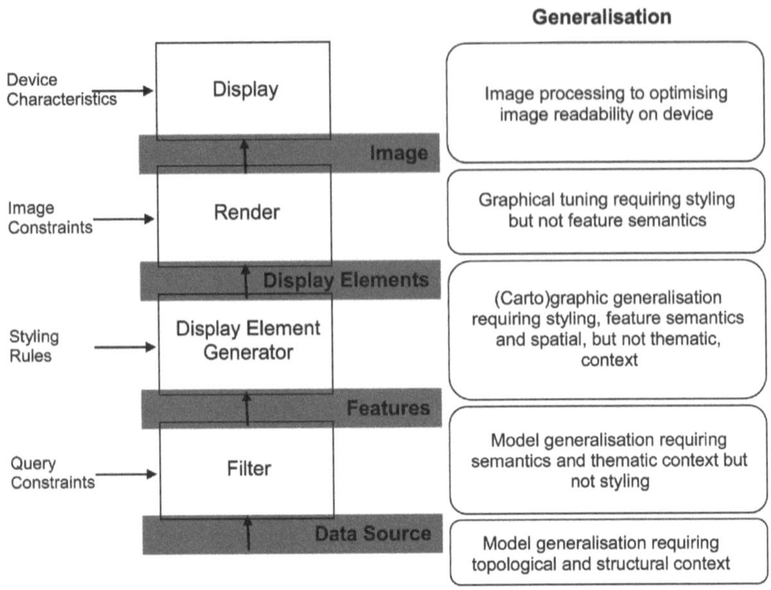

Fig. 8.2. The Portrayal model and generalisation (adapted from Cuthbert, 1998 and following the generalisation typology of Weibel and Dutton, 1998).

ated styling information. Data enrichment can be undertaken to support this by modelling complex multi-scale spatial relationships (Neun et al., 2004). Multiple representations of features and multi-resolution data can also be managed at this point (Weibel and Dutton, 1999; Hampe et al., 2003; Cecconi, 2003). Filtering allows control over the map level-of-detail and theme. Layers modelling different phenomena can be instantiated by selection according to scale, semantics, spatial relationships and relative importance. The display element generation step provides the most suitable point for graphical generalisation. Information is available on both feature semantics and styling, allowing graphical legibility constraints to be effectively analysed. At the rendering stage, only styled graphical primitives without semantics exist. Techniques based on coordinate transformations can be applied, for example variable scale projections (Harrie et al., 2002b; Rappo et al., 2004). Tuning (to enhance the impression of graphical variables such as brightness and colour contrasts (Bertin, 1973)) can also be performed. Text placement (Petzold et al., 2003) might also be applied here, particularly if the language of text is localised dynamically on a client. At the display stage, there is no knowledge about the image content so generalisation is not performed. To some extent, legibility related to the client's display might be improved, for example by applying anti-aliasing using super-sampled images or automatic adaptation of display brightness and contrast according to background light conditions.

8.3.2. *"Publish-Find-Bind" – Registry for a research platform*

In an open research model the idea is that every researcher can deploy their own generalisation service. Through the Internet and the use of platform independent technologies such services

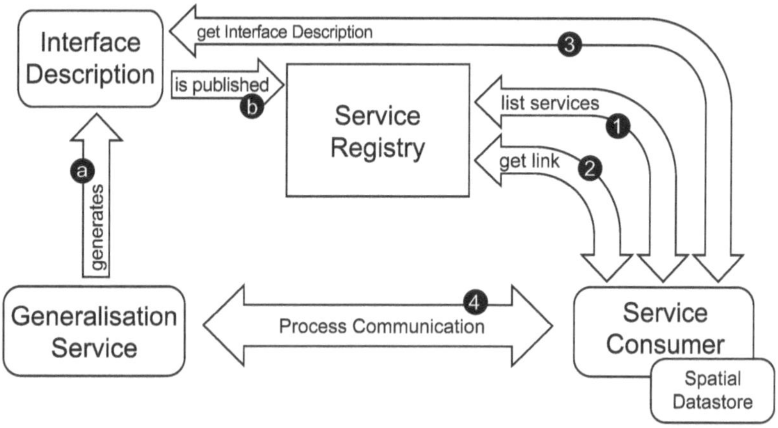

Fig. 8.3. Registry for Generalisation Services.

can reside on servers all over the world. To discover these services a "Yellow Pages" is needed, indicating which services are available, where they are located and what algorithms they offer (Burghardt et al., 2005). This is the "Registry" model for generalisation services. The registry offers a single access-point where all further information can be found. Whilst services can change and move they can always be found again through the registry.

This model for sharing and discovering generalisation services can be summarized by the "publish-find-bind" paradigm (UDDI, 2004), shown in Figure 8.3. The "publish" step is carried out by the service provider, e.g. a researcher who wants to make their generalisation operation available. They must create (a) an interface description containing the parameters of their generalisation service and the service endpoint, (an URL where the service can be accessed). Once the interface description is published (b) in the registry database, the community can access the service. The "find" step is done by a service consumer, e.g. a researcher who wants to test an operation. Having selected the desired service from those available (1), the link pointing to the interface description (2) is followed and the interface description itself retrieved (3). Using the interface description, the consumer can "bind" to a service and establish communication with the service-endpoint directly.

Accessing the interface and operations of a generalisation service can either be through a form-based webpage or a plug-in for mapping software. Only a webpage client has the potential to upload a file via HTTP. A plug-in in a GIS integrates seamlessly within a cartographic application. The user can access the service without needing to export and re-import data because features are encoded and decoded directly within the application. Features sent to the service are encoded in GML and embedded within a Simple Object Access Protocol (SOAP) message. The use of these XML formats (GML, SOAP) makes the approach very open and flexible. However, the costs of data conversions and transporting of large amounts of data can become a bottleneck in real-time applications (Gbei et al., 2003a), though for a research platform this is a minor problem.

Another way of supplying data to a service is with the URL of the data source (e.g. WFS) directly. The service then accesses the data source itself, processes the data, and sends it to the user (Sester et al., 2004). In this context, the service-consumer is a simple client controlling the process without uploading cartographic data. Similar concepts to this, e.g. using Common Object Request Broker Architecture (CORBA), have been previously implemented in mapping platforms (Hardy et al., 2003).

8.4. Implementation of the Two Architectures

Systems were implemented to demonstrate how openness could support research and help identify problems that this might entail. The first example implemented a generalisation service that dynamically generated points-of-interest maps for a location based service, as part of the EC project WebPark (Edwardes et al., 2003a). The second implementation demonstrated how a web services architecture could ease the sharing of techniques and results amongst researchers.

8.4.1. *On-the-fly generalization in the WebPark project*

a) Strategy
The aim of the project WebPark (2004) was to develop a mobile information system for visitors to natural and recreational areas. A particular emphasis in this research was the presentation of wildlife and "points of interest" information on small screen devices, using different forms of portrayal, scales and symbolisation (Edwardes et al., 2005). A two-pronged approach to generalisation was taken to achieve this. Analysis of user needs for information indicated that some data was relatively static and some very dynamic. This meant data could be split into background (or base map) and foreground (or thematic) types. Background data helped to orientate users within the information context (e.g. topographic maps and maps of general animal distributions or ecology). Foreground information delivered dynamic content in response to user queries (e.g. points of interest such as restaurants and recent animal sightings).

b) Implementation
A generalisation service was constructed using Deegree software (Fitzke et al., 2004). This was found to be a highly modular and adaptable framework. The generalisation of background information was performed offline using semi-automated approaches. This was organised in a database using multi-resolution data structures, and configured for access through a WMS using static, scale-dependent, layer definitions. Handling of dynamic data was enabled using mechanisms in the SLD. This allowed data definitions and data sources to be passed to the service at runtime. Data definitions could either be encoded using as FES (Filter Encoding Implementation Specification) constraints related to a WFS or data could be encoded in GML and passed in directly as a layer definition.

The style definition in the SLD was used to state when and how a layer should be generalised. This was indicated through the semantics of the style element which could be configured using the `SemanticTypeIdentifier`. Customisation of the Deegree framework was required so these identifiers could be interpreted by the system. The framework permitted the behaviour of the "GetMap" operation to be configured at runtime using a custom map request handling class. Thus modifications could be made separately from the core framework

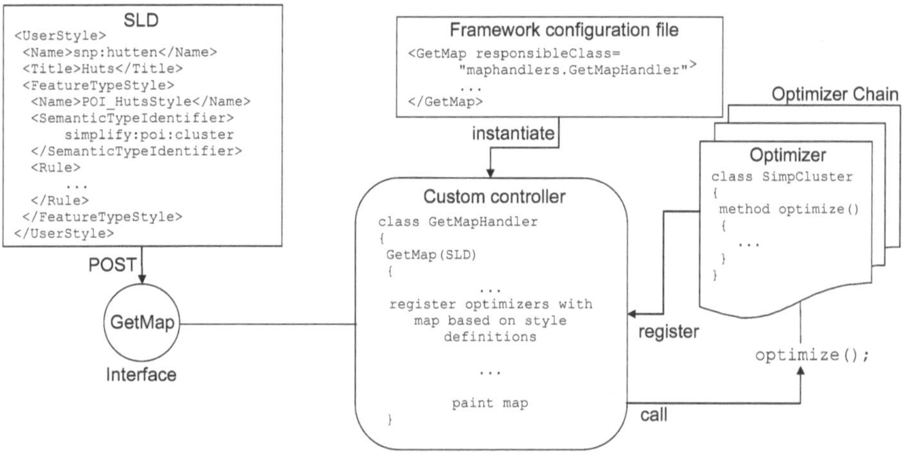

Fig. 8.4. Configuration and customisation of Deegree to add generalisation functionality.

code. Atomic generalisation operations were added using framework base-classes called display element *Optimizers*. Several operations could be applied sequentially using *Optimizer Chains*. Optimizers were created to perform on-the-fly typification of point-sets (Burghardt et al., 2004), simplification of point-sets and lines (Visvalingam and Whyatt, 1993), point displacement (Edwardes, 2004) and density surface generation from points. Figure 8.4 illustrates the places in the framework where configuration and customisation was made to add generalisation functionality.

c) Results and discussion

Figure 8.5 shows examples of point set generalisation. The first picture illustrates the overlapping of symbols (mountain huts in the Swiss National Park) without generalisation. The second, the same situation after applying an icon optimizer based on a quadtree approach (Burghardt et al., 2004). The final picture shows the result of icon optimisation using a hierarchical clustering approach. The size of an icon is changed dependent on the number of underlying features that the icon represents.

The approach was straightforward to implement and relatively fast (around 100 milliseconds per operation). Layers provided meaningful groupings for features and free access to geometric, and semantic and styling information of display elements was possible. This provided a two level view of the features (grouped by layer and individually) which was useful for simple generalisation but less adequate when considering thematic relationships such as semantic hierarchies amongst layers, and spatial and topological relationships amongst features in different layers. The styling definitions and optimizers gave flexibility in map portrayal, though the approach of chaining optimizers had limitations for complex generalisation. The sequence of processing (for example, first selection then displacement), had to be predefined, which limited how procedural intelligence could be incorporated in the system.

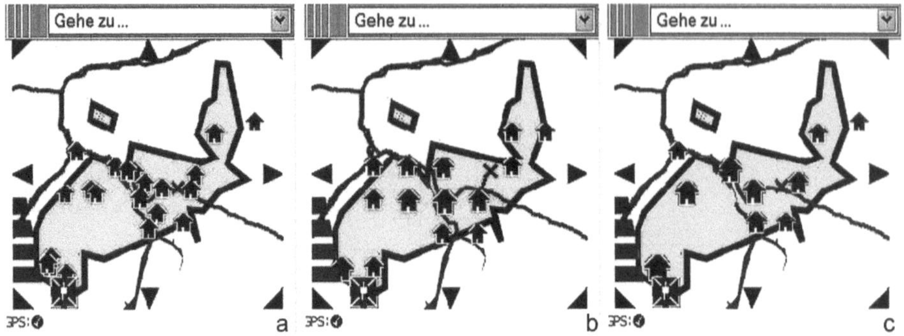

Fig. 8.5. Examples of graphical generalisation: (a) original, (b) quadtree typification, (c) hierarchical cluster simplification.

8.4.2. *Implementation of an interactive generalisation platform*

a) Strategy

The development was made as part of ongoing efforts by the ICA Commission on Map Generalisation and Multiple Representation to improve testing and sharing of generalisation algorithms (Badard and Braun, 2003; Edwardes et al., 2003b). The goal of the platform was to help researchers share their techniques through the ICA web site (ICA, 2005a), at the level of basic operators and measures, in a way that was independent of coding language and didn't require intellectual property to be exposed.

b) Implementation

The two usage scenarios, by web page and by mapping software plug-in (see § 8.3.2), were implemented, using Java Servlets. The JUMP provided tools for working with Shapefiles in the browser example. For the handling of Shape and GML geometries JTS tools were used. The plug-in was developed for use with JUMP as the client. Other plug-ins, for platforms such as ArcView®, are planned as future project work. The client was implemented as an easy-to-use, interactive graphical user interface. The web page example uses standard HTML pages accessible by any standard browser. The user accesses the service through a start page containing all available services. This page is dynamically created with information from the service registry. After selecting a particular Generalisation Service the user was presented with a new, dynamically created page which allowed parameters for the algorithm to be entered and a Shapefile containing source data to be uploaded from the local system.

The JUMP plug-in has the same functionality as the browser solution but integrates seamlessly into the software, so that the user does not have to quit the application and does not notice a significant difference between using a local or remote algorithm. The plug-in integrates into the JUMP menu bar. It automatically checks every time it is started for all available generalisation services. The result of this search is displayed in a selection list to the user. Figure 8.6 illustrates this idea in the plug-in.

The user selects the desired operation from the list. They are then presented with an entry form for the corresponding algorithm's parameters. These entry forms are dynamically created

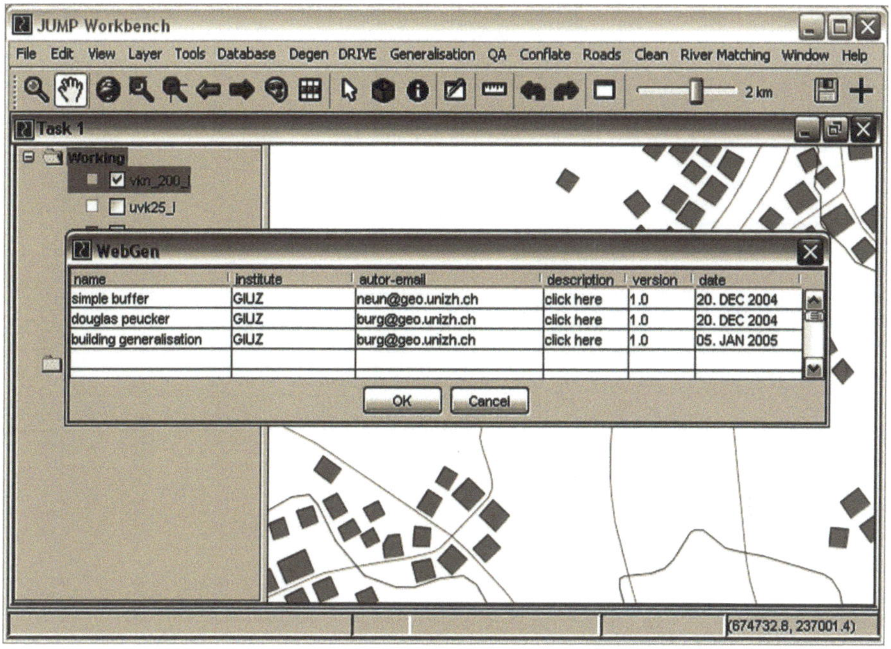

Fig. 8.6. List of available services (from the Generalisation Service registry).

```
<?xml version="1.0" encoding="UTF-8" ?>
<webgen xmlns:gml="http://www.opengis.net/gml" xmlns:SOAP-ENC="http://schemas.xmlsoap.org/soap/encoding/">
  <name>building simplification</name>
  <method>buildingSimplification</method>
  <endpoint>http://server/soap-endpoint</endpoint>
  <description>simplifies all buildings in a layer and returns resulting buildings</description>
  <config>
    <layer>
      <schema>
        <attribute name="geom" type="GEOMETRY">
          <allowed>gml:Polygon</allowed>
          <allowed>gml:MultiPolygon</allowed>
        </attribute>
      </schema>
    </layer>
    <param name="min edge length" type="SOAP-ENC:double">
      <description>Minimum Edge Length</description>
    </param>
  </config>
</webgen>
```

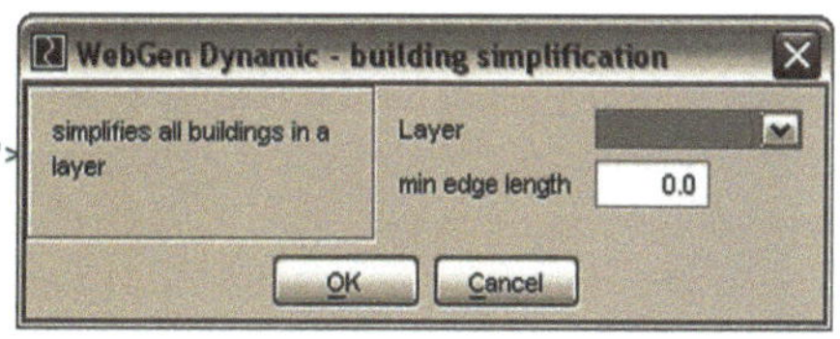

Fig. 8.7. XML Interface description.

using the interface description. An example of a simple interface description for a building simplification algorithm is shown in Figure 8.7. The data format for the interface description is XML, accessed via the Registry. This XML format extends the Web Services Definition Language (WSDL) with schema definitions for the simple features which are required by every generalisation operation. The registry uses the "publish-find-bind" concept from the Universal Description, Discovery and Integration protocol (UDDI) but is implemented as a

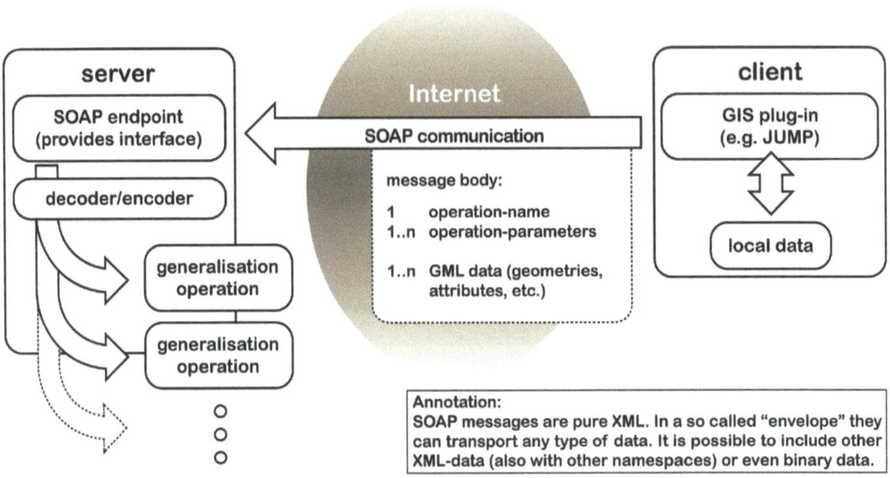

Fig. 8.8. SOAP client–server communication.

more lightweight database front-end which can be queried for available services and then returns the XML descriptions.

After all the parameter information has been entered and the operation started, the plug-in establishes communication with the server. The communication process is summarised in Figure 8.8. It consists of passing SOAP requests and responses. The client sends the name of the desired generalisation operation, the operation parameters and the data to generalise to the server. The generalised data or an exception report is sent back in response.

c) Results and discussion

The implementation of client and server components effectively demonstrated an open generalisation service with a registry. The use of standard protocols, such as GML, and SLD, to define parameter data types meant that geographic phenomena and map specification (e.g. scale and styling) could be reasonably represented. In particular, they gave good assistance for sharing research techniques for geometric generalisation of independent features or amongst features of the same feature-type. Additional input and practical experience with the system by the research community is now needed to understand how the service can be improved. Open issues are: (1) how protocols can better express more contextual inter-object and inter-class spatial relationships, necessary for complex generalisation, (2) what generic methods can be found to represent semantic relationships between features, e.g. importance ordering and attribute similarity, and (3) how differences in terminology for feature-types and generalisation operation descriptions can be best resolved. There are various other possibilities for extension and improvement. In the context of managing an entire generalisation workflow or accessing a completely server-based generalisation system, solutions for extended session-management and data persistence on a server are being considered. This stateful behaviour could also be a way for flexibly accessing more complex agent based systems as well as overcoming data transfer bottlenecks.

8.5. Conclusion

The research demonstrated two approaches to addressing the issue of improved access to research. In each case the use of standards and frameworks from the existing body of knowledge on architectural design was found to be highly desirable. This was partly for practical reasons, as it helped to focus research efforts by supplying auxiliary tools, but also because they provided technology neutral, standardised design concepts and protocols that made it easier to express where and how functionality could be deployed. However, it was clear that the standards also lacked concepts and methods required to undertake and share very complex generalisation research, for example modelling highly contextual geometric relationships or constructing sophisticated control structures for managing sequencing and orchestration of operations. For instance, simple constraints were handled implicitly by parameterising algorithms accordingly. Ideally, these would be specified independently in a more formal way (c.f. Edwardes and Mackaness, 1999; Hardy et al., 2003; Sarjakoski et al., 2005b). There is also wider semantic issue that needs to be addressed. Generalisation research often needs ontological descriptions about geographic phenomena. For example, the algorithm of Rainsford and Mackaness (2002) for generalising rural buildings needs to make explicit definitions about what constitutes a *rural building* to their algorithm. They "define" this geometrically using template matching. Without such descriptions, assumptions must be made which lead to difficulties in reusing techniques if the two parties (designer and re-user) do not share a common understanding over the definition of feature type. These interoperability issues can only be resolved through consensus formalisation within the generalisation research information community (OGC, 1999b) and by practical experience in sharing research. In this regard the registry platform has strong contributions to make. Its relative simplicity, language neutrality, and ease for publishing and accessing algorithms means the barriers to researchers interoperating are very low. Thus such experiences can be shared fairly rapidly. In addition, because the owners of each algorithm remain in complete control of its code and the responsibility for its maintenance, intellectual property obstructions and administration issues are fewer and cooperation more feasible between commercial and non-commercial parties.

Acknowledgements

This work was in part carried out through the EC-IST Framework 5 project "WebPark: Geographically relevant information for mobile users in protected areas" (IST 2000-31041). The authors gratefully acknowledge the financial support of the Swiss Office of Education and Science (OFES) within the scope of this project (BBW Nr. 01.0187-1).

Chapter 9

A Data Warehouse Strategy for on-Demand Multiscale Mapping

Eveline Bernier, Yvan Bédard

Centre de recherche en géomatique, Pavillon Louis-Jacques Casault, Université Laval, Québec G1K 7P4, Canada
e-mail: {eveline.bernier,yvan.bedard}@scg.ulaval.ca

Abstract

With the advent of web mapping applications, today's users of geographic information are more aware of their requirements and are now asking for customised products specifically suited to their needs. These applications must, among other things, be able to present the same phenomenon at different levels of abstraction and according to different points of view. As with any web application, they must also provide fast response times. Accordingly, generalisation processes as well as any other processes must take place on-the-fly. In spite of considerable advances in automatic generalisation, there is still no complete solution that can provide this. This chapter presents a data warehouse strategy aimed at supporting multiscale on-demand mapping. It is based on experimentation to create an integrated multiple representations data warehouse from eight existing data sources. It introduces the idea of UMapIT (Unrestricted Mapping Interactive Tool), an on-demand web mapping application offering maximum flexibility to the user and multiscale functionalities supporting intuitive navigation at different levels of abstraction. This application is based on a specialized multi-representation datamart and is developed according to existing interoperability standards.

Keywords: on-demand mapping, datawarehousing, multiple representation database, VUEL, UMapIT, XML, WFS, on-the-fly generalisation, web mapping, data integration

9.1. Introduction

To fully satisfy users' needs, on-demand web mapping applications should offer the capability to display spatial information at different levels of detail. Automatic generalisation mechanisms should form part of these applications. Given that such applications are Web-based, response times must be very fast and generalisation must take place on-the-fly. Unfortunately, there is no complete solution to automatic generalisation (Cecconi, 2003; Ruas, 2002b). Although some operators require little control, complex ones still require human intervention. As an alternative, multiple representation databases have been proposed (Vangenot, 1998; Kilpeläinen, 2001; Weibel and Dutton, 1999; Sarjakoski, this volume (Chapter 2)) but they suffer from difficulties in populating and updating. This chapter presents an innovative strategy for multiscale on-demand mapping. This strategy is akin to datawarehousing architecture where automatic cartographic generalisation (CG) and multiple representation (MR) are distributed between the MR-centric warehouse and specialised CG-centric datamarts (optimised subsets of the warehouse). After presenting the requirements of web-based on-demand mapping, we explain the underlying philosophy and the data warehousing architecture. Our experiment in building a multiple representation data warehouse from eight data sources in the context of the GEMURE (Generalisation and MUltiple REpresentations for on-demand map production and delivery) project is also discussed. Finally, this chapter introduces UMapIT (Unrestricted MAPping Interactive Tool), a web-based on-demand mapping tool offering flexible seamless generalisation.

9.2. Web-Based On-Demand Mapping

Nowadays, the term *On-demand* is being applied in several contexts, for example: on-demand printing, on-demand publishing, on-demand weather, and on-demand business. As IBM (2004) stated "On the surface, on-demand is everywhere". Given the digital era and the power of the Internet, customers can demand instant and customised services. They are more aware of the opportunities and conditions associated with the Internet (van der Steen, 2000). Cartography has also been influenced by the on-demand era. Some web sites allow users to select specific regions, to pick a map scale from a list, to choose between a number of map styles, to turn on or off map layers, to add symbols, and so on. We define *on-demand mapping* as "a *customer-driven* map production process that enables creation of *tailor-made maps, often in a ready-to-print unique form*". As opposed to traditional map production processes driven by the fixed specifications of map producers, on-demand mapping is driven by the capability of customers to define their own specifications. This is a major shift in the map production philosophy as the customer becomes the map's co-producer (van der Steen, 2000). Thus, on-demand mapping assumes that the user has the ability to specify their needs with regards to map content, area and resolution, to put emphasis on individual objects, to select graphic properties and symbols for the map, the layers and the individual objects at will, and to receive the result instantly.

This approach operates irrespective of the form of output (paper or screen) and does not necessarily imply that it should be created on-the-fly (Weibel et al., 2002). On-demand mapping emphasises user-selection instead of immediacy and user interaction with data (NOAA, 1997). These latter requirements are associated with *on-demand web mapping* which is a subset of the general trend towards on-demand mapping and implies both interaction and immediate delivery through the web.

Though in recent years we have witnessed a major growth in web mapping applications, only a few of them may be called *on-demand.* Today's web mapping applications are mostly based on predefined maps where the user has little control over map content, how it is displayed or what individual objects are emphasised. Some of these applications allow the user to select themes to display and/or the scale to be used. Depending on the user's selection from a limited choice, different pre-produced maps are loaded. In summary these "first generation" web-based personalised mapping tools support limited customisation and functionality. The expectation is that future generations will support greater levels of customization and interaction, and richer choice in the map specification including scaling, and emphasizing particular elements in the map. Such mechanisms will include on-the-fly cartographic generalisation as well as individual and thematic "drilling" (retrieving deeper, more detailed information about map entities). Though important progress has been made in automatic generalisation and multiple representation databases, they are still not effective enough for on-demand web mapping. In this chapter we present a solution based on a datawarehousing architecture that combines automatic generalisation and multiple representation. The next section gives an overview of datawarehousing principles and presents the architecture used in a project called GEMURE.

9.3. A Datawarehousing Architecture Combining Cartographic Generalisation and Multiple Representation

9.3.1. *Data warehousing architectures and principles*

Data warehouses stemmed in the early 90s from the need for large organisations, which are typically the home of several independent legacy databases, to have a unified view of their organisation for the purposes of strategic decision-making. Although independent departmental databases are useful for day-to-day operations, they are not suited to support decision-making that requires integrated, aggregated and summarised data. The term "data warehouse" can be defined as an enterprise-oriented, integrated, non-volatile and read-only collection of data imported from heterogeneous sources and stored at several levels of detail to support decision-making (Inmon et al., 1996).

Though data warehousing architectures can differ widely (e.g. two/three/"*n*"-tiers), they all rely on the same idea: retain the legacy databases for data acquisition and update, then copy subsets of data as required, integrate and summarise them to produce more global data, and finally restructure them for new data storage and dissemination systems specifically tailored for decision-support. According to Kimball (2002), there are four separate and distinct components to successful exploitation of data warehouse architectures:

1. Operational source systems: i.e. legacy systems where the main priorities are the processing performance and the availability of detailed data (usually at a low level of abstraction – very similar to the acquired data);
2. Data staging area: i.e. the stage where you Extract, Transform and Load your data (ETL); this includes integration, aggregation and summarisation of data;
3. Data presentation for decision-support: i.e. the location where data are organised, stored and made available for direct querying. Typically, this is the data warehouse;
4. Data access tools: i.e. the tools that query the data in the data warehouse's presentation area; they include OLAP (On-Line Analytical Processing), data mining tools, "dashboards" that typically embed an OLAP interface in a preset display for specific decision support (Turban et al., 2005).

Some architectures make use of datamarts. Usually, these datamarts refer to databases containing data specific to a particular domain or subject that derives from the data warehouse (Figure 9.1). These datamarts are accessed by the users with the same types of querying tools

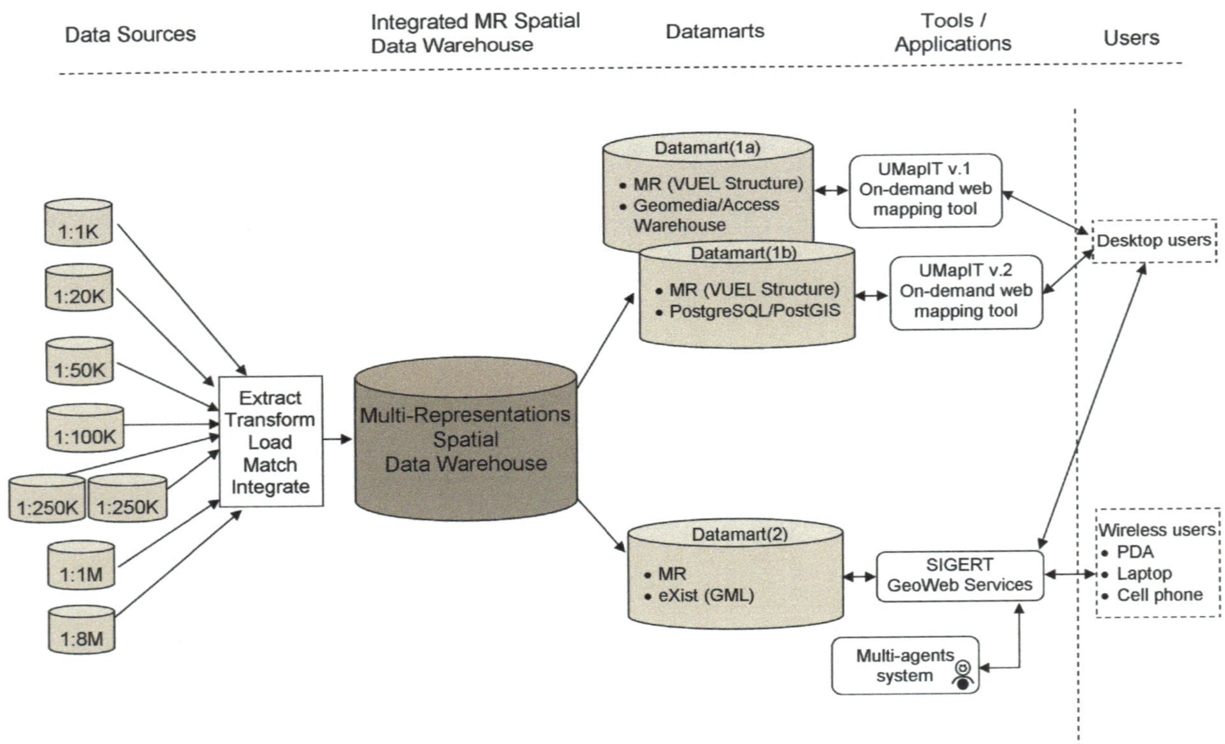

Fig. 9.1. Two-tiers Data Warehousing Architecture combining automatic CG and MRData warehousing in the context of GEMURE.

as the users of the warehouse. While the data warehouse is usually unique and intended to support global decisions involving strategic indicators for the whole enterprise, datamarts are more sector-based or domain-based. They are sometimes called "subject warehouses" or "mini warehouses" and are seen as more focussed subsets of the warehouse. In an organisation, typically there are several datamarts fed by a unique warehouse. This is the approach used in GEMURE.

The GEMURE project sought to develop new methods and tools that better support the combination of cartographic generalisation operators with multiple representation to enable easier on-demand cartographic information to delivery at different scales, whether for Web mapping or paper products. The project architecture is based on a central data warehouse and topic-driven or scale-centered datamarts (Figure 9.1). This warehousing strategy allows for the separation of acquisition databases (legacy systems) from specialised multiscale databases (datamarts) designed for dissemination. The GEMURE data warehouse lies between the legacy systems and the datamarts. It stores multiple representations of geographic objects which are selected and extracted from eight independent legacy systems (the eight data sources presented later in this chapter). The data warehouse provides the database link between each individual geographic object and its multiple representations for a large range of map scales and themes. Semantic coherence among objects has been achieved using traditional database conceptual schema integration techniques (ontological analysis) (see Mustière and van Smaalen, this volume (Chapter 6)), while spatial coherence has been achieved using traditional map projection algorithms and map merging techniques (all datasets used the same datum, although the spatial precision remains dependant on the source). To facilitate the population of the warehouse, we used the Feature Manipulation Engine (FME), a data translation and conversion tool from Safe Software (Safe, 2004) together with manual operations.

Once population of the warehouse is completed, each topic-oriented datamart imports a selected subset of data from the warehouse. A datamart is built for a predefined range of scales (e.g. from 1 : 10 000 to 1 : 250 000), so the user can perform seamless zooms within this range and see the map properly generalised automatically. In other words, each datamart stores and uses the required multiple representations – provided as "almost-ready" seamless-scaling maps that need minimal on-the-fly graphic generalisation.

Such database optimisation technique can be fine-tuned to obtain the best combination of MR and automatic CG. A balance must be found between: (1) importing and transforming the data from the warehouse into the datamart, (2) extracting the data from the datamart and transforming them automatically via generalisation. In the first step, easy-to-generalise data can be imported without transformation from the warehouse to the datamart while harder-to-generalise data can be processed once extracted from the warehouse and the results stored in the datamarts. For the remaining data that cannot be generalised automatically, the most appropriate representations of each geographic object can be extracted and imported directly into the datamart for later use. In the second step, (conveying the information from the datamart to the user), MR is used when automatic CG cannot be performed in the time available. In summary the datawarehouse provides multiple representations with which to build a datamart. The end user application uses on-the-fly cartographic generalisation where time permits, or equivalent multiple representations.

Such a solution allows the creation of applications that are guaranteed to work within the datamart topic and across a range of map scales. We call the "pivot-scale" the scale that represents the average between the minimum and maximum map scales. For example 1 : 225 000 would be the pivot-scale for a datamart that supported a range from 1 : 50 000 to 1 : 500 000). We call the "pivot-theme" the central theme associated with a given datamart. As an example Tourism would be the pivot-theme for a combination of geographic objects that included road networks, parks, public attractions and stores. Thus each datamart is built around a pivot-theme and a pivot-scale supporting the representation of a subset of geographic objects at a range of map scales.

Such a "divide-and-conquer" database approach to map generalisation allows datamart users to quickly derive topic-oriented maps at scales compatible with the pivot scale of the datamart, mostly using automatic cartographic generalisation processes (e.g. producing a road map at 1 : 125 000 from the transportation objects from the 1 : 100 000 datamart). All the heavy processing and manual manipulation is done beforehand (i.e. when building the warehouse and the datamart) and is done only once which is a major cost/benefit consideration. Thus the warehouse acts as a federated database supporting multiple representations of individual objects and storing the results of labour intensive data integration and generalisation procedures. These datamarts can then be accessed by different applications such as UMapIT (see §9.5). The following sub-sections present in more detail the components of the developed architecture with a focus on where cartographic generalisation and multiple representation are involved.

9.3.2. *The data sources*

In the GEMURE project, data sources were provided by three partners (Natural Resources Canada, Natural Resources Quebec, and Defence Canada). In total, there were eight independent data sources at the scales of 1 : 1000, 1 : 20 000, 1 : 50 000, 1 : 100 000, 1 : 250 000, 1 : 1 million, 1 : 8 million. At 1 : 250 000 there is two data sets from separate organisations. The pilot zone included the greater Quebec City area and covered 266 km^2. These sources were produced according to different standards. They included topographic data (over 110 object classes, including road and hydrographic network elements), administrative boundaries, 2D and 3D data. Thematic, non-spatial data were also added for a tourism-oriented datamart. These datasets used different ontologies, were in different formats (Shape, ArcInfo and MapInfo files), based on various map projections and coordinate systems, and met varying spatial precision standards.

In order to integrate them into a data warehouse, we first transformed them into the same format (Shape) and used the same spatial reference system (3°TM projection, NAD83). We assumed that the spatial precision was appropriate for the level of detail required and kept the original data "as is" within the warehouse. Nevertheless, we had to compare and select the most appropriate data for each datamart. Temporal discrepancies between datasets could afford to be left unsolved in the context of the GEMURE project but would have to be solved for industrial projects. Semantic differences between the datasets were numerous and subtle, requiring skilled personnel to select and merge the data properly.

9.3.3. *The multiple representation data warehouse*

The data warehouse is the centerpiece of the architecture. It has been populated by integrating the data imported as independent files from the legacy systems (our data sources) and it supports multiple representations. During the ETL (Extract, Transform, Load) phase typical of data warehousing, explicit links have been created between corresponding geometries for the same objects (see §9.4 for details). These links are used to extract the multiple representations of geographic objects in order to simulate cartographic generalisation when it cannot be done automatically on-the-fly. For example, when users zoom out on a specific region, these links enable display of the generalised view of the region by showing the less detailed representations of the objects. Similar solutions are already implemented in existing web mapping applications through the use of multi-layer structures. When users zoom in or out, these applications simply display another layer with more or less detail respectively. Although this solution is very fast, it is not sufficiently flexible as often all objects on the same layer change at the same time, without considering their specific characteristics (geometric or semantic). Using today's GIS intelligent zooming where objects appear between two predefined map scales (e.g. buildings appear between 1 : 1000 and 1 : 50 000 scales), one could refine such approaches and create tens of layers (e.g. Houses between 1 : 1000 and 1 : 10 000; Stores smaller than 1000 m^2 between 1 : 1000 and 1 : 20 000; Larger stores between 1 : 1000 and 1 : 20 000 represented as polygons and between 1 : 20 000 and 1 : 50 000 represented as points). However, this rapidly becomes unmanageable as complexity increases, redundancy is created and updating becomes tedious because the geometry of non-changing objects is repeated for every layer that remains the same. Moreover the contextual coherence between objects would be difficult to ensure and nearly impossible to maintain.

On the other hand, taking account of the characteristics of individual objects allows some occurrences to remain unchanged while others from the same object class are generalised (e.g. collapsed). For example, a zoom out could show the occurrences of buildings having the attribute value "commercial" still displayed with a polygonal geometry while other occurrences of buildings having the attribute value "residential" being generalised to become a point geometry. The generalisation could also be based on a combination of attribute values or on geometric criteria, such as the building area or minimum dimension, leaving the larger buildings unchanged while generalising the smaller ones. Of course, this also depends on the zoom factor since at smaller scales buildings will simply not be displayed. Such flexibility does not depend on map layers, but on the characteristics of individual objects. This allows an application to move away from a "fixed-layers/fixed-scales" generalisation approach towards a seamless continuous generalisation where every small change to a map scale can produce a different map without changing layers.

The approach used in the GEMURE warehouse uses multiple representation links at the occurrence level, making it possible to navigate through the different geometries of the same object based on any attribute or geometric value (Mustière and van Smaalen, this volume (Chapter 6) present different alternatives for multiple representation modelling). This allows the user to select only one object and display its detailed, or inversely, its generalised geometry. This is called "drill down" and "drill up" respectively in warehousing and OLAP terminology (On-Line Analytical Processing) (Thomsen et al., 1999) and we applied these operations to the geometric characteristics of individual objects (object drills) as well as object classes (class drills). Such advanced capability is important for on-demand web mapping applications since users want immediate results and increased flexibility. Whether changes in the geometry of an object result from a query involving the multiple representation links or from on-the-fly generalisation, it remains transparent to the user. These operations are further illustrated in §9.5.

Whilst we have discussed traditional multi-scale geometric multiplicity: an object has several geometries that differ according to the scale (Bernier and Bédard, 2002), it may also be the case that we have multiple geometries at the same scale for the single object. Since there were two data sets at the scale of 1 : 250 000, it was sometimes necessary to keep two geometries of the same object. The data warehouse therefore supports single-scale geometric multiplicity. These geometries share the same level of granularity and may look like unwanted redundancy. In reality they represent two different ways of representing the reality for different contexts. Furthermore, they already fit with different combinations of other cartographic features (Figure 9.2). Keeping them as alternate geometries for a given level of detail facilitates both their generalisation and layer integration processes when automatic generalisation cannot produce these geometries on-the-fly. Changing from one geometry to another geometry at the same level of detail is referred to, in OLAP terms, as "geometric drill-across" (Rivest et al., 2001) and provides extra flexibility to the user. Thus storing alternate geometries to support single-scale geometric multiplicity is a database optimisation choice.

Similarly, we also support semantic multiplicities (Bernier and Bédard, 2002). Accordingly, one shape may have different meanings or labels at the same level of detail (i.e. synonyms and quasi-synonyms) or for different levels of detail (i.e. generalisation and aggregation relationships within the database schema). This increases the flexibility for users since they can make semantic drills up, down and across. In addition, this allows one to recreate the original source products taking into consideration geometric as well as semantic aspects. From a technological point of view, the present version of the data warehouse uses a relational structure within Integraph's GeoMedia (Access Warehouse) while the datamarts have been developed using a variety of technologies. Maintaining the MR links in the warehouse and datamarts allows the user to customise their maps using drills up and down and generalise at will and on-the-fly individual occurrences as well as groups of objects (e.g. a class) within the scope of the datamart.

9.3.4. *Datamarts*

The GEMURE architecture includes specific datamarts. These datamarts differ in their content, structure and in the way they combine CG and MR methodologies. This is intentional since GEMURE is a research project whose objective is to find and test different approaches. We currently have two applications using their own datamarts from the same warehouse. The first datamart is being used by UMapIT and is described in §9.5. This datamart relies entirely on MR and is based on a very flexible concept called Vuel (Bernier and Bédard, 2002). This datamart is currently implemented in a relational database (Microsoft Access Technology) using a snowflake schema typical of datamart applications that simulates a multidimensional structure (Inmon, 2002; Kimball, 2002; Poole et al., 2002). Multidimensional structures, as opposed to relational ones, are denormalised to obtain faster answers as required in complex decision-making processes. The multidimensional paradigm is used today by most decision-support solutions. Turban et al. (2005), Inmon (2002), Kimball (2002), Poole et al. (2002) provide a useful overview of these technologies.

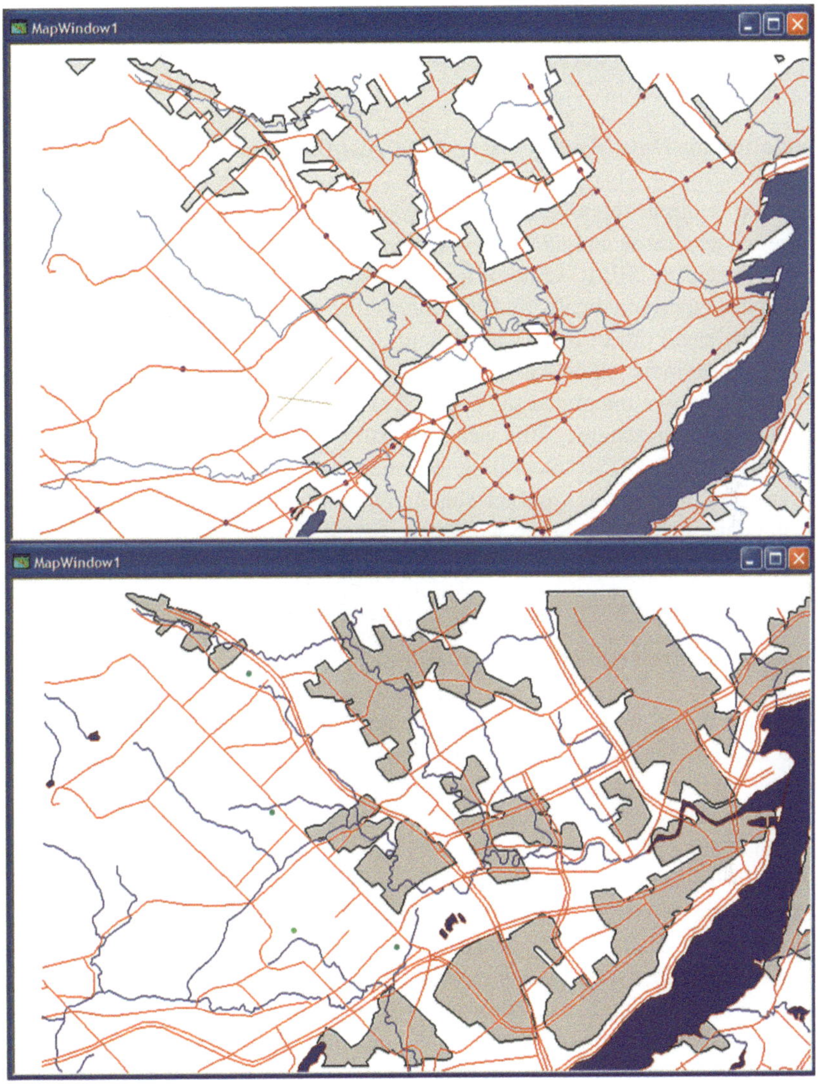

Fig. 9.2. Example of two datasets at the same scale (1 : 250 000) coming from two different data producers.

A second datamart, called SIGERT, has been built to support geo-located web services (Gbei et al., 2003b). This multi-modal application supports tourist and recreational activities at different levels of detail. This datamart includes data for downtown Quebec City, a subset of the GEMURE pilot zone. SIGERT focuses on solving crowding problems that can occur when displaying a map, especially for small devices such as PDAs. A multi-agent system has been developed to use generalisation operators (exaggeration, simplification and displacement) coupled with artificial intelligence techniques to detect conflicts and resolve them. The reader is invited to consult (Gbei et al., 2003b) for additional information.

9.4. Populating the Data Warehouse and the Vuel Datamart

Populating the multiple representation data warehouse was done in three stages. First, we selected the object classes among those in the data sources to be stored in the MR data warehouse. In the context of GEMURE, only a subset was needed (a total of twenty object classes). We analysed the data semantically and spatially and selected classes that had a multi-scale behaviour (i.e. the same object class was present in different data sources at different scales). The selected subset included typical object classes such as buildings, roads and hydrographic networks. Secondly, we integrated the conceptual schema of each data source into a unique multiple representation schema. This was done using the multiple representation capabilities of the spatially-extended UML modelling tool called Perceptory (Bédard et al., 2004). ETL – extract, transform, load – was the third step and allowed us to integrate copies of the source data into the data warehouse. Some work was carried out automatically with the FME tool while other work had to be done manually using ArcGIS (ESRI). These three steps are now detailed.

9.4.1. *Semantic analysis and database schema integration*

Having selected the desired data warehouse content, we proceeded with a semantic analysis of the selected object classes. Given that the data sources were coming from separate organisations, they were not produced with the same specifications and ontologies. These specifications varied in class definition (semantic), the acquisition rules (e.g. minimum size to be digitised), and data type (point vs. surface). A conceptual database schema with data dictionary existed for some of the sources while others had only a detailed textual specification. Most sources were very well known by the research team. Such thorough understanding of the specifications of each dataset was necessary to make proper correspondences, aggregations and generalisations of concepts in order to map them to the MR database schema of the warehouse. It is beyond the scope of the present chapter to describe in detail the steps of semantic analysis and schema integration, a topic widely addressed in the database literature.

The integrated conceptual model of the MR data warehouse has been built using Perceptory, a free spatially-extended UML-based visual modelling tool that provides mechanisms to describe multiple representation and generalisation (Bédard et al., 2002; Proulx et al., 2002). Figure 9.3 shows the Building class resulting from the integration of the Building classes from three different sources. In the 1 : 20 000 source, buildings can be represented by any geometry (point, line or polygon) while in the 1 : 50 000, they are represented by points or polygons. Finally, large buildings are represented in linear form in the 1 : 100 000 source dataset.

To keep the model as simple as possible, the details are stored in a data dictionary. This dictionary includes all the information regarding an object class (such as the class definition, attribute definitions, spatial and temporal definitions). For example, the 1 : 20 000 building in Figure 9.3 is defined in the dictionary such that we know when buildings are represented by a polygon or when they are represented by points or lines.

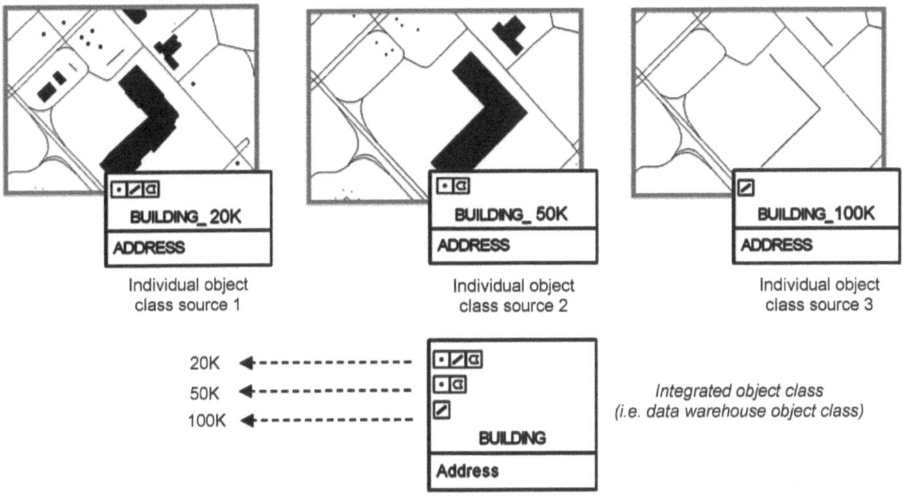

Fig. 9.3. Modelling multi-scale data with Perceptory.

9.4.2. *Data integration*

To have an integrated MR data warehouse where all the geometries of an object can be linked to that object, we must first identify which geometries in the different data sources represent that object. This operation has been done automatically with FME for non-network data and manually (with ArcGIS) for network data.

9.4.2.1. *Non-network data*

Discontinuous data, such as buildings, were matched automatically by using an operation called "NeighborFinder" available in the FME Tool, from Safe Software. Figure 9.4 shows the FME workbench being used to match *Dams* from two data sources (1 : 20000 and 1 : 50000). First, some semantic manipulations were performed: (1) creating a new attribute (source) that will be used to identify, in the data warehouse, from which data source the object came from, and (2) apply a semantic filter such that all dams are classified in the same manner. The NeighborFinder operation was applied to find the closest *candidate* feature within a maximum distance (buffer) of each *base* feature. If no *candidate* feature was found, then the *base* feature was output as unchanged. If a *candidate* feature was found, then all the attributes from the closest *candidate* feature were added to the *base* feature. *Candidate* features not close enough to any *base* feature were output as unchanged.

Several matching possibilities exist. Figure 9.5 shows different cases of data matching using only two data sources (A and B). The first case represents a polygon in source A that is represented by 3 points in source B. The second one represents an object that has the same geometry type in both sources. The third case shows an object represented by a line in source A and by a point in source B. The fourth one depicts an object that is represented by a polygon in source A and by a point in source B while in the fifth case, the same object is represented by a point in both sources.

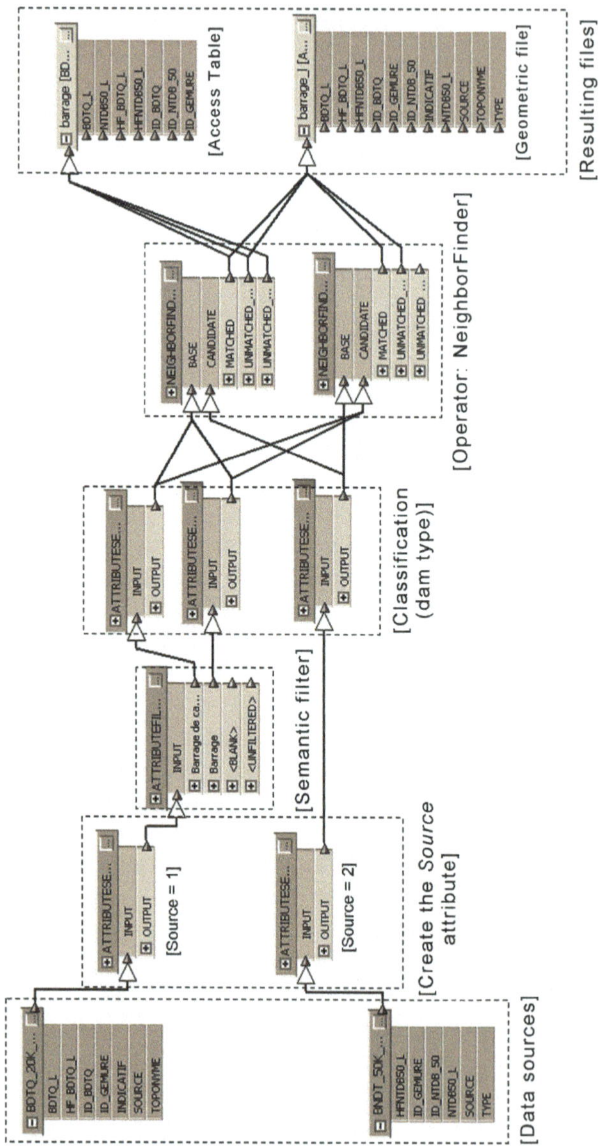

Fig. 9.4. Data matching with FME Workbench.

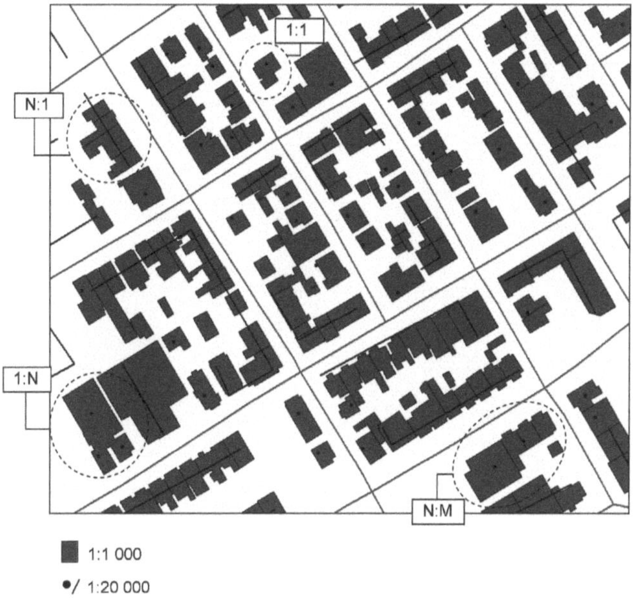

Fig. 9.5. Different data matching cases.

NeighborFinder did not work in every instance but once these exceptions had been handled, the end result of this matching process was two files: a semantic one (MS Access table) and a geometric one (Shape file). The latter included all the geometries, whether they have been matched or not (Figure 9.6). All the resulting shape files were then exported into the warehouse along with their attributes. The resulting semantic file shows some attribute information and indicates the source of the corresponding geometries for multiple representations (Figure 9.7).

9.4.3. *Network data*

Network data, such as hydrographic and road networks, are more complex to match as the FME software does not take into account the connections between vectors. For example in Figure 9.8, if we want to match *River A* (grey plain line) with *River B* (grey dashed line), FME will create a buffer around the base feature (*River A*), then match it with the candidate feature river B *plus* all the other river segments that intersect the buffer. This provides an incorrect result. Furthermore, since the river segmentation criteria differ among sources, the complexity of this process increases to a point that it could not be described in formal rules nor fully automated.

Consequently, data matching for the hydrographic network was done manually using ArcGIS. Data matching for the road network used a semi-automated approach based on the spatial join function of ArcGIS. First we automatically matched the large scale data ($1:20000$, $1:50000$ and $1:100000$), then the smaller scales ($1:250000$, $1:250000_1$, $1:250000_2$, $1:1$ M), and then manually matched these two results.

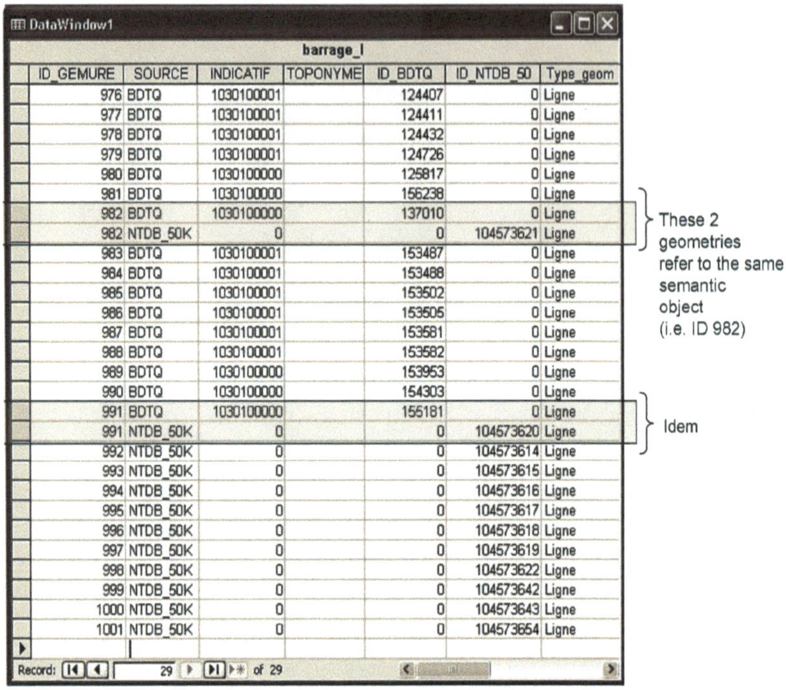

ID_GEMURE	SOURCE	INDICATIF	TOPONYME	ID_BDTQ	ID_NTDB_50	Type_geom
976	BDTQ	1030100001		124407	0	Ligne
977	BDTQ	1030100001		124411	0	Ligne
978	BDTQ	1030100001		124432	0	Ligne
979	BDTQ	1030100001		124726	0	Ligne
980	BDTQ	1030100000		125817	0	Ligne
981	BDTQ	1030100000		156238	0	Ligne
982	BDTQ	1030100000		137010	0	Ligne
982	NTDB_50K	0		0	104573621	Ligne
983	BDTQ	1030100001		153487	0	Ligne
984	BDTQ	1030100001		153488	0	Ligne
985	BDTQ	1030100001		153502	0	Ligne
986	BDTQ	1030100001		153505	0	Ligne
987	BDTQ	1030100001		153581	0	Ligne
988	BDTQ	1030100001		153582	0	Ligne
989	BDTQ	1030100000		153953	0	Ligne
990	BDTQ	1030100000		154303	0	Ligne
991	BDTQ	1030100000		155181	0	Ligne
991	NTDB_50K	0		0	104573620	Ligne
992	NTDB_50K	0		0	104573614	Ligne
993	NTDB_50K	0		0	104573615	Ligne
994	NTDB_50K	0		0	104573616	Ligne
995	NTDB_50K	0		0	104573617	Ligne
996	NTDB_50K	0		0	104573618	Ligne
997	NTDB_50K	0		0	104573619	Ligne
998	NTDB_50K	0		0	104573622	Ligne
999	NTDB_50K	0		0	104573642	Ligne
1000	NTDB_50K	0		0	104573643	Ligne
1001	NTDB_50K	0		0	104573654	Ligne

Fig. 9.6. The geometric file resulting from the data matching process.

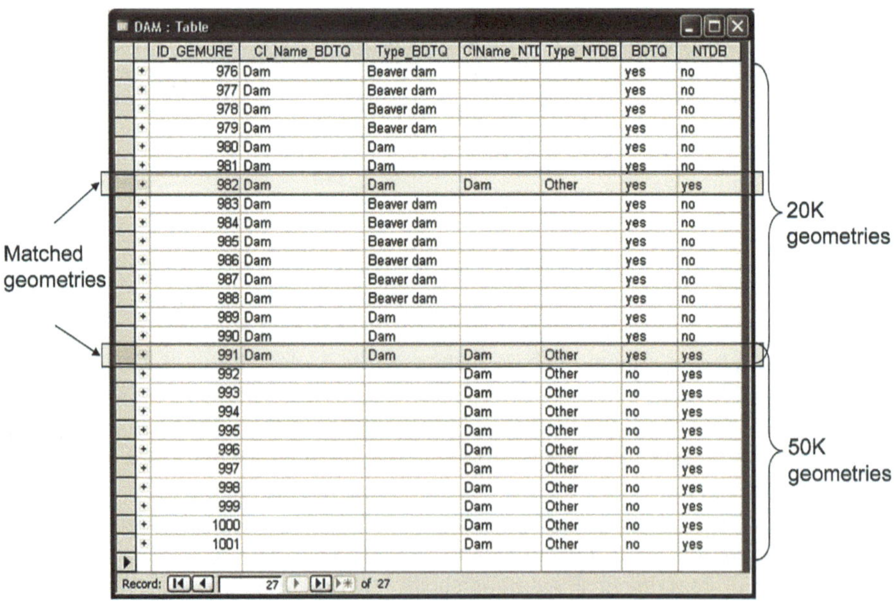

ID_GEMURE	Cl_Name_BDTQ	Type_BDTQ	ClName_NTI	Type_NTDB	BDTQ	NTDB
976	Dam	Beaver dam			yes	no
977	Dam	Beaver dam			yes	no
978	Dam	Beaver dam			yes	no
979	Dam	Beaver dam			yes	no
980	Dam	Dam			yes	no
981	Dam	Dam			yes	no
982	Dam	Dam	Dam	Other	yes	yes
983	Dam	Beaver dam			yes	no
984	Dam	Beaver dam			yes	no
985	Dam	Beaver dam			yes	no
986	Dam	Beaver dam			yes	no
987	Dam	Beaver dam			yes	no
988	Dam	Beaver dam			yes	no
989	Dam	Dam			yes	no
990	Dam	Dam			yes	no
991	Dam	Dam	Dam	Other	yes	yes
992			Dam	Other	no	yes
993			Dam	Other	no	yes
994			Dam	Other	no	yes
995			Dam	Other	no	yes
996			Dam	Other	no	yes
997			Dam	Other	no	yes
998			Dam	Other	no	yes
999			Dam	Other	no	yes
1000			Dam	Other	no	yes
1001			Dam	Other	no	yes

Fig. 9.7. The semantic file resulting from the data matching process.

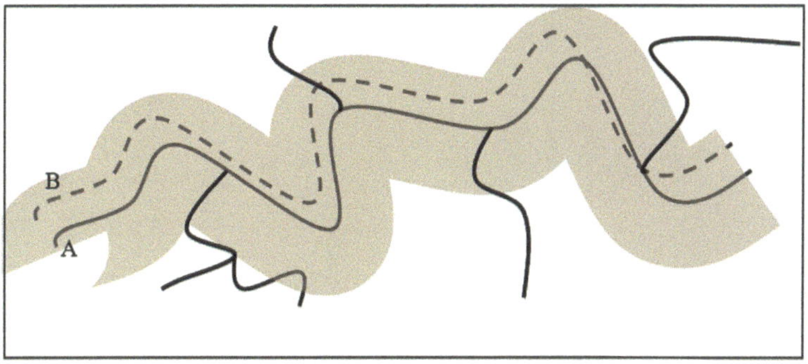

Fig. 9.8. Data matching on network data.

9.4.4. *Populating the Vuel datamart*

The Vuel datamart is based on a structure that differs from the warehouse structure. The latter is based on the relational paradigm while the datamart structure is based on a multidimensional paradigm. This structure is based on the Vuel concept (Bernier and Bédard, 2002) which separates the geometric properties from the semantic and the graphic properties. These three aspects are combined through a central fact table typical of multidimensional structures which contains Vuel objects (i.e. a unique combination of a geometric representation, a semantic and a graphic representation). Programs, developed in Visual Basic, have thus been written to extract the data from the warehouse and to restructure them to fit within the datamart architecture. For academic purposes, we built two versions. The first version of this datamart, used by UMapIT v.1, stored the data in an Access-Geomedia warehouse while the second version, used by UMapIT v.2, stored them in a PostgreSQL/PostGIS format.

9.5. Exploiting the Vuel Datamart Using the UMapIT Prototype

9.5.1. *General architecture*

UMapIT is a prototype that seeks to provide very flexible on-demand web mapping that supports seamless occurrence-based generalisation and drilling (Bernier et al., 2003). We have introduced this *occurrence-based* approach to add extra flexibility when one wants to customize only one or a few features of a map. For example, we may want to display a few selected individual buildings with their detailed geometry while leaving the other buildings with their simplified geometry (as seen on several downtown tourists maps, or to indicate a route from point A to point B). Furthermore, when zooming, such occurrence-based geometry selection provides each feature of a class (e.g. each individual building) with the capability to display a more or less generalized geometry for a given map scale depending on its individual properties such as area or ownership (public/private) information. With this occurrence-based behaviour for zooming, the choice of geometry for each feature is made independently from the other occurrences of the same class rather than changing simultaneously the geometry

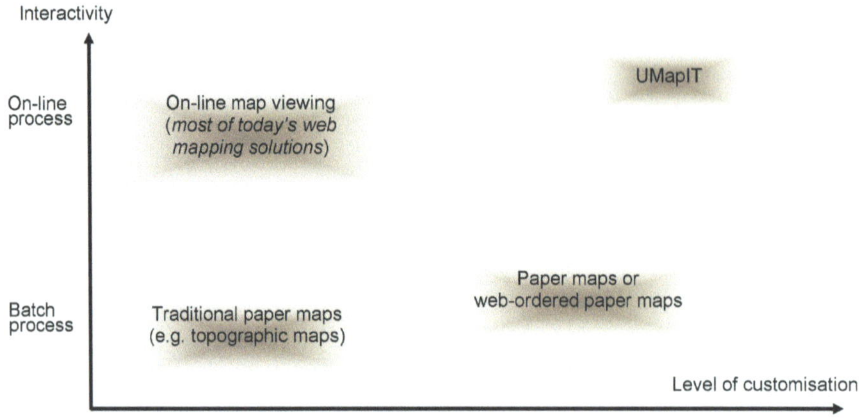

Fig. 9.9. UMapIT in a map production context.

of every occurrence of the same class (which is the typical layer-based GIS approach using minimum and maximum map scales for display).

It supports the intuitive creation of customised maps. Based on an occurrence-based approach, it allows requests to be applied to a specific object (e.g. drill-down or drill-up, changing the geometry or graphic properties of a single object or a group of objects) in order to produce highly personalised maps. For example, a user can produce a map at a given scale, represent all buildings as points, and drill-down on selected buildings to increase their level of detail (e.g. in polygonal form). As with most web mapping applications, map creation and navigation is mouse driven. Users do not need to know any query language such as SQL. Thus, UMapIT offers high levels of interactivity and customisability (Figure 9.9).

Currently there are two versions of the UMapIT application. The first one has been developed using Intergraph's technology GeoMedia Webmap (Figure 9.10). It is connected to a specific datamart (GeoMedia-Access Warehouse) of the GEMURE data warehousing architecture based on the Vuel structure.

The desire for increased interoperability and flexibility as well as academic needs led to development of a new architecture that complied with emerging international standards, such as those promoted by the World Wide Web Consortium (W3C) and the Open Geospatial Consortium (OGC). The second version of UMapIT relies on an architecture composed of a data server (the datamart), a geographic services component (geographic server and Vuel server) and a Web client (Figure 9.11).

The datamart of this second version still uses the Vuel structure but the data are stored in a PostGIS/PostgreSQL database. PostGIS represents the spatial data stored in PostgreSQL according to the "Simple Features" specification (OGC, 1999a). Tests have included the use of Oracle.

9.5.2. *Geographic services*

The geographic server is based on the WFS (Web Feature Service) standard, which gives access to feature-level geospatial data. As opposed to traditional map servers, the client now ob-

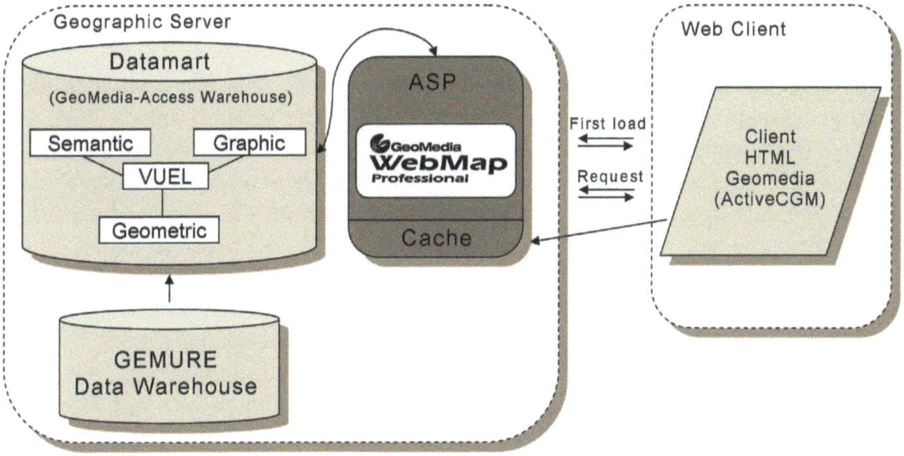

Fig. 9.10. UMapIT v.1 architecture.

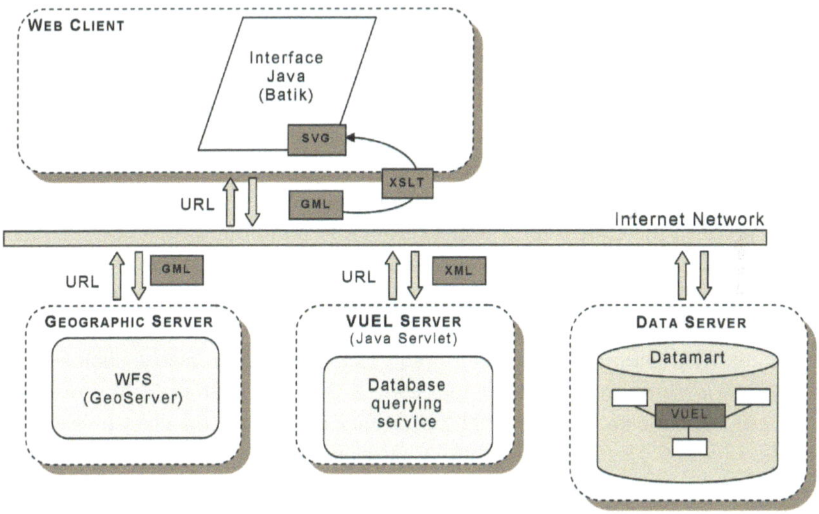

Fig. 9.11. UMapIT v.2 architecture.

tains fine-grained information about geospatial features, at both the geometry and the attribute levels (OGC, 2002b). This information is sent to the client in the GML (Geography Markup Language) format, which is an XML-based format for spatial data (OGC, 2003b). This geographic server has been developed using the GeoServer open source project (GEOSERVER, 2004). The Vuel server is a Java servlet offering access to the Vuel datamart via HTTP/XML requests through an Internet/Intranet network.

9.5.3. *Web client*

The web client was used to visualise and manipulate geographic data. The GML data obtained from the geographic server are transformed into SVG (Scalable Vector Graphics; W3C, 2003), using XSL (Extensible Stylesheet Language) transformations (W3C, 2001). This step is essential as GML files separate the content from the display. There is no information about how to display the data in a GML file. The same GML file can thus be displayed according to different rules. The web client was developed upon the Batik tool (Apache, 2005). This Java technology based toolkit offers a SVG viewer and some basic navigation functionalities. It is possible to extend the existing solution using Java. This web client interface can be embedded in a HTTP page (via an Applet) or can be deployed using JWS technology.

The main interface, illustrated in Figure 9.12, can be divided in three parts: the selection and customization part (left), the viewer (center) and the navigation pane (right). The first part is composed of a dynamic tree which presents the available object classes at different levels of granularity. Using this tree, users select the geographic objects they want displayed. Users also have the possibility to define the geometric granularity of selected object classes as well as their graphic properties. The first button in Figure 9.12 displays the available geometric representations of the selected object class. Users can choose between a detailed geometry (e.g. a polygonal representation) or a simplified one (e.g. a point representation) and apply it either to an entire object class or to an occurrence. For example, the user can decide to display all the residential buildings with a point geometry and use a polygonal geometry for their house and other landmarks that they want to highlight. It is also possible to define some graphic properties (e.g. fill colour, line weight and style) using the palette button. Again, these properties can be applied to a specific occurrence or to the entire object class.

UMapIT offers some navigational tools to navigate among the geographic objects once they are displayed in the cartographic area (viewer). In addition to the traditional navigation functionalities such as zoom in, zoom out, zoom extend, pan or fit, there are other tools from the OLAP domain. These operations are called Drills. They allow the user to navigate among the different geometric granularities of objects. The spatial *drill-down* operator allows us to go from a simplified geometry, such as a point, to a detailed geometry, such as a polygon, while the spatial *drill-up* operator does the reverse. These operators (drill down/up) can be applied once again at the occurrence-level (occurrence drills) or at the class-level (class drills). For example, Figure 9.13 shows the result of an occurrence drill on a single object which is the polygon building in the right bottom corner of Figure 9.12. The result shows that this polygon is in fact an aggregated polygon, composed of two distinct objects. Accordingly, a drill-up on one of these two objects would simulate an aggregation operation in cartographic generalisation. Figure 9.14 illustrates the result of a class drill-down on the Building class. All the occurrences of this class are thus displayed with a detailed geometry.

There are other operations that transparently combine zoom operations with drills operations to offer the user seamless generalisation while zooming. These operations are called Gen Zooms (for Generalisation Zooms) and they adjust the content of a map individually for each occurrence according to the zoom operation. This is made possible through the explicit links stored in the datamart between the geometries of an object and its semantic record. Because the management of geometric multiplicities is done at the occurrence level, a Gen Zoom out operation will not affect all the occurrences of an object class in the same manner. For exam-

Fig. 9.12. UMapIT web client interface.

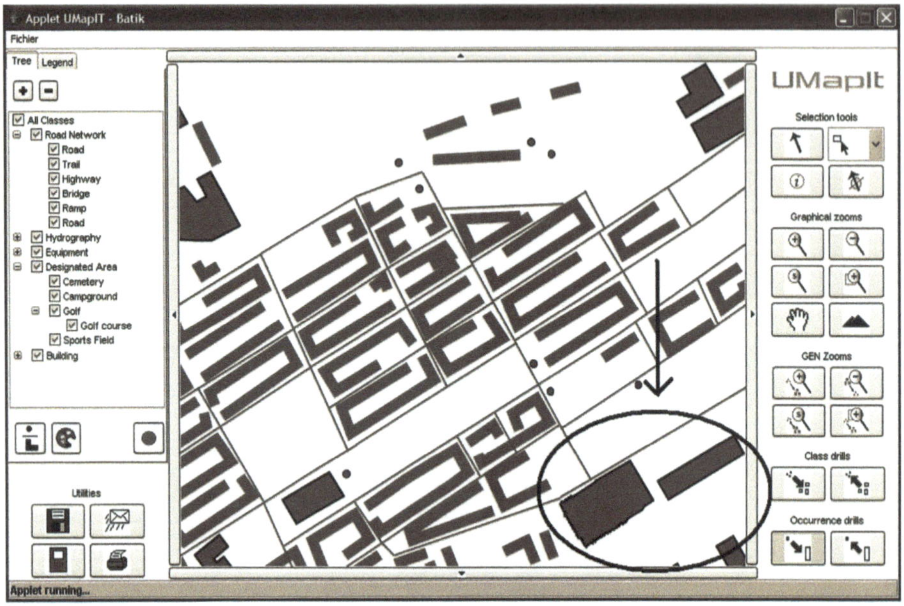

Fig. 9.13. Result of an occurrence drill-down.

Fig. 9.14. Result of a class drill-down (on buildings).

ple, smaller ones may be generalised while larger ones may remain unchanged. This is quite different from what we have seen so far with existing applications which are based on fixed layers and where a zoom operation will simply replace a layer by another one, thus affecting all the occurrences.

9.6. Future Improvements to UMapIT

UMapIT relies solely on multiple representations and simulates on-the-fly generalisation by exploiting the Vuel structure. This offers very fast response times but it also results in a voluminous database. In order to reduce the data volume and facilitate updates, the plan is to include some automatic generalisation capabilities through the concept of Self-Generalising Object (SGO), an agent-based concept developed in GEMURE. This new concept combines generalisation algorithms with geometric patterns at the occurrence-based level. SGO are defined as a special type of object that are able to generalise a cartographic object or a geometric primitive using: (1) one or more geometric patterns, (2) processing patterns (simple generalisation algorithms), and (3) spatial integrity constraints. These SGO will be tagged to exact geometries stored in the datamart and will be used during operations such as zooms and drills. SGO are being investigated in a PhD thesis while geometric patterns have already been defined (Cardenas, 2004).

Among the improvements and enhanced customisation capabilities of the next version, is the intention to implement geometric and semantic drill-across operations. These operations will allow the user to navigate between the different geometries of the same object (for the same level of detail, and different semantics of the same object but still at the same level of detail). For example, the geometric drill-across operation could be used to navigate between the geometry of source A to the geometry of source B, for the scale 1 : 250 000 (as mentioned earlier, there were two data sources, sourced from two data providers, at the scale 1 : 250 000).

Currently the Gen Zoom operations are based solely on an object's size but additional criteria such as semantic properties are planned. Semantic information could be used to control the threshold of a geometric change during a zoom operation. For example, a user could give more importance to commercial buildings than residential ones. During a zoom out, commercial buildings would keep their detailed geometry longer than the other types of buildings as long as they did not conflict with others. So, the user will potentially end up with a small number of customised Gen Zooms.

9.7. Conclusion

As we enter an on-demand era, we are facing new challenges in cartography, including on-the-fly generalisation and occurrence-level customisation. Recent years have witnessed a major spreading of web mapping applications allowing users to easily access maps. Today, users are becoming much more aware of map production opportunities. Though existing applications offer some functionality to personalise maps and produce them at varying scales, they mostly rely on predefined maps or layers and present very limited customisation capabilities. As a first step towards the ultimate On-Demand Web Mapping application, we have proposed in this chapter a solution that relies on a datawarehousing architecture that combines automatic generalisation and multiple representation. This architecture is composed of a "multiple representation-centric" data warehouse and of specialised "on-the-fly generalisation-centric" datamarts. Inspired by the "divide-and-conquer" philosophy, our approach to map generalisation allows datamart users to gain quick access to maps at different scales and on different themes and to easily navigate within the controlled scope of the datamart. In order to show the potential of such an approach, we presented UMapIT, a web-based on-demand mapping tool based on a specialised datamart extracting data from a data warehouse and offering flexible seamless generalisation.

More research and development is needed to optimise our solution. Integrating the new concepts of Geometric Patterns and Self-Generalising Objects (SGO) in the data warehousing architecture will bring a significant performance gain and reduction of storage space. Although several papers regarding MR-DB management have been published, one of the main challenges remains one of populating such databases. The authors believe that such a process could be facilitated by slightly modifying the way data are acquired. We are currently working on adapting the photogrammetric data acquisition process, and the map digitising process, so that they include geometric patterns and SGO.

It is only when these issues are completely solved that we can expect to achieve the ultimate on-demand web mapping application that supports both on-the-fly automatic generalisation and full customisation at the occurrence-level. On-demand mapping must allow the user to produce a map at the desired scale and content without worrying about generalisation complexity or performance. In the specific case of web-mapping, response times must stay within the 10 seconds cognitive band defined by Newell (1990) in order to keep the user on-line. Thus, web-based on-demand mapping represents a major challenge for map generalisation as it requires on-the-fly and fully automatic generalisation as well as transparent operations for the user.

Acknowledgements

The authors wish to thank Canada GEOIDE Network of Centers of Excellence for funding the GEMURE project (2002–2005) along with the following organizations: Natural Resources Canada, Research & Development Defence Canada, Natural Resources Quebec, Intergraph.

Generalisation of Geographic Information: Cartographic Modelling and Applications
W.A. Mackaness, A. Ruas and L.T. Sarjakoski (Editors)

Chapter 10

Relevance of Generalisation to the Extraction and Communication of Wayfinding Information

Monika Sester, Birgit Elias

Institute of Cartography and Geoinformatics, University of Hannover, Germany
e-mail:{Monika.Sester,Birgit.Elias}@ikg.uni-hannover.de

Abstract

Wayfinding is the process of efficiently finding and following a route between a starting point and a destination. It relies on spatial information that has to be communicated to the user in order to solve this specific task. This can result in delivery of highly abstracted and sparse but efficient spatial information in terms of route descriptions or depictions. This information is typically generated from an information-rich environment. The abstraction process requires application of generalisation to spatial descriptions and depictions. Generalisation is needed for two tasks: firstly, in the selection of relevant wayfinding information; secondly in the visualisation of that information such that it can be immediately understood by the user. Since landmarks play an important role in wayfinding tasks they first have to be identified from the data set and then visualised appropriately, for example through emphasis (using other objects only as background information). In this chapter first an overview of the wayfinding task is given, focusing on cognitive aspects. This leads to the identification of a class of objects that play an important role in wayfinding. Methods to characterise and automatically determine landmarks are presented, followed by generalising methods that can be used to communicate wayfinding descriptions. Finally, applications utilising landmarks for navigation and wayfinding are described.

Keywords: spatial cognition, wayfinding, automatic feature extraction

10.1. Aspects of Wayfinding

10.1.1. *Wayfinding tasks and wayfinding aids*

The skill of orientating oneself in spatial environments is a fundamental human ability and a prerequisite to wayfinding. It has been studied intensively by psychologists and geographers in order to understand the underlying processes and to exploit this knowledge in the creation of effective navigation applications, such as car navigation systems that assist users in wayfinding or robot systems that gather information and navigate space autonomously (Davis, 1986). Navigation and wayfinding are closely related concepts that are often used interchangeably, leading to different definitions in the literature: In general, *navigation* is defined as the process of orientating oneself in a geographic environment in order to reach a specific location. The navigation process consists of three activities: positioning (i.e. establishing location and orientation in the environment), route planning (i.e. calculating which transport network segments to follow) and locomotion (i.e. implementing physical movement). From the spatial cognition research perspective, we can divide navigation into two components. The first is locomotion, the movement of one's body around an environment coordinated with respect to one's immediate surrounding. The second is wayfinding – regarded as the goal-directed and planned movement of one's body through an environment in an efficient way (Montello, 2005). Golledge (1999), defined *wayfinding* as the process of determining and following a path or route from a starting point towards a destination. It is a purposive, directed, and motivated activity. Collectively then, both positioning and route planning information is required for wayfinding. A further classification of wayfinding considers two different aspects:

1. Wayfinding tasks such as travelling to a previously known or unknown destination or exploration of the environment with the purpose of returning home (Golledge, 1999), and
2. Wayfinding techniques including the different methods of navigating through space, such as tracking, dead reckoning, piloting and navigating (Hunt and Waller, 1999; Mallot, 2000): Tracking is defined as following exocentric, local cues. Dead reckoning is based on egocentric cues, such as the physical turns one makes or the mental sense of distance covered. Piloting is the process of navigation combining reference to exocentric objects (such as landmarks) with egocentric bearings. Finally navigating is defined as moving among a set of landmarks configured in such a way that the user is able to calculate distance and bearings. It requires the user to know where they are, both at the outset and during navigation.

The process of navigating can be supported using a variety of different kinds of wayfinding aids. Examples include:

- *Cognitive maps (mental maps):* internal spatial representations of the environment, gradually learned by exploration or by maps. Because of their systematic distortions they are also known as "cognitive collages". They consist of basic spatial relations, relations between elements and relations between an element and a reference frame (Tversky, 1993);
- *Directions:* conveying route instructions, mostly in verbal form. They are a composite of descriptive discourse (nature and position of landmarks) and instructions (actions at critical points along the route) (Tversky and Lee, 1998; Tversky and Lee, 1999; Denis et al., 1999);
- *(Topographic) maps:* traditional wayfinding aids, can be used to acquire configurational knowledge (such as the spatial relations between a set of landmarks) (MacEachren, 1995);
- *YAH-maps:* You-Are-Here-maps are usually vertically fixed (e.g. against a wall) clearly labelled, and have an indicator showing the location of the viewer (A "You are here" annotation). The optimal placement of the map is governed by the need to support structure matching (relating the map to the terrain) and orientation (map alignment) (Levine, 1982; Richter and Klippel, 2002);
- *Route-maps:* route maps portray specific routes, either in the form of a sketch map or as a strip map (with no fixed scale and orientation) (Tversky and Lee, 1998; MacEachren, 1986);
- *Schematic maps:* are simplified general maps derived by relaxing spatial constraints, with a task-specific focus. They can be considered as a hybrid of maps and verbal descriptions

(Casakin et al., 2000; Freksa, 1999). Popular examples are plans of transportation networks (Avelar, 2002);
- *Sketch maps:* portrayals of space using map elements, visualisation of verbal directions, neglecting spatial constraints and completeness, and often less accurate than schematic maps (Freksa, 1999). They can also be used for mere localisation in an unknown environment (Kopczynski and Sester, 2004).

In summary, there are a range of techniques used to move through space, supported by a range of map forms. A variety of navigational techniques (searching, aiming, piloting) depend on the description and identification of landmarks and routes in order to move through that geographic space (Mallot, 2000).

10.1.2. *Cognitive aspects of wayfinding*

The spatial representation of the environment in the human mind is formed from a variety of geographic features. According to Lynch (1960) these can be divided into paths, districts, edges, landmarks and nodes. They include the locations of these features in the environment, the distances among them and the knowledge necessary to orientate oneself in the environment. The primary function of this spatial representation is to facilitate location and movement within the larger physical environment and to prevent oneself from getting lost (Siegel and White, 1975). According to Thorndyke (1981) there are three different types of spatial knowledge:

- *Landmark knowledge:* memory of prominent geographic features.
- *Procedural knowledge:* knowledge of route representation, action sequences that connect separate locations, and procedures for navigating between points.
- *Survey knowledge:* knowledge of the configuration of features and the global organisation of those objects and the relationship between different routes.

It is generally felt that these three types of knowledge build sequentially; that a person's knowledge typically progresses from landmark to procedural to survey knowledge with increasing familiarity with the environment (Thorndyke, 1981). Siegel and White (1975) proposed that landmarks and routes are the necessary and sufficient elements for "minimal" representations that allow "wayfinding" to occur. Also Agrawala (2002) judged that this information was essential, whereas other structural information such as localised and broader contexts was of minor importance.

10.1.3. *Generalising wayfinding descriptions*

Generally speaking, topographic maps have been used as a basis for wayfinding. However, when the task is to convey a specific route or access to a specific object or place, adapted descriptions are more appropriate to provide a quick and unambiguous aid in solving the problem. At this point, information in the real environment has to be abstracted whilst taking into account the nature of the task. Research in cognitive psychology indicates that effective route descriptions in wayfinding need to convey all turning and decision points along the route. However, precise depiction of lengths, angles and shape of the route are less important (Denis et al., 1999). The preservation of the topology of the route or the network is of utmost importance. This requires presentation of all turning points, whereas other geometric properties can be relaxed or even omitted in favour of better readability and immediate understanding. Due to their sparseness these representations provide perceptual stimuli that support and facilitate the formulation of mental descriptions without overpowering the users' imagination with the completeness of the concrete visual world or overly detailed maps (Avelar, 2002).

Cartographic generalisation can be used to achieve this abstraction, as it provides techniques for determining these important objects by selection and omission, simplification, enhancement, classification, typification, and displacement. This idea of abstraction relates to both pictorial and verbal descriptions. For example turning instructions tend to be given as "turn right", even when no exact right angle occurs. Only in cases where confusion could occur, is more specific information needed. Similarly, in verbal communication, different levels of detail are given according to the generality of description. For example Appelstraße – Hannover – State Lower Saxony – Germany – Europe is a hierarchy that moves from fine to coarse detail. Such hierarchical descriptions can be taken from existing taxonomies or ontologies for spatial information (Jones et al., 2003). Claramunt and Mainguenaud (1997) propose a combined data model for the storage of partial navigational knowledge representing different levels of hierarchy.

10.2. Characterisation and Extraction of Objects Important to Wayfinding and Navigation

Effective route descriptions consist of both landmarks and routes. Providing the route information automatically is a key feature of current route planning systems. However, the use of landmark information has not yet become part of commercial navigational systems and services. This is in spite of the fact that research findings confirm the important role of landmarks for the effectiveness of route directions and descriptions (Deakin, 1996; Denis et al., 1999; Michon and Denis, 2001; Tom and Denis, 2003). Though prominent landmark objects such as monuments and important buildings can be identified manually, a more efficient method is to identify them automatically – based on their relative size, shape and other non-geometric attributes. The automatic identification of landmarks along a route is an interpretive process, and wayfinding can be considered as a typical generalisation task. As such, it consists of the selection (semantic abstraction) of relevant objects for the given task and their adequate presentation and communication using cartographic generalisation. The cartographic portrayal of route information including landmarks requires application of generalisation techniques, and includes exaggeration of landmarks in order to give them increased prominence in the map.

10.2.1. *Characterisation of landmarks and their salient features*

A landmark may be any element in an environment that supports recognition and confirmation of the user among a set of features and landscapes. An early attempt to characterise landmarks was made by Lynch (1960): A landmark is singular in its environment, because it has a sharp figure-background-contrast, is perhaps located at a road junction, and it differs in size, shape, position or age relative to the objects around it. Good candidates would be those landmarks that have particular visual characteristics, unique purpose or meaning, or a central or prominent location (Appleyard, 1969). Sorrows and Hirtle (1999) propose three categories of landmarks: visual (visually standing out, in contrast with surroundings), cognitive (important meaning or prototypical, cultural significance) and structural landmarks (highly accessible, prominent location, or important intersections or plaza). Landmarks that match all these categories have the greatest effect.

The position of the landmark relative to the route leads to different types of landmarks, called distant/global and local landmarks (Lynch, 1960; Steck and Mallot, 2000) or off-route and on-route landmarks (Lovelace et al., 1999). Some researchers suggest that landmarks are not required uniformly along a route (Denis et al., 1999) but principally close to critical nodes or decision points along the route (Michon and Denis, 2001). Lovelace et al. (1999) propose a more detailed classification of landmarks: those at critical points along a route (perhaps associated with a turning point), and those at non-critical points (used to confirm that the user is on the right path). In their study Lovelace et al. (1999) found that the quality of route directions was dependent on both forms of landmarks being provided. Steck and Mallot (2000) investigated the role of global and local landmarks in an experiment using a virtual environment. This experiment revealed that both types are used in wayfinding decisions – the optimal mix varying depending on personal preferences. Removing one landmark type revealed that humans were able to switch between the landmark types. Elias et al. (2005) made the observation that the type of objects useful as landmarks depends on the mode of travelling. For example, for in-vehicle navigation purposes "road furniture" landmarks (such as traffic lights, pedestrian crossing and petrol stations) seem to be appropriate (Burnett et al., 2001). In contrast, for pedestrian navigation in urban areas, buildings and facades have to be conveyed (Winter et al., 2005).

10.2.2. *Methods for automatic extraction of landmarks*

The automatic identification of landmarks is an ongoing research topic. One approach is based on the concept of Sorrows and Hirtle (1999) and provides measures to formally specify the landmark saliency of buildings. The strength or appropriateness of landmarks is determined by parameters relating to visual, semantic, and structural attraction (Raubal and Winter, 2002). This measure of saliency can be extended to include a measure of visibility determined from the point of approach (Winter, 2003).

Another approach has been to extract the building landmarks from existing GIS databases using a two-step procedure, summarised in Figure 10.1. The starting point is to investigate the surroundings of junctions, where each intersection is a possible decision point in a route description. All buildings visible from a decision point are used in the analysis. The next task is to select those buildings that are unusual or unique in their local environment. Attributes of the buildings that can be taken into account include geometric, topologic, and semantic properties. A selection of possible attributes derived from the German cadastral map are proposed in Figure 10.2. Spatial Data Mining (Han and Kamber, 2001) methods or modified Machine Learning algorithms (ID3 from Quinlan, 1986) can be used to create a decision tree that shows the discriminating attributes that characterise an object as a landmark (Elias, 2003b).

In the next stage, the selection of potential landmarks is further reduced by assessing the degree of visibility at specific decision points. Tracking the degree of visibility whilst approaching the decision point can also form part of the analysis – those buildings that can be seen for the greatest amount of time during the approach being more valuable (Elias and Brenner, 2004). A digital surface model is required in order to determine the visibility of objects

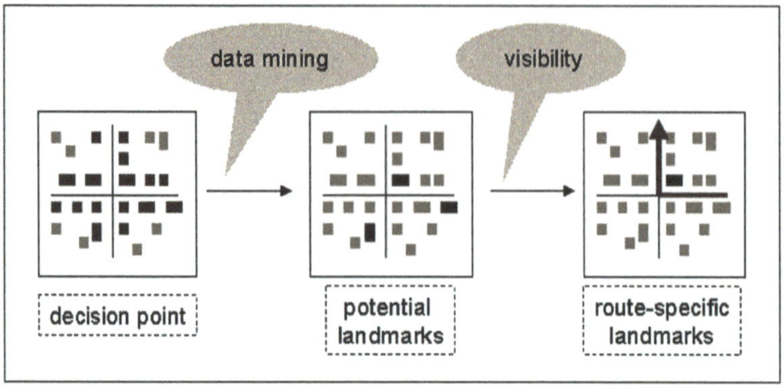

Fig. 10.1. Extraction procedure for landmarks.

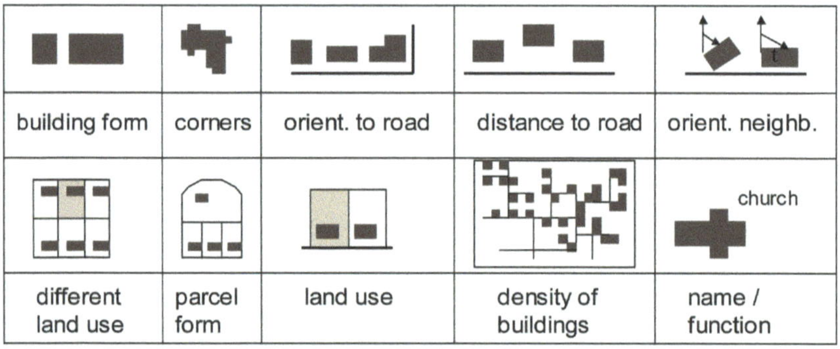

Fig. 10.2. Building attributes: geometric, topologic and semantic properties of buildings.

at the various decision points and can be created by laser scanning data in combination with cadastral ground plans. A virtual panoramic view of the decision point can then be calculated. The degree of visibility can be used to rank the landmarks. The portion of coverage in the virtual panoramic view is measured in pixels.

In Figure 10.3, an example of landmark extraction is shown. A road intersection at the University campus of Hannover is investigated. First, all visible buildings are analysed using data mining techniques, reducing the choice to three potential landmarks. After that, a further investigation of the visibility (see Table 10.1) reveals that only object number 2 (as labelled in Figure 10.3) is a suitable candidate. This landmark is the only object that is clearly visible and perceivable from the decision point. The most discriminating attributes that were relevant to the selection process are given in Table 10.1. This table shows the visibility related to the amount of pixel in the virtual panoramic view and the distance of the object to the road intersection. Both measures are used to determine the suitability of objects as candidate landmarks. Furthermore, the description of the object used in the cadastral map is given. This label can be used to reference the object in a wayfinding description.

Fig. 10.3. Example of landmark extraction, (a): candidate buildings (dark grey) for the decision point (black triangle), from which potential landmarks are selected (b).

Table 10.1. Degree of visibility, taken from Elias and Brenner (2004).

No.	Visibility (pixel)	Distance (m)	Description	Discriminating attribute
1	4	216	high voltage transformer building	different usage than neighbours
2	1278	60	cafeteria of university	singular building function
3	3	118	kindergarten	singular building function

Future research aspects will examine the quality of the route description, taking into account the reliability of the description (i.e. do users consistently achieve their goal when following the route description). This requires that unambiguous instructions are given in terms of landmarks used (Elias, 2003a). This includes descriptions that are "fault tolerant" – that where the user makes a mistake in route execution, they can still find their way back onto the route and to their destination (Duckham et al., 2003).

10.3. Methods of Communicating Wayfinding Information

10.3.1. *Different forms of communication*

Communicating information efficiently is a goal that can be achieved using Intelligent Multimedia Presentation Systems (IMMPS) (Roth and Hefley, 1993). Such systems strive to automatically combine a set of alternative media and presentation methods to optimally achieve a given goal. Communicating wayfinding information needs to be set within the broader challenge of conveying geographical information. In order to reduce the cognitive effort in finding a desired route between a start point and a destination, specific ways of communication are used, either verbally or graphically (Freksa, 1999). Using verbal descriptions requires a language. Language is a very powerful medium and can describe features, relations and structures. However communication is linear, sequential, and somewhat one-dimensional. Complex 2D or 3D spatial relations are therefore difficult to describe verbally. In contrast, using graphical spatial representations, e.g. in the form of maps, 2D and 3D features, relations and structures can be better described. The spatial relations become visible through their implicit representation.

Whilst we can code a great deal of information into the map, it requires the user to have skills sufficient to then extract and interpret that information in the context of a particular task. The general form and relatively high information content of topographic maps makes them less effective for wayfinding tasks. On the other hand, strip maps, a very abstracted form of map, are more appropriate for wayfinding – largely ignoring issues such as orientation, scale and detail (Agrawala, 2002). The emphasis on linear features and process descriptions in strip maps matches the human cognitive mapping process, so potentially the user can understand these types of maps more quickly. Other have suggested that strip maps are less useful or may even confuse users who already have a detailed mental map of the situation (MacEachren, 1986). Thus adequate abstractions using generalisation methods are needed that are adapted both to the task at hand and for different user groups with varying spatial knowledge (Sarjakoski and Sarjakoski, this volume (Chapter 7)). It has been suggested that one way of determining the appropriateness of the level of abstraction is to undertake eye-movement studies (Castner and Eastman, 1984; Brodersen and Andersen, 2001) but determining the link between cognitive processes and map design methodologies has always proved problematic.

10.3.2. *Generalisation methods*

Cartographic generalisation offers techniques for abstracting spatial information. In this section different wayfinding aids are described, together with possible generalisation methods appropriate for their creation. Table 10.2 lists the possibilities for conveying route descriptions, together with examples and corresponding generalisation operators. The different visualisations describe general ways of communicating route information; some of them are especially suited for conveying landmarks:

- *Route highlighting:* the desired route is overlaid on a conventional map. Although this presentation mode gives a good overview of the whole route, it fails to provide details about junctions and turning instructions. Typically, no generalisation is applied;
- *Overview and focus maps:* in addition to the route highlighted on the map, detailed focus maps are presented that depict each individual junction at larger scale. Such descriptions are typical of route planning programs on the Internet, for example Map24 (2005). For a user it can be difficult to put these detailed maps together, as scale and orientation may change. Sometimes the level of detail of the description in the focus maps is too high to be really useful (Denis et al., 1999). In car navigation systems, this problem is overcome by providing both the overview map and detailed pictograms of the junction, and accompanied by verbal descriptions including real distances ("Turn right in 300 m"). In this way, both general and detailed information are available. The multimodal communication of the turning instructions helps to ease the cognitive load the driver is exposed to. The pictograms can be generated from the information about the junction: either it is transformed to a prototype junction from a given taxonomy (e.g. Klippel et al., 2003) or schematised to the closest angles in a given qualitative description. The different description options have decreasing levels of accuracy: quantitatively, in terms of exact angles, or qualitatively using 8 main directions, 4 cardinal directions, or no orientation at all (Barkowsky and Freksa, 1997);

Table 10.2. Map examples and corresponding generalisation techniques.

<table>
<tr><th>Type of map</th><th>Example</th><th>Used generalisation techniques</th></tr>
<tr><td>Route highlighting</td><td>
</td><td>No generalisation as such; mainly symbolisation using graphical variables</td></tr>
<tr><td>Overview/Focus map</td><td>
</td><td>Overview map: no generalisation; Focus map: symbolisation, simplification, schematisation, typification</td></tr>
<tr><td>Sketch map</td><td>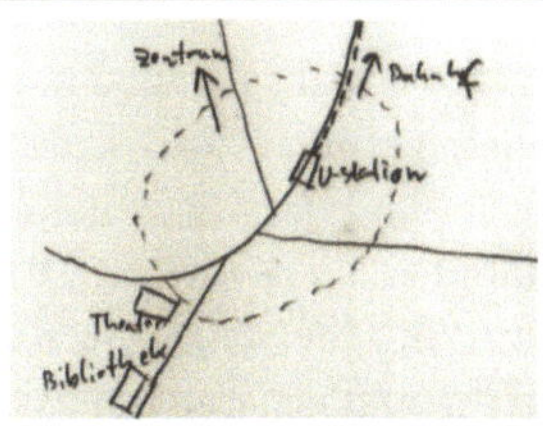</td><td>Manually produced sketch: human generalisation
Automatically produced maps (Agrawala, 2002): selection, simplification, schematisation, vario-scale, typification</td></tr>
<tr><td>Schematic map</td><td>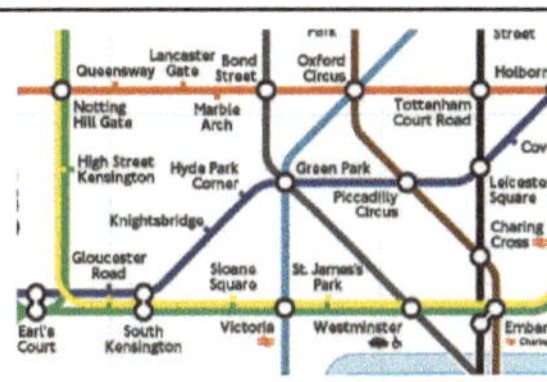
</td><td>Selection, schematisation, enhancement, typification, displacement</td></tr>
<tr><td>Vario-scale map</td><td>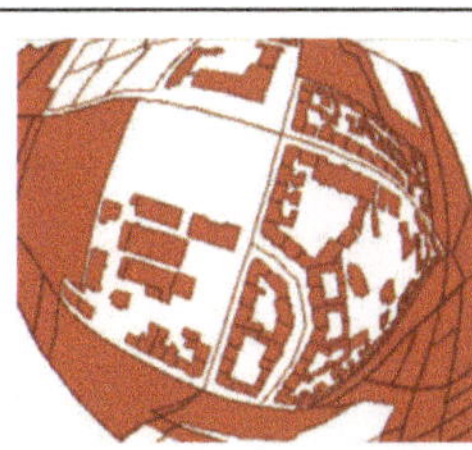</td><td>Transformation function to produce different scales at different map locations In addition: more details in focus area using selection, emphasis, simplification, aggregation</td></tr>
<tr><td>3D-representation</td><td></td><td>3D-visualisation and 3D-generalisation</td></tr>
</table>

- *Sketch maps or hand-drawn maps:* sketch maps emphasise relevant features and relations by adaptively accentuating turning angles and route lengths and simplifying the road elements between turning points. There is no uniform scale, as short distances are typically enlarged and long distances are shortened. However, the relations between adjoining short and long route sections are preserved (Agrawala, 2002). This means that the properties and relations can be relaxed to enhance aspects relevant to the task. Thus certain spatial relations will not be faithfully represented. For example, the spatial orientation of outgoing roads can be abstracted as described above; the same holds for the distances between the turning points for which different scales can be applied;
- *Schematic maps:* schematic maps serve a similar purpose as sketch maps, but are useful not only for one particular route, but for navigating in a larger environment and across different routes. They are well known for visualising public transportation networks (Garland, 1994). In this case, the schematisation of routes between decision points is even more enhanced – representing them with straight lines (simplification) and imposing orientations in fixed angles such as 45 or 90 degrees or 30, 60, or 90 degrees (typification) (Avelar, 2002). Furthermore, close lines are separated by minimal visual distances (displacement);
- *Vario-scale maps:* in order to allow for a more detailed presentation of start- and endpoints of a route, as well as important decision points, vario-scale (or polyfocal) maps can be used. They have a magnifying glass-effect, leading to an enlargement in the areas of interest and a reduction in scale in the background areas (Lichtner, 1979b; Keahey, 1997; Harrie et al., 2002b). Besides continuously modifying the scale, discrete scale changes are possible by highlighting important objects and de-emphasising the background objects using appropriate generalisation operators, such as simplification or aggregation (Sester, 2002) (for an example see Figure 10.4);
- *3D-representations:* 3D-presentations of the route provide a perspective view (car navigation prototype systems are currently being tested in Japan). In these presentations, landmark buildings are presented in 3D. In this way, the visibility of objects can be taken into account

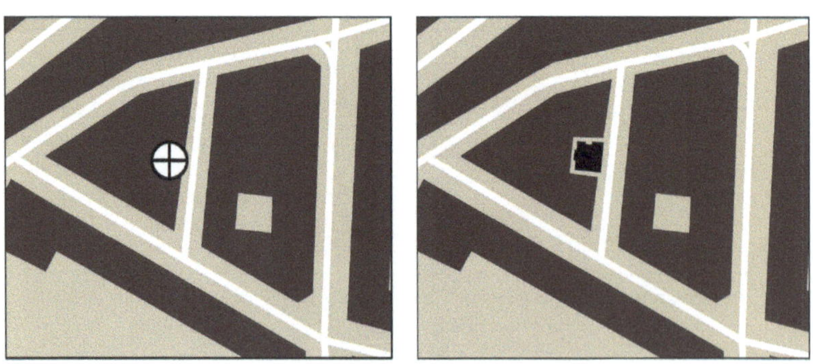

Fig. 10.4. Two ways of visualising landmarks: in the small scale representation (left), all buildings are aggregated, and therefore the landmark building cannot be separated from the others. Using discrete generalisation (right) shows the landmark building in full detail and aggregates the background objects.

– something that is not possible when looking at 2D representations. 3D-representations have also been tested for LBS applications (Schilling et al., 2005). Techniques have been developed for the generalisation of 3D objects in order to reduce both the amount of data to be visualised and the cognitive load on the user (see Meng and Forberg, this volume (Chapter 11)).

The choice of representation (Table 10.2) and the generalisation methods applied need to take into account the preferences of the user, and the task they are performing. The user needs to understand the link between the reality they are navigating through and the schematic form of the map.

Aggregation and abstraction has also been used in the generalisation of verbal descriptions, by grouping several potential decision points – a process called spatial chunking (Klippel, 2003; Klippel et al., 2003). An example would be "Turn right at the third road intersection". This description can be generated by counting the number of junctions where there is no deviation from the route. This grouping of information tends only to work for a limited number of objects.

Studies have shown that schematic maps and verbal descriptions result largely in equally effective wayfinding. In some studies a slight preference for map presentations was reported (Burnett, 2000b; Schlender et al., 2000). This equivalence in effectiveness is due to the fact that the content of depictions and descriptions has the same underlying structure and semantics, and can thus be transformed automatically into each other (Tversky and Lee, 1999). Their research showed that depictions were superior to verbal descriptions because of the iconicity that facilitates identification in the real environment, and the fact that spatial relations among groups of objects are easier to determine from maps.

10.4. Application of Landmarks for Wayfinding and Navigation

Today's route guidance and navigation systems for vehicles provide information in the form of "turn-by-turn" instructions. Distance-to-run information can be conveyed, either in absolute terms ("Right turn in 100 metres") or in non-absolute, time-based terms ("Right turn soon") (Burnett, 2000a). There is empirical evidence to suggest that the use of these systems as compared with paper maps leads to less navigational errors, shorter journey times and reduced mental workload (Streeter et al., 1985), but that further improvements can be made to make the interface more "intelligent" and naturalistic. It is likely that the integration of landmarks will improve navigation systems (see also Burnett, 2000b). Within the field of Location Based Services there is growing interest in delivery of navigational systems for pedestrians (Bartie and Mackaness, 2006). The level of detail required is greater than that for car navigation systems, since the pedestrian is not constrained by the network, and simple distance-to-turn instructions are not adequate. In these applications, the wayfinding instructions must also be adapted to the mode of transport and the speed of the pedestrian (Maaß, 1995; Elias et al., 2005).

In the previous sections the importance of landmarks for wayfinding was stressed; in §10.3.2 different ways of communicating routes have been reviewed. We now present a scenario in which an appropriate visualisation of a landmark was chosen based on the process described in §10.2.2. We focus on a 2D presentation, where the task is to visualise a landmark object, in this case a building. The landmark could be visualised just by highlighting it with a certain colour or marking it with a symbol. This makes it difficult to identify the object, as it might be "hidden" between other objects. Presenting only the landmark object without the environment, however, would lead to a loss of context. Therefore, both landmark object and background are presented in one graphic: the visualisation of the landmark in the local environment is achieved by representing the landmark in full detail, whereas the background information is given in a highly generalised form. Thus the overall spatial context can easily be recognised, as well as the local context the specific landmark is embedded in (Figure 10.4). The presentation can be generated using different generalisation operators: simplifying or aggregating background objects, and enhancing the landmark object. In order to achieve an optimal presentation, the characteristics of the buildings are conveyed. These are identified via a Machine Learning process that discriminates them as landmarks for visualisation (see §10.2.2). For example, if the colour or shape of an object has been an indication of its uniqueness, then these attributes can be used in the visualisation by enhancing this property and a combination of colour and exaggeration of form (Hampe and Elias, 2003).

10.5. Conclusion and Outlook

It would appear that highly abstracted information is required for wayfinding. Generalisation offers the means by which spatial information can be abstracted and visualised for specific tasks. Generalisation methods are vital for deriving wayfinding descriptions and depictions both for the initial selection of relevant and important objects through to their appropriate visual communication. Current research in wayfinding theory are focused on the formal analysis of the wayfinding process. In order to achieve this, theories and methods from different disciplines have to be integrated (notably cognitive psychology, human computer interaction, cartography, and computer science). These theories need to be evaluated and further expanded via empirical studies. Collectively work in this field will lead to the specification of a generic wayfinding ontology (Timpf et al., 1992) which will include optimal cartographic portrayal of route instructions. Automatic generation of enhanced route instructions can be applied to car or pedestrian navigation systems to improve their efficiency and effectiveness. In fully automated systems, reliability of the derived solution is of critical importance. Therefore, methods for describing the quality of a wayfinding description have to be developed in order to characterise good and bad descriptions.

W.A. Mackaness, A. Ruas and L.T. Sarjakoski (Editors)

Chapter 11

3D Building Generalisation

Liqiu Meng[a], Andrea Forberg[b]

[a]*Chair of Cartography, Technische Universität München, Arcisstr. 21, 80333 München, Germany*
e-mail: meng@bv.tum.de
[b]*Weiglstr. 9, 80636 München, Germany*
e-mail: Forberg.Andrea@GMX.de

Abstract

Starting with an introduction on 3D city models and their popularity among users, this chapter begins by arguing the need for automatic 3D generalisation. An overview of on-going research efforts is provided, including theoretic concepts and assessment of the feasibility of methods. Implementation work is presented, elucidating ideas in automatic segmentation, recognition, simplification and viewer-dependent visualisation of 3D building structures. The examples seek to illustrate the methodological and pragmatic challenges in modelling and generalising 3D structures. The authors attempt to provide a basis from which new techniques can be developed as well as highlight the need to extend 2D operations such as aggregation, typification and landmark exaggeration to cope with 3D generalisation problems.

Keywords: scale space, level of detail, description model, recognition of building structure, squaring, parallel shift, simplification, segmentation

11.1. Background

The rapid development of multisensor and multimedia technologies has made it possible to construct and visualise detailed 3D city models. As illustrated in Figure 11.1, 3D city models are typically rendered as central perspectives with rich depth cues and a self-explaining character. They offer an intuitive organisation of spatial objects that replicates or reflects the real world, thus utilising the viewer's natural perception and memory of space as well as spatial relationships (Mallot et al., 2002; Germanchis and Cartwright, 2003). Indeed 3D city models have been increasingly applied as communication languages and working tools in a growing number of fields such as architecture, construction, archaeological reconstruction, urban planning, tourism, civil engineering, mobile telecommunication, energy supply, navigation, facility management, disaster simulation, spatial cognition and computer game industries. Although these application fields share the common demand for 3D information, their special requirements considerably differ with regard to precision, actuality, spatial coverage and interoperability. In other words, what is needed is not one single solution, but rather a number

Fig. 11.1. Examples of 3D city models: Zurich (top) (CyberCity AG, 2005), Philadelphia (bottom) Marlin Studios, 2005.

of 3D city models, which can be (1) different resolutions of a city model, (2) different updates of a city model, or (3) interoperable models of different cities spread over a large region. While case (2) and (3) deal with the research issues of spatial-temporal data acquisition and modelling, case (1) focuses on the study of 3D objects in the scale space, which is the main topic in this chapter with an emphasis on 3D buildings.

11.1.1. *3D buildings and their levels of detail*

The scale space of 3D buildings is essentially a linear continuum, along which an arbitrary number of milestones can be said to exist referred to as Levels of Detail (LoD). Each LoD corresponds to a certain degree of generalisation. Unlike the 2D topographic maps that have standard official scale series, there are no generally agreed LoDs for 3D buildings. As exemplified in the following list, the currently available LoDs are mainly determined in relation to the resolution of sensor data, the precision of semantic information and the relevant application:

Thiemann (2004) summarises three LoDs for settlements and buildings:

- LoD1 = aggregated settlement blocks with a uniform height,
- LoD2 = block of the individual buildings without roof form,
- LoD3 = LoD2 enhanced with a simplified roof form.

The Netlexikon (akademie.de, 2005) suggests five LoDs for individual buildings:

- LoD1 = Popping up of the ground plan to a uniform height,
- LoD2 = LoD1 enhanced with a texture,
- LoD3 = External hull of the building with a roof form and small surface elements,
- LoD4 = LoD3 enhanced with external textures,
- LoD5 = LoD4 enhanced with internal structures.

Gröger et al. (2004) defines five LoDs of 3D landscapes:

- LOD 0 = A digital terrain model with draped orthophoto and classification of land use,
- LOD 1 = Popping up of the ground plan to a uniform height,
- LOD 2 = LoD1 enhanced with roof textures, roof structures and vegetation features,
- LOD 3 = Architecture models with vegetation features and street furniture,
- LOD 4 = Indoors architecture models.

In Schilcher et al. (1998) three LoDs for individual buildings are explained:

- LoD1 = Popping up of the ground plan to a uniform height,
- LoD2 = LoD1 enhanced with a standard roof form,
- LoD3 = LoD2 enhanced with photorealistic textures and small surface features.

Similarly, three LoDs can be found in Kolbe (2004):

- LoD1 = Popping up of the ground plan to a uniform height,
- LoD2 = LoD1 enhanced with a standard roof form and simulated wall texture,
- LoD3 = LoD2 refined with a detailed roof form, small surface elements and photorealistic textures.

In spite of the differences in the aforementioned definitions of LoD, they all follow the logic of successive refinement from lower LoDs with coarser resolutions to higher LoDs with finer

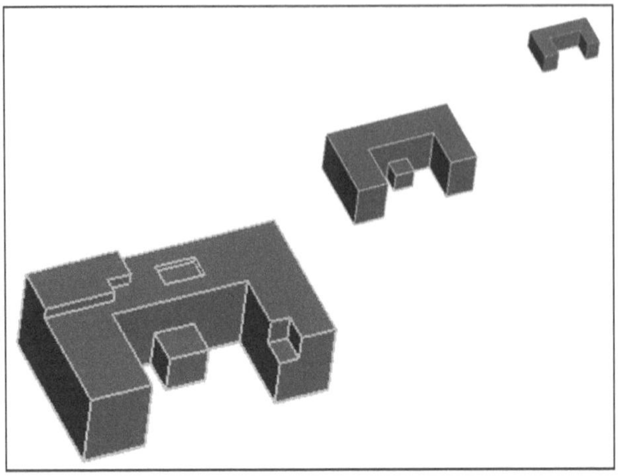

Fig. 11.2. Successive refinement of 3D building structures depending on the viewing distance: the nearer the building objects, the more details they reveal.

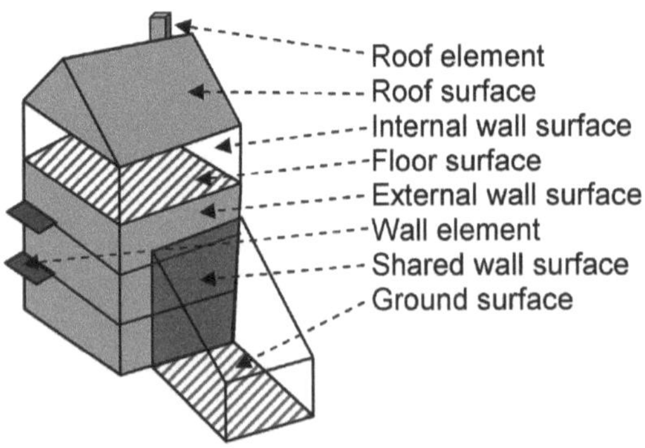

Fig. 11.3. Structural components of a typical building.

resolutions. Figure 11.2 demonstrates the visual effect of three successive LoDs of building structures.

At its finest LoD, a building object can be typically described by its external components, roof surface, roof element, external wall surface, wall element, and internal components, floor surface, internal wall, shared wall surface, ground surface, and ceiling surface as shown in Figure 11.3 (Ramos et al., 2004). Typically associated with its cadastral footprint, a building hull can be assigned a number of general attributes describing the various qualities of the building. Each storey level can be further attributed for example by information about the occupancy. Finally, every internal or external component can be integrated with application-specific attributes such as incident solar energy, temperature or building construction material.

11.1.2. *Acquisition of 3D building models*

3D building data can be acquired using a variety of terrestrial and non-terrestrial techniques. Among others, aerial photogrammetry, aerial laser scanning, terrestrial measurement, close range photogrammetry, terrestrial laser scanning and official cadastral information have been widely applied.

(1) Aerial photogrammetry
Aerial photogrammetry is able to economically capture the roof landscape and ground texture of a large built-up area. The limited resolution of aerial images, however, does not allow the detection of small roof elements. Neither are façade structures acquirable as they are mostly invisible from the air.

(2) Aerial laser scanning
In use since the 1990's, aerial laser scanning based on LIDAR (LIght Detection And Ranging) technology can be used for direct acquisition of 3D building surfaces. LIDAR scanning can take place day or night, as long as clear flying conditions are present. The cost of laser scanning is usually more expensive than photogrammetric methods, but the directly available 3D surface characterised by a point cloud allows for straightforward data processing (Ding, 2000).

(3) Terrestrial measurement
Terrestrial measurement is a complementary method for the acquisition of fine details, especially the individual structure points that cannot be observed from the air. The high precision of this method requires laborious field work since terrestrial details are usually selected and measured on site.

(4) Close range photogrammetry
Close range photogrammetry is an economic method for the geometric documentation of complex buildings and texture registration of façades. The result of stereophotogrammetric analysis is usually a precise 3D line drawing composed of the visually characteristic edges and points on building surfaces. Areas between the edges, however, can hardly be interpreted in fine detail.

(5) Terrestrial laser scanning
Terrestrial laser scanning or ground-based LIDAR technology is used to capture 3D models of complex and irregular buildings (Früh, 2002). It is relatively expensive and requires large storage capacity since the footprint contains many measure points that do not belong to the building structure. The 3D scanning does not reach as high a precision on structural edges of buildings as close range photogrammetry or terrestrial measurement, but its surface-based working principle allows a precise interpretation of the surface areas between characteristic edges (Boehler and Marbs, 2004).

(6) Derivation from official cadastral databases and maps
The geometric and semantic attributes of buildings documented in cadastral databases and maps provide rich sources for the derivation of building models of different LoDs. Information such as ground plan, the number of storeys and the hypothetical assumptions about the average storey height can easily lead to a block model. Further information such as ridge and eave lines and their terrestrially measured heights can extend the block model to include roof forms (Schilcher et al., 1998). An important advantage of seamlessly avail-

able cadastral data is that individual building models from different cities can be easily sewn together to form a value-added 3D model covering a large region (Averdung, 2004).

All these existing methods can be combined to construct high-fidelity and photorealistic 3D building models. An image-recording camera can be integrated with the aerial laser scanning system so that orthophoto mosaics can be directly created. The combination of image data with the point cloud makes the interpretation of a 3D scene easier and more reliable (Brenner et al., 2003). Likewise, a terrestrial laser scanning system allows the embedding of a camera to record terrestrial images, thus enabling the precise acquisition of both structure lines and surface details of complex buildings. Furthermore, the superimposition of aerial images and cadastral maps supports the reliable reconstruction of standard roof forms and the determination of height for individual buildings. Similarly, the footprint of an aerial laser scanner can be superimposed on the ground plans of buildings to speed up the detection of 3D buildings.

For the time being the accessibility of 3D building models worldwide is rather variable. Some regions have been completely covered with redundant 3D data of more than three LoDs, some regions have only access to a very coarse LoD or to a fine LoD for a limited number of buildings. Elsewhere 3D building models are entirely missing. The interoperability between existing 3D building models is problematic since they are still constructed in an isolated manner and in the absence of shared standards.

11.2. The Necessity for 3D Building Generalisation

Although methods for the acquisition of 3D building geometries have been constantly improving with regard to precision, reliability, degree of automation and processing speed, a fully automated procedure for constructing high fidelity 3D building models is not yet in sight. The existing approaches, even in combination, are too time intensive and expensive for 3D data acquisition and updating (Brenner et al., 2003). The lack of access to actual and extensive building models in various LoDs has hindered the integration of thematic information of different granularities. Moreover, the existing different LoDs of 3D buildings within a limited spatial scope are often captured separately, using different methods. The missing linkages result in a high maintenance cost and difficulties for the user in conducting multi-scale spatial analysis. One of the possible remedies is to reuse and value-add to the existing datasets (Hampe et al., 2003; Thiemann, 2004). This can be realised in three different ways:

- A 3D building model with the finest possible LoD is used as a source model for the derivation of coarser LoDs;
- Links are established between different LoDs of an existing 3D building model, that is, a Multiple Representation Database (MRDB) (Mustière and van Smaalen, this volume (Chapter 6)) is constructed;
- Existing 3D building models from adjacent areas are made interoperable by transforming them to a uniform spatial reference system and LoD.

Generalisation is an essential ingredient in all three approaches. For the derivation of a coarse LoD from a fine one, the geometric and semantic attributes of individual 3D buildings as well as the relationships among them must be simplified, aggregated or typified, while for the establishment of links between corresponding buildings at two different LoDs, generalisation can serve as a support for efficient object matching. Finally, the harmonisation of spatially adjacent 3D building models often requires an adjustment of content density which can only be realised by means of generalisation. In addition to reducing the acquisition and maintenance cost of 3D building models, generalisation is an essential strategy for real-time visualisation of 3D building models. With the additional third dimension, a 3D building consists of far more geometric primitives than a 2D footprint, therefore, it is more computing-intensive to render on a computer screen. A real-time 3D visualisation free of graphic conflicts and without quality deterioration is not possible unless the generalisation mechanism, that allows object selection or hiding, conflict handling and perception-constrained enhancement is embedded in the visualisation pipeline. Analogous to the terminology applied in 2D, 3D generalisation can be divided into model generalisation for the purpose of data reduction, and graphic (or cartographic) generalisation for the purpose of visualisation (Hake et al., 2002; Sarjakoski, this volume (Chapter 2)). Model generalisation is a data-to-data transformation that deals with model objects, their geometric and semantic precision and their topological consistency in the scale space. The generalised building models are either distributed as data services for integration with thematic information of comparable resolution or regarded as a prior stage to graphic generalisation which transforms the data to a graphic presentation. Graphic generalisation focuses on the visual impression of model objects together with 3D graphic artefacts in relation to the selected projection, visualisation style, camera position etc. Many applications, especially 3D the spatial analysis needs the results of graphic generalisation as an interface to the underlying object attributes.

11.3. Challenges of 3D Building Generalisation

Adding the third dimension has dramatically increased the complexity of a building model in both a geometric and a semantic sense. While a large-scale 2D building is represented by its cadastral ground plan, the appearance of a 3D building is characterised by a lot more surface elements. Consequently, it can take many possible forms. The meaning of a 2D building is usually expressed by a number of semantic attributes attached to its ground plan. In 3D space, however, every surface element of a building can be described by special semantic attributes in addition to its more general attributes. As an example the relationship between two individual 2D buildings can be judged by relative location, form and relative size of ground plan, orientation, proximity, and horizontal alignment, whereas two individual 3D buildings are related additionally by vertical alignment, roof form, relative height or surface texture. Bearing in mind the complexity of a 3D building model, cartographers are confronted with the challenging task of deriving constraints for model generalisation from the interdependencies among building parts, neighbourhood relationships among individual buildings, and spatial structures of settlement blocks. So far the knowledge and conceptual models necessary to support these structures are still largely missing.

With regard to the graphic generalisation needed for 3D visualisation, cartographers face the challenge of acquiring knowledge of users and their tasks. In comparison to 2D visualisation which is traditionally constrained to a plan view, 3D visualisation is inherently more user-oriented in terms of viewing point, eye level and vision field. Although the central perspective view of a 3D building model gives a naturalistic impression, the observer experiences it differently from reality. For example, an observer typically has difficulties in estimating distances and orientations in the virtual space due to the varying scale of the presented model objects.

In computer graphics, the surface of an irregular 3D object can be generally represented by a wire frame that consists of triangles or quadrilaterals as typical mesh elements. By successively simplifying the mesh elements, a coarser resolution of the wire frame can be derived. However, this approach is very inefficient, if not impossible at preserving the semantic and structural characteristics of building objects. In current 3D visualisation systems, building objects are typically rendered at three pre-determined LoDs. During a fly-through, the three LoDs are dynamically switched over from one to another according to predefined distance thresholds. This leads to two main problems: (1) each building regardless of its size and position in the vision field preserves a uniform LoD, so discontinuity is inevitable between adjacent buildings, and (2) abrupt changes or "popping" effects in the shape of the building occur simultaneously to many objects in a rather unforeseen manner. In order to achieve a smooth graphic transition while preserving the visual clarity of important objects, a building needs to be examined at its geometric primitives, with each being rendered at a LoD according to the distance to the viewer, whilst a sufficiently large number of intermediate morphing steps are inserted between adjacent LoDs. Such an approach requires both a thorough understanding of the meaning of the individual primitives and their relationships in the context of use and, the availability of a 3D MRDB. Both requirements are difficult to satisfy. Moreover, too little is known about how many LoDs an observer really desires for their personal convenience, how well they will recognise the characteristics of a 3D building model at different abstraction levels, what kind of impacts their task will have on their perception and cognition and what kind of interactions make sense.

Another challenge lies in the relative immaturity of existing 3D systems. Although many GIS vendors tend to expand their tool kits to include 3D interactions, users are usually only allowed to change a subset of the visualisation parameters (camera position, light source, texture and colour). Often the individual 3D objects or object parts and their associated attributes are inaccessible due to the absence of necessary analytical and interaction methods in spite of their acknowledged importance (Hedley, 2003). Over recent years, efforts have been made to develop a range of intuitively operable 3D widgets that allow the direct manipulation of 3D data, such as selection by virtual pointer, modification and deformation of various spatial entities (Leiner et al., 1997; Rahman and Khuan, 2003; Yang et al., 2004). However, these 3D widgets are often incompatible with each other (often due to a lack of standards). Without the support of 3D interactive functions, both developers of generalisation methods and users have to invest considerable effort in understanding the behaviour of 3D objects and their generalised forms.

11.4. Description Models of 3D Buildings

Regardless of their complexity, individual 3D buildings can be described in one of five ways (Lang, 1999; Thiemann and Sester, 2004; Forberg, 2005):

(1) Voxel model
A 3D building can be organised as a matrix composed of voxels (volume elements) which are small cubes of a uniform size. Each voxel can be attached to one or more semantic attributes. The voxel model shares the same advantages and disadvantages with 2D raster data. On the one hand, it can model arbitrarily complex 3D buildings and allow direct access as well as simple image processing operations (Ayasse and Müller, 2001). On the other hand, it requires large storage capacity and long rendering time even if data compression methods such as octree have been applied.

(2) Parametric description
This is analogous to the geon-based theory for human image understanding proposed by Biederman (1987). He believed that the recognition of spatial objects and their perceptual organisations (e.g. curvature, co-linearity, symmetry, parallelism and co-termination) relied upon the detection of the generalised sub components of the structure called geons. Similarly a 3D building can be partitioned into a number of primitive bodies such as cuboid, sphere, cylinder, cone and pyramid which can be completely described by a few simple parameters such as side length, width, radius and height. The absolute position of the building is defined by six further parameters of rotation and translation. Parametric description is suitable for the description of simple buildings which are characterised by their planar roof surfaces and the orthogonal relationship between walls and ground plans.

(3) Constructive solid geometry (CSG)
The solid geometry of a 3D building is constructed through Boolean operations such as intersection, difference, union or inversion of elementary building parts. The sequence of operations is stored in a CSG tree. Usually the length of the sequence reflects the relative complexity or irregularity of the corresponding 3D building. Figure 11.4 shows a CSG tree of a building constructed with the "union" operation.

(4) Boundary Representation (BRep)
The geometry of an arbitrarily complex 3D building is described by its boundary surface using topological elements such as mesh, edge and vertex. The most popular form of a BRep is a Triangulated Irregular Network (TIN), which is often based on the principle of Delaunay triangulation. Curved parts can be approximated with small triangles. For certain applications a constrained TIN is constructed in which characteristic points and structure lines of the building serve as vertices and edges of triangles. Another popular form of 3D boundary representation is provided by the Virtual Reality Modelling Language (VRML). VRML allows the creation of interactive 3D scenes that can be networked and hyperlinked in the Internet. Each scene is composed of various nodes which describe shapes, texts, textures, lights, sounds and backgrounds. The geometries of virtual

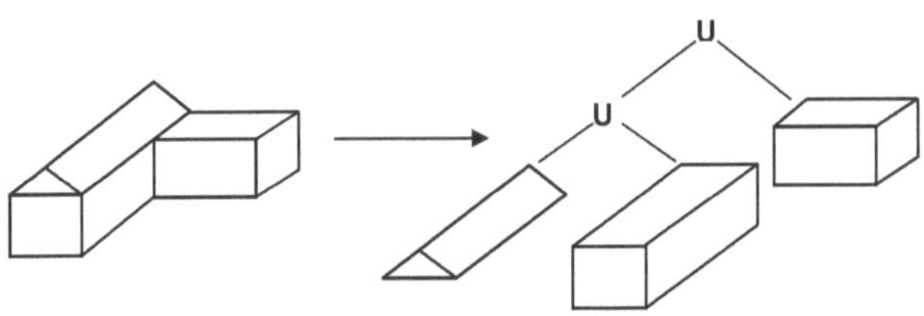

Fig. 11.4. A CSG-tree of a building (Rottensteiner, 2001).

objects in the scene are built from rudimentary shapes or primitive points, lines and faces. They can be reused, translated, rotated and grouped into more complex geometries.

(5) Solid representation (SRep)
The geometry of an arbitrarily complex 3D building is described by a Tetrahedral Network (TEN) composed of the regular or irregular topological elements: tetrahedron, triangle, edge and vertex (Song et al., 2004). Analogous to TIN, a TEN is usually based on the principle of Delaunay Tetrahedral Tessellation (DTT). Similar to TINs, constrained DTTs can be built in which characteristic points, structure lines and planar facets of the building serve as vertices, edges and meshes of tetrahedrons.

No single approach was found to support optimal interactive visualisation but through experimentation a combination of these techniques proved promising. This chapter reports on a model that used a combination of parametric description, CSG and BReps.

11.5. Methods of 3D Building Generalisation

Current research efforts in 3D building generalisation are concentrated on the following main tasks: (1) segmentation of building structures, (2) recognition of building structures, (3) derivation of coarser LoDs from the recognised structures (model generalisation), and (4) visualisation of different LoDs of building structures in a legible way free of graphic conflicts (graphic generalisation).

11.5.1. *Segmentation of building structures*

On the basis of the polyhedron segmentation proposed by Ribelles et al. (2001), Thiemann and Sester (2004) have developed a "feature-finding" algorithm. A feature is defined in this context as a connected region that can be easily separated from the rest of the surface. In this approach, a planar-structured complex building with BRep is recursively partitioned into a set of convex parts which are then stored as a cell complex and a CSG-tree. The cell complex represents the topological adjacency of the convex parts, while the CSG-tree logs the partitioning history.

Figure 11.5 illustrates two splitting planes used for segmentation of the same building. Each splitting plane divides the space into two half spaces, defining the space behind the plane as solid and the space in the front as empty. The intersection of the polyhedron (here the building surface) with one or more half spaces leads to the detection of extruding features (e.g. bumps, peaks) and intruding features (e.g. holes, notches). The remaining part of the polyhedron is formed either by cutting away the extruding features or by filling the intruding features. As long as the detected features or the remaining part of the polyhedron are concave, they will be further segmented until no further features can be detected, which means that all resulting parts are convex.

The intersection process starts with one splitting plane. Only where no good results can be found – for instance where small features touch more than one facet of the polyhedron boundary – is a second, third or even fourth plane added. A 3D building can be segmented in many alternative ways – each resulting in the detection of different features. Therefore a control mechanism is required. Ribelles et al. (2001) defined a quality measure based on the size between the detected feature and the remaining part lying in the splitting plane(s). The

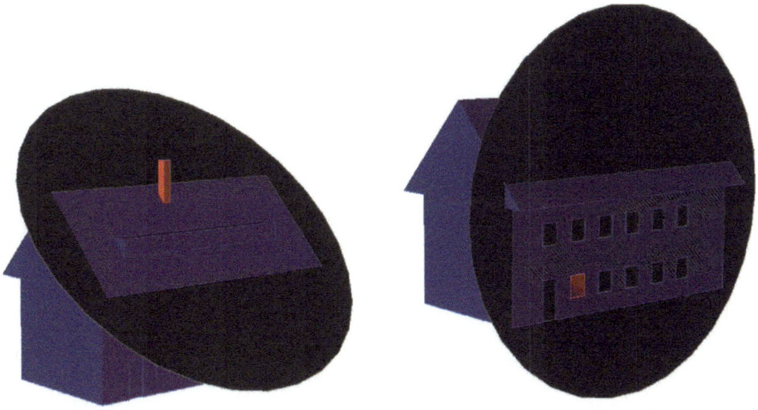

Fig. 11.5. Two possible splitting planes in grey with one intended for the detection of bumps on the roof (left) and the other for the detection of both bumps and holes on the wall (right) (Thiemann and Sester, 2004).

Fig. 11.6. Segmentation of a sample building with 34 split planes. The resulting cell complex contains 19 parts (Thiemann, 2004).

smaller the size of the detected feature, the better its "quality" is considered to be. Splitting plane(s) that produce the best feature quality are selected first. Thiemann and Sester (2004) added a further constraint whereby only the features with a relative size smaller than 1 are considered valid. This heuristic leads to a considerable reduction in computing time. If some detected features show the same best quality, they can be segmented at the same time. This can occur where buildings have a repeating pattern in their shape – such as the existence of many equally large windows lying in the same splitting plane.

Figure 11.6 shows the segmentation result of a sample building with 34 split planes. The segmentation was realised using the ACIS 3D Geometric Modeler software. ACIS uses the BRep which is essentially a hierarchical decomposition of object topology. The cell complex yielded by the segmentation contains 19 cells. Its corresponding CSG tree has 19 leafs and 8 inner nodes, reflecting the recursive nature of the tree. Since small features are segmented early in the process, they are located close to the root while the largest feature is stored in the deepest leaf.

In a subsequent processing step, the individual convex parts need to be identified as meaningful features. Examples of such features include chimneys, balconies, bays, windows, doors and roofs. These are ranked in terms of their relative significance by using methods of 3D structure recognition with input parameters such as relative size and position, inclination, form, orientation and the available generic knowledge. The CSG tree can thus be reorganised as a hierarchical tree composed of the detected features with successively declining significance from the root to the leaves. In this sense, the generalisation problem of an individual building is reduced to a data retrieval task. By keeping all the features, the original building can be reconstructed, while pruning away the features below a certain threshold of significance results in a simplified building with preserved core characteristics such as parallelism, rectangularity, horizontality, verticality or co-planarity. This approach is comparable to the creation of the BLG-tree (Binary Line Generalisation) and the GAP-tree (Generalised Area Partitioning) in 2D modelling. This stores the features (discrete points along a curved line in the case of BLG-tree and non-overlapping area parcels within a region in the case of a GAP-tree) in a hierarchical order based on their relative importance (van Oosterom and Schenkelaars, 1995; van Oosterom, 1995).

11.5.2. *Recognition of 3D building structures*

One of the fundamental requirements for the development of 3D building generalisation methods lies in the preservation of core characteristics of building objects. A systematic analysis is given by Lal and Meng (2003), in which 3D building structures are parameterised at three different abstraction levels: individual buildings, adjacent buildings and building clusters. The study attempted to find recurring templates and the essential parameters describing them. Figure 11.7 illustrates a set of planar-structured building templates characterised by the type of ground plan in combination with the roof type.

In order to automatically recognise the planar-structured building types, Lal and Meng (2004) implemented an algorithm based on a hierarchical neural network. In a first step, the algorithm identifies the type of ground plan using input parameters such as 3D coordinates of

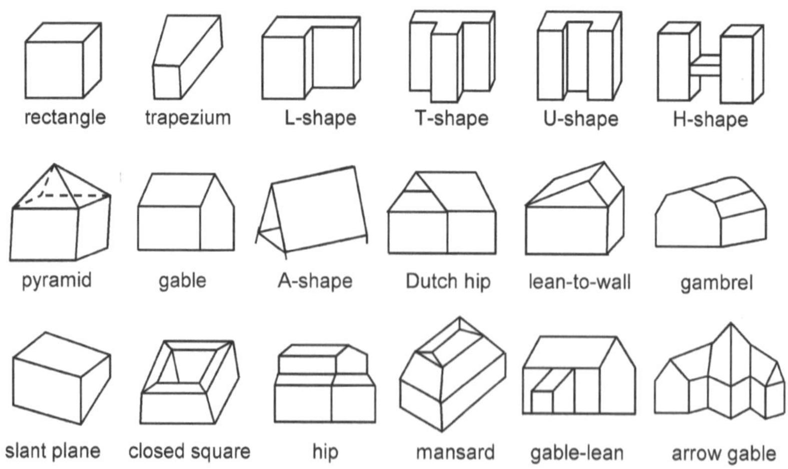

Fig. 11.7. Types of ground plan in the top row; Roof types in the middle and bottom row.

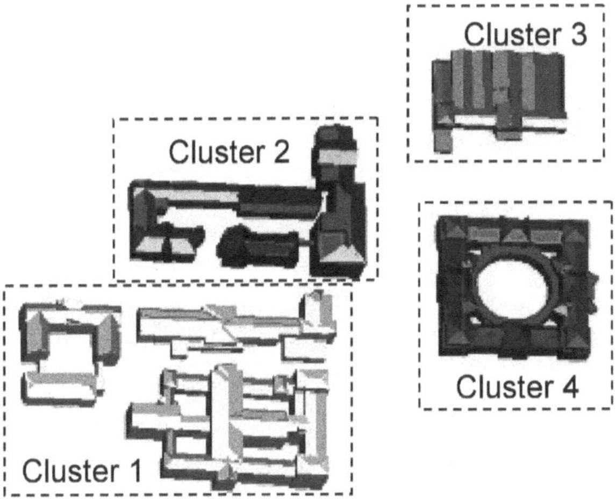

Fig. 11.8. Automatically recognised 3D building clusters with test data from the Nussallee area of city Bonn (constructed by the Institute of Photogrammetry, University of Bonn). The clusters are encircled by dash lines (Lal and Meng, 2004).

vertices, the number of edges, the number of closed facets, and the orientation of the longer axis. In the second step, roof types are recognised with input parameters such as 3D coordinates of vertices, the number of edges, the number of facets and the angles between the neighbouring facets. Finally, the recognised type of ground plan and roof type are fed into the algorithm together with parameters describing the component parts such as 3D coordinates of vertices, the number of edges, the number of facets and the number of parallel edges. The recognised building type is further used as one of the input parameters for the classification of neighbourhood relationships and the detection of building clusters. Figure 11.8 illustrates the recognised 3D building clusters based on this hierarchical approach. The performance of the algorithm can be improved by feeding the algorithm first with an intuitively selected small set of parameters. Further parameters are incrementally added if the learning process tends to be unstable or error-prone.

11.5.3. *Model generalisation of 3D buildings*

In this section, an approach to model generalisation, particularly simplification of building data, is introduced. In this context, a building is considered to have three generic surface types: a ground plan (horizontal facets), a number of walls (vertical facets), and a roof (inclined or horizontal facets). With regard to the semantic differentiation between walls and roof, the process of simplification of an individual building is split into the simplification of parallel structures (§11.5.3.1) and the squaring of inclined facets (§11.5.3.2).

11.5.3.1. *Simplification of parallel structures*

In Forberg and Mayer (2003) an approach to the simplification of 3D building data is proposed, which extends the idea of scale spaces applied in image analysis. Two scale spaces,

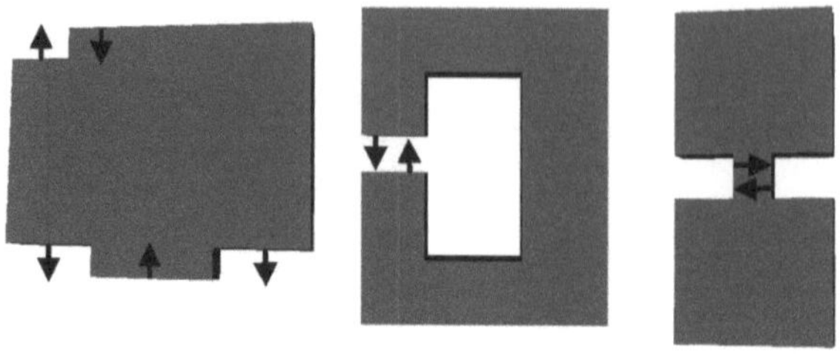

Fig. 11.9. Parallel facets under a certain distance are shifted towards each other, until the facets of the building merge.

a 3D version of mathematical morphology and the so-called 3D curvature space are applied separately. However, a rather complex analysis is needed for curvature space, which may dramatically slow down the generalisation process. In a later approach (Forberg, 2004a), the advantages of mathematical morphology and curvature space have been unified into one process. In this approach, if the distance between two neighbouring parallel facets falls below a predefined threshold, one or both facets are moved towards each other until they merge into the same plane (Figure 11.9). Such a "parallel shift" may lead to the simplification of all parallel structures including the splitting or merging of different object parts, and the elimination or adjustment of local protrusions, such as step and box structures.

The parallel shift is realised using the ACIS Geometry Modeller of Spatial Corp. (Spatial Corp., 2005), which allows specific offset distances to be defined. The selection of a facet pair and the specific shift distances are based on the analysis of the relations between the facets. The decision if one or both facets of a facet pair are shifted depends on a threshold that is chosen intuitively. Here, both facets are shifted half of the distance, if the area of the smaller facet is bigger than a third of the area of its partner's facet. Otherwise, only the smaller facet is shifted for the whole distance. Figure 11.10 illustrates the use of these two kinds of weighted movements. The area-dependent shift has the advantage that a shape simplification and adjustment takes place simultaneously.

If there are several pairs of parallel facets with the same smallest distance, one pair is selected randomly for the parallel shift. In Figure 11.11, a building with a symmetrical box structure is shown. Within the symmetrical box structure, one of three pairs of parallel facets (marked in dark grey) is chosen at random to be shifted. If both facets of a pair are shifted by the same amount, three possible results can occur (shown in white), i.e., the result is not predictable. Additionally, the symmetry is lost. This problem can be avoided by shifting only the smaller facet by the whole distance (as described above). Since the result remains the same in spite of random choices, the symmetry of such box structures can be preserved. This means, for box structures the random choice does not matter. But for other structures, it can lead to unexpected results. In Figure 11.10, the fourth building from the left (marked with a rectangle) shows a structure where one facet has the same distance to two parallel facets. The random selection leads to the elimination of one of two nearly symmetrical building parts. A human would most likely have closed the gap instead.

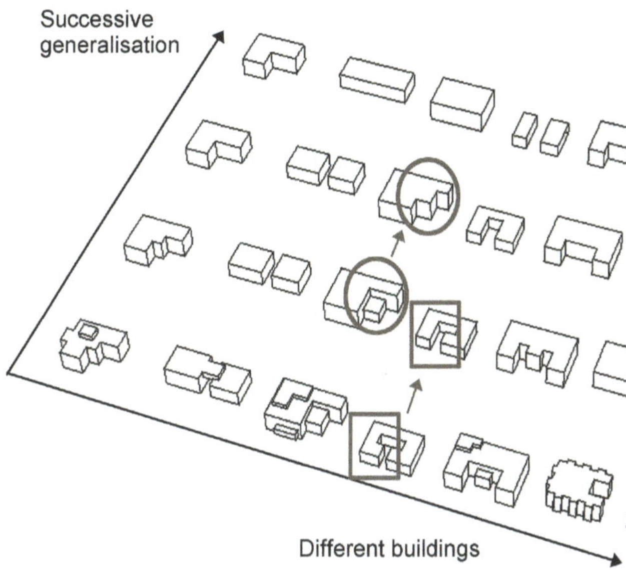

Fig. 11.10. Results for the simplification based on parallel shifts. Some object parts are removed and adjusted such that the characteristic shape is preserved and slightly emphasised (circled). Random selection of facets causes objects to change randomly (rectangled).

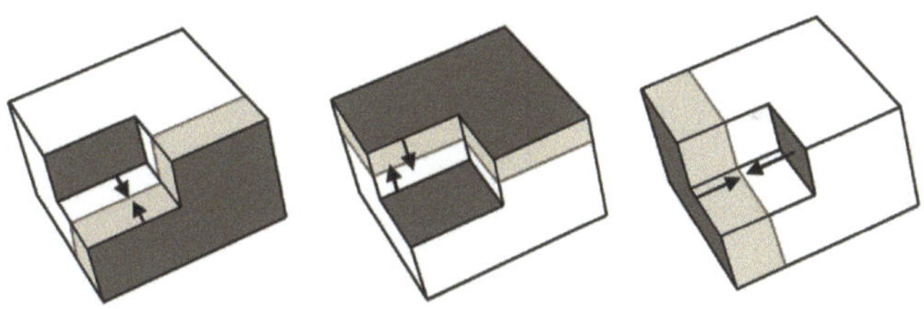

Fig. 11.11. Three possible choices of equidistant parallel facets. Facets to be shifted are marked in dark grey. Parts marked in light grey are removed by the generalisation.

11.5.3.2. *Squaring of inclined roof-facets*

Simplification using the parallel shift works only for parallel structures. In order to simplify object parts with non-orthogonal structures, particularly roofs, a method for enforcing right angles has been developed (Forberg, 2004b). In this solution the inclined roof-facets (either ridges or eaves) are forced to become horizontal or vertical by rotating them around one of their edges. The choice between the eave and the ridge (the taper-edge) as well as between the horizontal and vertical direction of flattening (the taper-direction) depends upon the combination of two neighbouring facets, the facet sharing its eaves and the facet sharing its ridges (Figure 11.12).

A reasonable generalisation requires that the contextual information of the inclined facets, – i.e. the facets belonging to the same roof unit, defined by connected ridge lines, – be treated

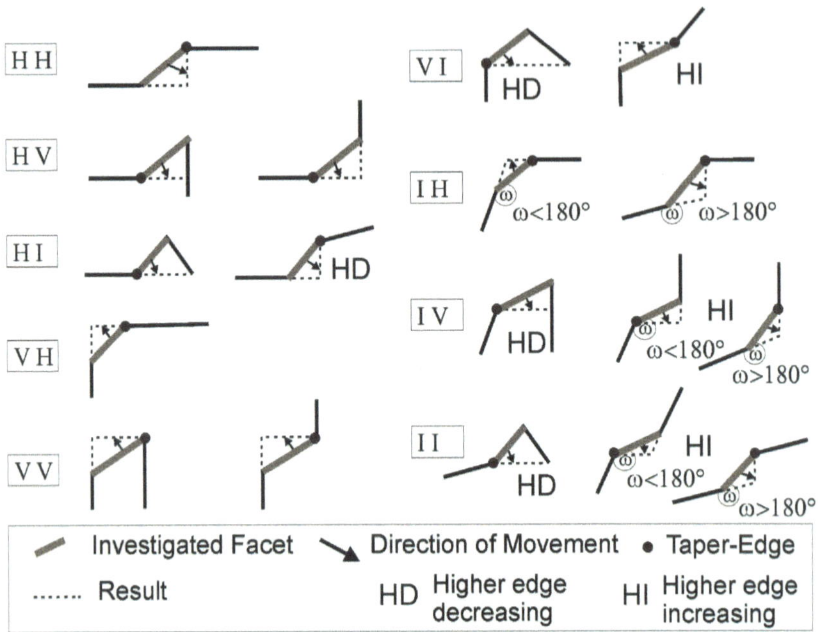

Fig. 11.12. Taper-edge and taper-direction depend on the relation of the inclined facet to its neighbouring facets (horizontal – H, vertical – V, inclined – I). In some cases additional information, such as the angle ω, is needed.

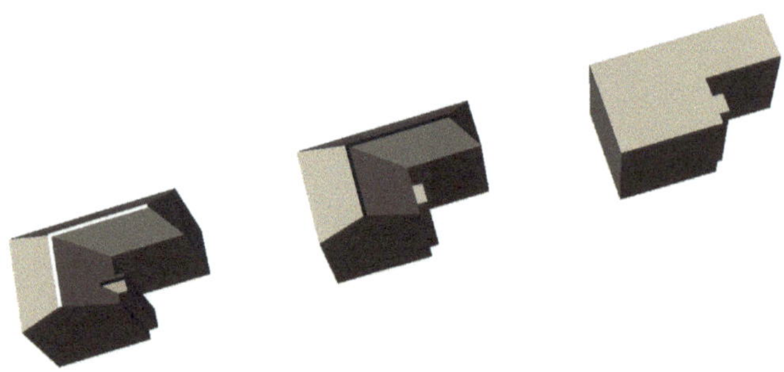

Fig. 11.13. Connected horizontal ridges (black and white) define two roof units. Depending on the average facet area, roof units are removed.

together. If the average facet area of a unit falls below a certain threshold, the individual facets are rotated, so the roof structure is reduced to a flat form. Figure 11.13 illustrates a building with two roof units. The unit with the smaller facet area (marked by a black ridge) is squared first. Some examples for the roof-squaring realised in ACIS are illustrated in Figure 11.14.

When squaring, the height of a building can be changed to the ridge height or the eave height. This then makes a scaling necessary. After flattening the roofs, it is theoretically pos-

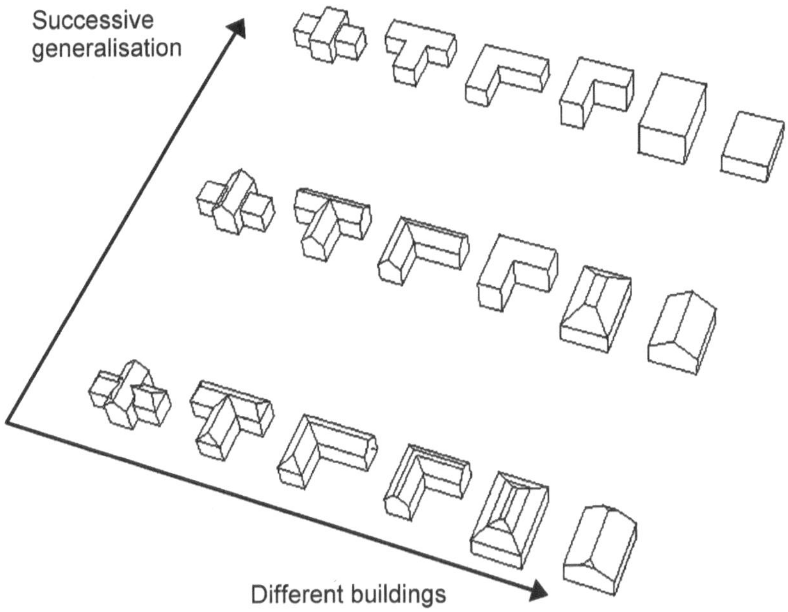

Fig. 11.14. Results of the roof-squaring procedure.

sible to aggregate parallel facets of adjacent buildings by applying the parallel shift method. When adjacent buildings have non-flat roof forms, the aggregation operation proves rather complicated and needs to satisfy many constraints concerning the roof type, orientation and texture. A method for height-scaling as well as the aggregation of buildings with different roof types has not yet been developed.

11.5.4. *Graphic generalisation of 3D buildings*

When 3D buildings are brought into view, the user expects appropriate graphic legibility and a real-time rendering of the scene whenever the viewing parameters are changed. This requires that the generalisation operations be embedded in the 3D data model and/or performed on the fly. Unlike the approach for model generalisation described in § 11.5.3, where the finest LoD is used as a source model for the derivation of coarser LoDs and the focus is mainly on data reduction, the following sections describe the existing approaches to graphic simplification that take into account users' visual perception and cognition.

11.5.4.1. *Invasive simplification of 3D buildings*

Invasive simplification techniques serve the purpose of continuously reducing and adjusting the geometric details of 3D buildings. With increasing viewing distance, the geometric primitives (vertices, edges or meshes) on building boundaries are progressively removed. At first, they are grouped into small clusters. For each cluster, a representative primitive is determined (via sampling) or a new primitive is calculated. The process terminates when the desired legibility level is reached (Heckbert and Garland, 1997). Algorithms of these types are targeted at

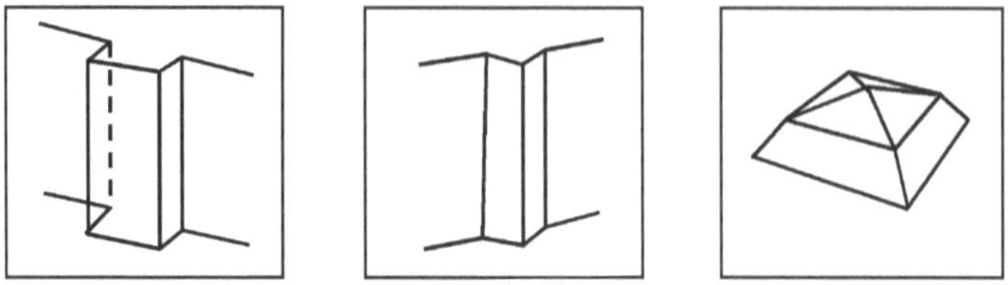

Fig. 11.15. Some typical small structures in U-, V- and peak form on a building surface (Kada, 2002).

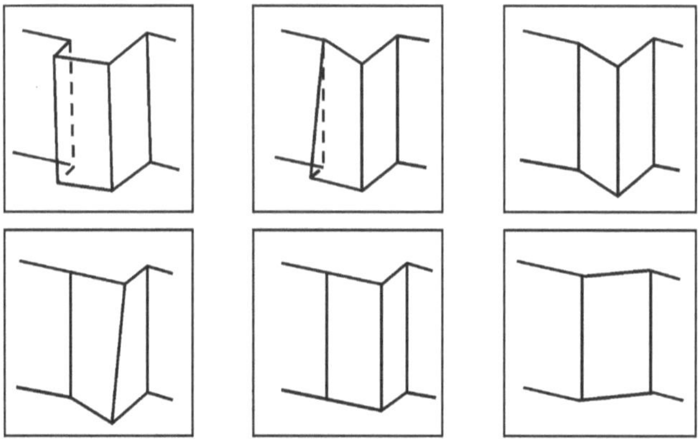

Fig. 11.16. Steps in the elimination of a small extruding structure (Kada, 2002).

individual buildings with redundance-rich BRep. The process is relatively computer intensive, but allows for a smooth graphic transition. With the decreasing viewing distance, the reverse process is applied to bring back the original details to the observer. The generic characteristics of building objects can be preserved by adding constraints. Kada (2002) reported an algorithm of constrained edge reduction. Depending on the characteristics of small surface structures as illustrated in Figure 11.15, different simplification functions are applied. In Figure 11.16, two unequally short lines are removed through edge collapse under the constraint of parallelism. Sester and Klein (1999) introduced a rule base which can guide the façade generalisation including aggregation of neighbouring windows, elimination, enlargement or displacement of small façade features depending on their relative importance.

11.5.4.2. *Non-invasive simplification of 3D buildings*

Non-invasive simplification techniques are used to accelerate the rendering process without modifying the underlying database (Plümer, 2002). In essence, they are data-hiding or optically illusive operations. The following strategies have been widely used in computer graphics:

- Back face culling: Back faces of the individual buildings from the actual viewing point are omitted.

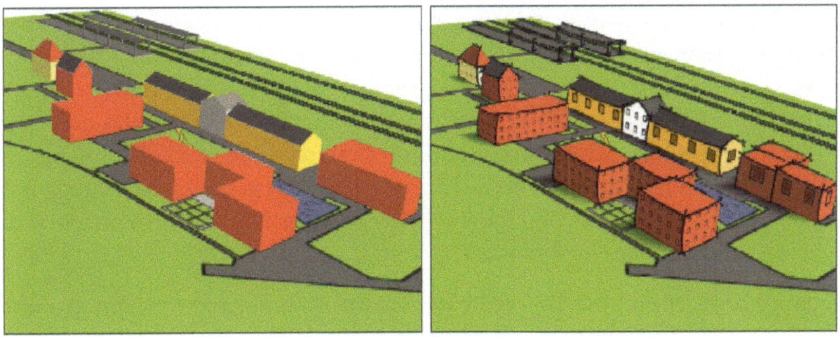

Fig. 11.17. Presentation based on standard illumination (left), presentation based on expressive line drawing (right) (Döllner and Walther, 2003).

- Frustum culling: Buildings or building parts that fall beyond the current vision field are omitted.
- Occlusion culling: Buildings or building parts that are covered by front faces from the actual viewing point are omitted.
- Impostor: A complex building or building group beyond a distance threshold is rendered as an impostor or cache which is a static image projected on a transparent quadrilateral. The same impostor can be reused in several consecutively changing frames (Haase et al., 1997; Kada, 2002).
- Coarse shadowing: The shadow area is rendered as a simplified bounding box of the corresponding building model.
- Expressive line drawing: This is a real-time non-photorealistic rendering technique which aims to exaggerate the boundaries of 3D building objects (Döllner and Walther, 2003). Two-tone or three-tone artificial shading can be added to building surfaces. Some complex building elements can be replaced by procedurally generated façades. Further, simulated shadows can be created to enhance the understanding of building structures. An example is illustrated in Figure 11.17, where the right presentation is an automatically created enhancement of the left one.

11.5.4.3. *MRDB and interpolating generalisation*

In an attempt to develop an interoperable 3D model of a large region, Kolbe and Gröger (2003) worked out a MRDB concept. Three cases that cover all the possible relationships are identified:

(1) A building has an identifiable representation in each LoD.
A building is represented as a block in LoD1, a textured hull with a simple roof in LoD2 and a life-like model with complex roof as well as small surface elements in LoD3. This approach states that for each 3D building, there exists exactly one corresponding model object in each LoD.

(2) A building becomes a part of a model object in the adjacent LoD.
Spatially neighbouring buildings are represented as identifiable blocks in LoD1, they are merged, e.g. to a row in LoD2. The objects in LoD2 are further merged, e.g. to a district in

LoD3. This approach states that there is a progressive aggregation hierarchy. For example, each building belongs to exactly one row, and each row to exactly one district.

(3) Buildings are partially contained in an aggregation of the adjacent LoD.
Due to independent data collection, a hierarchical relation between LoDs may be missing. For example, spatially neighbouring buildings are represented as identifiable blocks in LoD1, they are merged to a row in LoD2 and to a district in LoD3, but there is no relation between LoD2 and LoD3. A building row may only partly belong to a district. Unlike the second case, the structure of this scene is not a tree, but a directed acyclic graph.

With the availability of an MRDB, a real-time rendering speed can easily be realised. However, the abrupt transition between adjacent LoDs can lead to buildings "popping" as they suddenly change from one form to another. This is especially noticeable when there are only a few LoDs. Depending on the linking relationships, different interpolating strategies can be applied to create a smooth transition. In the case of a $1:1$ relation, an animated morphing technique can be applied. Starting from one LoD, the geometric details of each building are continuously deformed and hidden (or made visible if zooming in the reverse direction) until the adjacent LoD is reached. For example, a tetrahedron can be reduced to a triangle by successively collapsing one of its nodes as shown in Figure 11.18 (left). In the case of an $n:m$ relation, i.e. n buildings in one LoD correspond to m buildings in the adjacent LoD, where $n \neq m$, the fading technique is applied according to which source buildings continuously fade away in the background while the acuity of the target buildings increases smoothly in the vision field. Figure 11.18 (right) illustrates the fading principle.

Meng (2002) suggested that parallel perspective projections may provide a remedy to alleviating some of the drawbacks of the central perspective view. In a parallel perspective, there are no vanishing points at which objects "collapse". Without having to sacrifice their naturalistic appearance, all 3D building objects in a parallel projection are assumed to be equidistant to the observer. Therefore they can be displayed at a uniform LoD. As soon as the viewing distance changes beyond a predefined threshold, all objects will simultaneously alter their appearance to the adjacent LoD. Though a "jerky" impression is still inevitable, it is rather predictable and the observer will not feel uneasy. In addition, the preserved parallelism allows a good measurability of spatial distances and orientations. Moreover, the less intensive computing associated with rendering a parallel perspective means that strategies developed in 2D generalisation can be incorporated into the system. For example, the idea of decomposing gen-

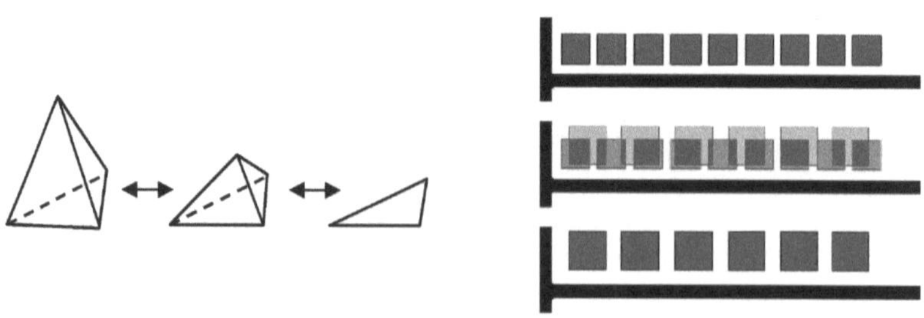

Fig. 11.18. Morphing from a tetrahedron to a triangle (left); Fading from 9 objects to 6 (right) (van Kreveld, 2001).

eralisation methods into a sequence of reversible elementary operations that gradually modify building ground plans between adjacent LoDs (Sester and Brenner, 2004) can be adapted to 3D building surfaces.

11.6. Concluding Remarks and Outlook

The development of 3D generalisation methods is still in its infancy. Most of the available methods have focused on the geometric simplification of a limited set of sample buildings. Experiments on large building databases have been rarely reported. Experiences so far have shown that many of the existing 2D generalisation algorithms as well as the MRDB concept can be extended to cope with 3D problems but not without substantial adaptation. Attempts in the future will include (1) the refinement of existing methods by considering additional constraints, and semantic attributes, (2) a feasibility study examining the application of other generalisation operations such as typification, landmark exaggeration, aggregation, integration of abstract 3D building symbols, displacement, texture and shadow generalisation, and (3) the development of 3D matching methods. One of the most important prerequisites for undertaking this work is a thorough understanding of user's behaviours and tasks. As early as the 1960s, Lynch (1960) suggested five elements that are especially significant in shaping people's image of the city. These five elements (landmarks, nodes, paths, districts and edges) govern the geometric appearance of the city. According to Haken and Portugali (2003), there are "semantic" elements – symbolic, cultural, personal –, that contribute to the interpretation and navigation of the city. Thus a landmark is governed by its geometry, its external appearance, as well as the meaning attached to it. A key goal of generalising 3D buildings is not merely to reproduce a physical form, but to enhance the process of spatial cognition by making characteristic parts more salient.

Acknowledgement

A substantial part of this chapter has arisen from the research project "Generalisation of 3D settlement structures based on scale-spaces and structure recognition" financed by the German Natural Science Foundation (DFG).

Chapter 12

Characterising Space via Pattern Recognition Techniques: Identifying Patterns in Road Networks

Frauke Heinzle, Karl-Heinrich Anders

Institute of Cartography and Geoinformatics, University of Hannover, Appelstraße 9a, D-30167 Hannover, Germany
e-mail: {frauke.heinzle,karl-heinrich.anders}@ikg.uni-hannover.de

Abstract

The chapter will introduce the subject of pattern recognition in road networks. It begins by describing the concepts of random and scale free graphs, since scale free graphs contain important information that can be used to control generalisation processes. It is argued that pattern recognition techniques can be useful in deriving scale free graphs from road networks. By describing typical design and lay-out of roads in urban areas, we will familiarise the reader with different patterns in road networks. Approaches to the automatic detection of these patterns in vector data are reviewed and different patterns of graphs are presented. The investigation of the topological structure of the graphs underpins the recognition of patterns. Techniques for detection of four of the patterns described have been implemented and are presented in detail. The chapter also presents methods for identifying the centre of cities using a combination of various pattern characteristics. It is argued that pattern recognition techniques are critical to the automatic characterisation and generalisation of higher order structures in geographical data.

Keywords: pattern recognition, city centre, star pattern, grid pattern, strokes, road network, vector data, single-source shortest path, Dijkstra algorithm, Hough transform, clustering, scale free graph, random graph

12.1. Introduction

The legibility of maps and their accuracy in respect of content depends to a great extent on the quality of the original data. However, the presentation and thematic fidelity of the data can vary considerably depending on the scale and the theme. The generalisation process seeks to highlight salient information whilst omitting information not pertinent to the particular theme. Different criteria are used to control elimination, presentation and emphasis of geometrical elements – criteria such as measures of object density and shape of geometric elements. Where an object of the same class occurs frequently, then only a subset of that group might be represented. For example minor roads might be pruned from dense, urban areas of the road network. Thematically insignificant elements might be omitted more often than important features, which themselves may be given additional emphasis.

In such cases the essential question becomes: which objects are important and which ones are not significant? It is possible to devise various rules that enable us to rank the relative importance of objects according to the intended theme and the level of detail. Underpinning those rules is the need for spatial analysis techniques that can make explicit the patterns and relationships that exist among the various objects. In this chapter we explore the role of pattern recognition techniques in map generalisation, focusing on their application in the analysis and generalisation of road networks.

Usually the design and lay-out of roads is based upon specific types of patterns. Their shape, connectedness and general form very much governed by the nature of transportation and the activities of humans. Humans conceptualise patterns in their minds, and interpret the role of the network, the design of which is driven by economic factors (such as fast, cost-effective and profitable connections between settlements), aspects of urban planning, and environmental factors. This chapter will present the concept of road network generalisation, exploring the role of pattern recognition techniques in analysing such structures and modelling changes in the structure through the application of map generalisation techniques. We begin by familiarising the reader with different road network patterns and approaches to their automatic detection (assuming vector data as input). A variety of examples illustrate various challenges in automating this process.

The chapter begins with a short overview of existing applications of pattern recognition (§12.2). In this section we also describe the concepts of random and scale free graphs. The idea being to derive scale free graphs from road networks by using pattern recognition techniques. These scale free graphs contain important information that can be used to control the generalisation process. Section 12.3 summarises the typical patterns found in road networks, namely node types, strokes, grids, stars, rings and city centres. We describe the properties of each of these patterns as well as our implemented algorithms used to detect these patterns, highlighting their application through examples. In §12.4, we provide a summary and perspective on the future development and application of pattern recognition techniques.

12.2. What is pattern recognition?

12.2.1. *Theoretical introduction*

In general, pattern recognition can be described as knowledge enrichment and decision making based upon the relationships inherent among a set of (spatial) objects. Patterns (and the associated information) can be extracted via a variety of techniques; techniques that have been developed to mirror the sorts of patterns and associations seen by the human eye. Pattern recognition techniques have been applied to a diverse range of applications, from engineering, medical science, astronomy, and hand writing recognition, through to automatic feature extraction and land classification techniques applied to remotely sensed imagery. Almost all of these applications involve the use of images as the information source. Therefore a central part of the pattern recognition process is identifying and extracting interesting features from the image. The complexity of the task depends on how clearly the feature can be separated from its surroundings. Hopefully it contains some unique set of properties (shape, size, reflectance value) that makes it distinguishable from surrounding features. Accordingly a variety of different image analysis techniques have been developed. For example techniques that

(1) separate the image foreground from the background, (2) that detect edges, (3) that match features against templates, and (4) that segment the image. This might include the use of Hough transform or various types of clustering algorithms. A good introduction to the theory of pattern recognition can be found in Duda et al. (2000).

The work presented here does not take as input raster based imagery, but instead solely focuses on pattern detection among vector data (the format used to store nearly all mapping data). The feature geometry and attributes are available in standard GIS formats. In some instances the data are poorly attributed, and not topologically structured – the term "Spaghetti" is often used to describe such data that is simply described in the Cartesian form of points, lines and polygons. This was the form of the data used in this project, with the specific objective of making explicit the patterns found solely within network data.

In the absence of topology and other structures, we used a graph theoretic network to represent the road network. As previous research has shown, this approach is useful in making structural and characteristic information of the network explicit (Mackaness, 1995b; Thomson and Brooks, this volume (Chapter 13)). The basic elements of the graph are nodes and edges. We used a graph with nodes representing start- and endpoints of lines, edges representing the lines. The degree of the nodes results from the outgoing edges, which in our case is equivalent to the number of lines linked with a node. Various texts provide an introduction to graph theory (Harary, 1969; Diestel, 2000; Thulasiraman and Swamy, 1992).

12.2.2. *Interaction of pattern recognition and generalisation*

Patterns can occur in multifarious ways, whereas qualities such as shape, orientation, colour, connectivity, density, distribution, regular repetition are characteristics that determine the appearance of a pattern (Mackaness and Edwards, 2002). Simple structures are usually limited to one object type with a particular arrangement of the individual objects. However patterns can be very complex. It is often the case that several types of objects play a prominent role in such structures. The intermodal arrangement of the individual objects of different types such as roads and buildings result in new patterns. The clear link between buildings and roads is reflected in the development of ribbon settlements that are band-shaped along traffic routes. The human eye immediately interprets this association and gives semantic meaning to such features. And where associations are not present elsewhere in the map, a different set of meanings are taken from the map. Another example of a complex pattern may be found in the crossings of railways and roads. This can take the form of a level crossing, or an over- or under-pass. One could refer to such complex patterns as meta patterns. This issue will be addressed in detail via the example in §12.3.6. In the following section we concentrate on characteristics and definitions of simple patterns in road networks. An introduction to the modelling of simple structures such as grids and stars (and their associated properties) is given by Zhang (2004).

In general, networks can convey different patterns, and these can be classified in different ways. The distribution of the node degrees in the network is one possibility to distinguish between different types of graphs. In this context we introduce the concept of the random graph (Erdös and Rényi, 1960; Bollobás, 2001) and the scale free graph (Bollobás and Riordan, 2004) as two types of representation that can be used in such classification. In a random graph the degree of the majority of nodes will statistically level off at a certain number. In

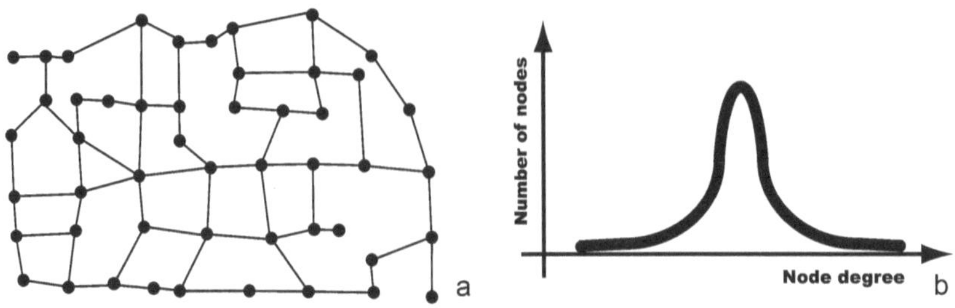

Fig. 12.1. (a) Example of a random graph and (b) a normal distribution for the degree of all nodes.

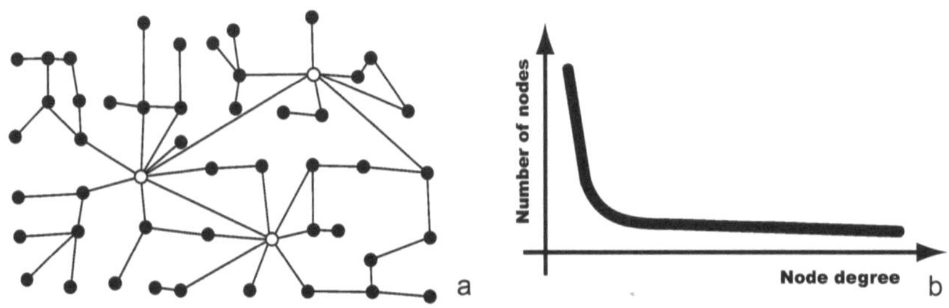

Fig. 12.2. (a) Example of a scale free graph and (b) its typical power law distribution.

the case of a road network we can assume that this number is three or four edges per node. Comparatively fewer nodes will show a lower or higher degree of connectivity. The number of nodes in relation to their degree resembles a bell shaped curve or a normal distribution (Figure 12.1).

A scale free graph does not mean that there is no map scale associated with that graph. The name "scale free" simply expresses the idea that the graph has no typical node degree. Typically scale free graphs contain nodes with a high node degree which are very important in summarising the connectivity of the graph (Figure 12.2). If such a node (a so-called hub) is removed from the graph, then the connectivity (reachability from one node to another) is significantly reduced. From a mathematical point of view in a scale free graph the node degree follows a power law distribution (Faloutsos et al., 1999). Popular examples of these types of graph are the world wide web or maps showing flight connections between cities.

In order to generalise road networks (assuming large changes in scale), it is necessary to identify the hubs (where many roads come together – typically within a city), and to preserve these hubs throughout the generalisation process. The roads connecting such hubs have significance – reflected in the volume of traffic and movement of goods between cities via major roads and highways. In some instances it is useful to attach a weighting to the edge in the graph in order to reflect the importance of that road in connecting major hubs. The node degree would count not only the number of edges, but the sum of the weighted edges. The result would be a transition of the random graph into a scale free graph. Such hierarchical ranking of edges and nodes can be used by generalisation algorithms: the lower the weight, the more

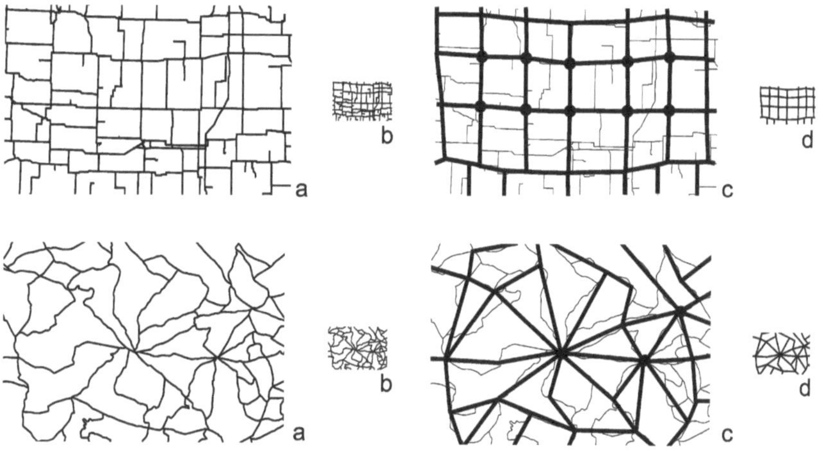

Fig. 12.3. Example of a road network generalisation. (a) Source. (b) Original data scaled down. (c) Generalisation with hubs. (d) Generalised data scaled down. (Source upper row: Highway (2004); Source lower row: Digital Chart of China (2005).)

probable that streets can be removed or moved. The greater the importance, the more they should be emphasised and their true location retained.

Hubs can be identified even without information about times and volumes of traffic. Shortest path algorithms, investigations of connectivity, analysis of the centrality of nodes and additional pattern searching can provide new information regarding the importance of separate nodes or regions in a graph. The two examples in Figure 12.3 show possible generalisation solutions that have retained the key edges and nodes after generalisation. It is likely that without the detection of such hubs, the structure of the road network would be destroyed.

Figure 12.3 illustrates the idea of condensing a set of crossroads into one hub. The centre point does not have to exist in the original dataset. This simplification of the network can lead to a new shared point of intersection. Pattern recognition algorithms can be used to detect structures such as those shown in Figure 12.3, and determine characteristics such as centre points and parallel edges. In this way, global structures and regularities can be recognised. The preservation of hubs and key links act as link points between the generalised form and the original and play a key role in the human recognition process (i.e. that the user understands that they are looking at the same region, albeit at a lower level of detail).

12.3. Patterns in Road Network

A graph theoretic representation is assumed to underpin all the patterns analysed and presented here. The basic graph contains all topographical points and lines as nodes and edges. Successive graphs contain only those nodes where three or more lines intersect (i.e. all nodes of degree two are redundant and removed from the graph). What remain are the "strokes" which are edges representing road sections between intersections. Each generalised graph is linked to the graph at the finest level of detail.

Fig. 12.4. Different node types (Sester, 1995).

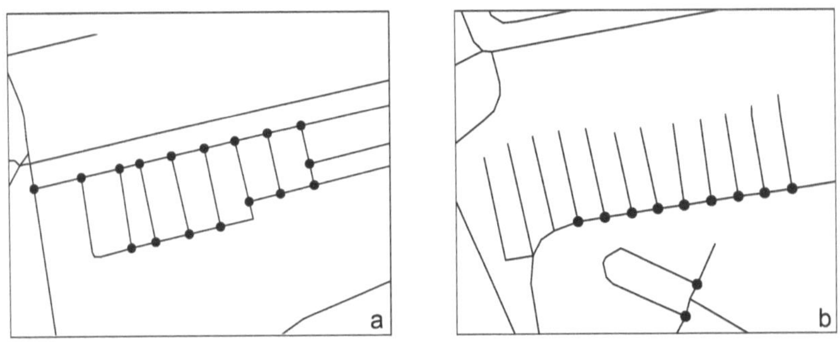

Fig. 12.5. (a) A "ladder" structure. (b) A "comb" structure.

12.3.1. *Different node types*

The nodes, i.e. the intersection points of roads, can vary greatly in appearance. Figure 12.4 for example, illustrates different types of road intersection based on the work of Sester (1995). Each node varies in the number of intersecting roads and their arrangement. The letter codings are used to classify the images in the following figures.

Knowledge of the type of node can be very useful in identifying patterns in the road network. In the examples presented in Figures 12.5–12.7, we can clearly see how degree, and the angles between edges can be used to describe different structures and types of patterns. For example a series of opposite "TEE" nodes often results in a ladder structure (Figure 12.5a); a series of single "TEE" nodes located at the foot of cul-de-sac roads reveals a "comb" structure (Figure 12.5b).

Very high numbers of "FRK" nodes can be found in the vicinity of motorway interchanges reflecting the large number of access roads leading onto important traffic routes (Figure 12.6). Bifurcations are typical in such areas (Figure 12.7).

12.3.2. *Strokes*

Within the scope of road network generalisation it is very common to talk in terms of "strokes". According to the degree of generalisation it is possible to reduce the road network by considering the length of a stroke. The idea of strokes (Thomson and Richardson, 1999) is based on the assumption that major roads should be as continuous as possible, and make gentle rather than abrupt changes in direction (akin to the continuous brush stroke of a painter). Major roads are very continuous and uninterrupted while minor roads are shorter and less continuous. By examining the angles at which edges "enter" and "exit" a node, it is pos-

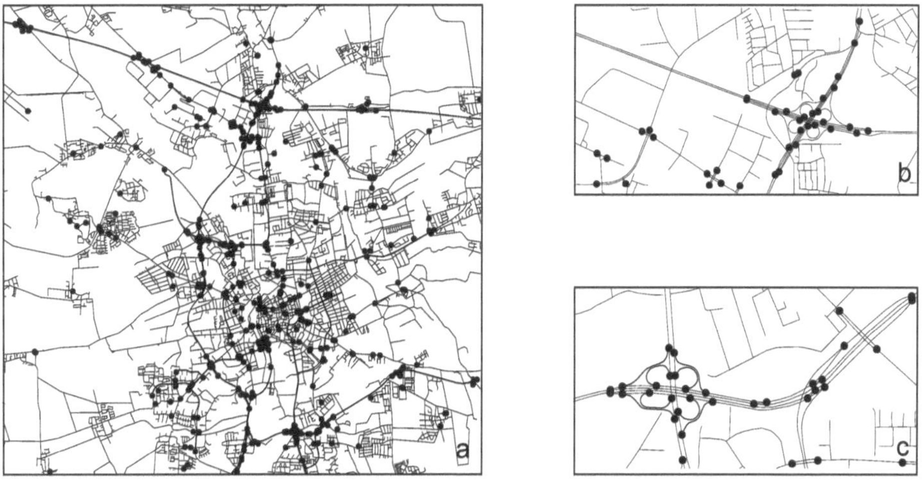

Fig. 12.6. (a) Example of "FRK" nodes. (b) Nodes in a motorway interchange. (c) Motorway interchange with additional access roads.

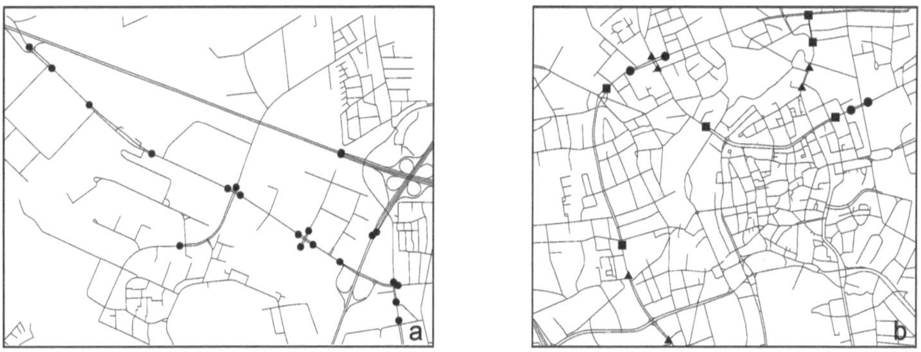

Fig. 12.7. (a) Typical splitting of high order roads into several lanes. (b) Detection of bifurcations.

sible to pair together major roads at intersections into single strokes (Thomson and Brooks, this volume (Chapter 13)).

The algorithm for tracing strokes in the graph is well known in principle. At each node an edge is selected that shows the minimum change in direction. However the splitting of a route into individual lanes creates problems since different strokes are computed when in fact it is essentially the same road. This idea is illustrated in Figure 12.8. Different strokes are generated depending on the sequence in which the edges are processed (approaching from a different direction may create different strokes). For this reason the pairs of branches (as shown in Figure 12.8) were investigated separately. Further discrepancies can arise between the entrance and exit roads at the intersection with highways, which can be overcome by considering the entire length of the stroke. Figure 12.9 shows the successful detection of an

Fig. 12.9. General principle of strokes.

individual stroke in a road network. Where the direction of the line changes gradually, it is straightforward to detect such main roads.

"FRK" nodes are particularly difficult to deal with when tracing the stroke. It is important to determine whether once a line has split, whether it rejoins later in the stroke. Figure 12.8 is an example of where "FRK" nodes split and then rejoin along the same length of road. By defining a single stroke for these dividing roads, we can generalise the line as if it was one stroke.

A further problem that can arise at "FRK" nodes, is when the access roads deviate slowly from the main road. An example is given in Figure 12.10, in which the algorithm has found the access road to be a better "stroke" than the main motorway (Figure 12.10b). However in the overall view it can be seen that the motorway route continues to run towards the lower edge of the image (Figure 12.10c). By investigating not only the preceding edge but the directional behaviour of all preceding edges such incorrect decisions can be avoided.

The following two illustrations (Figure 12.11) show a range of examples, in which only the longer strokes have been coloured. The basic configuration of the road network in the region can still be easily recognised. The second illustration shows, that the ring shaped road around the town centre has been explicitly identified – even where the road is represented as two separate lanes.

12.3.3. *Detecting grid patterns*

The grid structure is a common pattern in road networks. Artificially built settlements such as those found in new development schemes often contain straight-line roads, which enclose smaller internal building areas. Grid structures often occur in small regions and within urban areas, but rarely comprises all of a city.

Fig. 12.8. (a) A stroke is split in different parts at "FRK" nodes. (b) The parts of the strokes are combined at the "FRK" nodes.

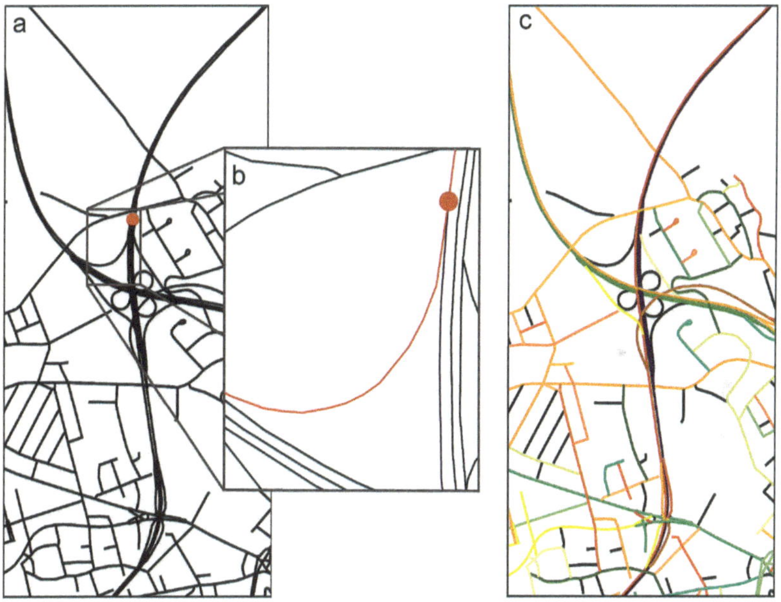

Fig. 12.10. (a) Pictures of an access road to a motorway. (b) The red coloured "FRK" node marks a wrong decision. (c) The right stroke is found, if the whole directional behaviour is taken into consideration.

The fundamental characteristics of a grid can be discerned from inspection of Figure 12.12. A grid is characterised by a set of mostly parallel lines, which are crossed by a second set of parallel lines. It is often, but not always, the case that the two sets of parallel lines intersect at right angle. The invariant characteristic of the grid is the parallelism of the groups of lines. The directions of the two groups of lines, and the angle of intersection between them remains invariant. What is variant, and distinguishes one grid from another, are the length and distance between the lines, and size of the grid overall. In turn, these values control the size of the

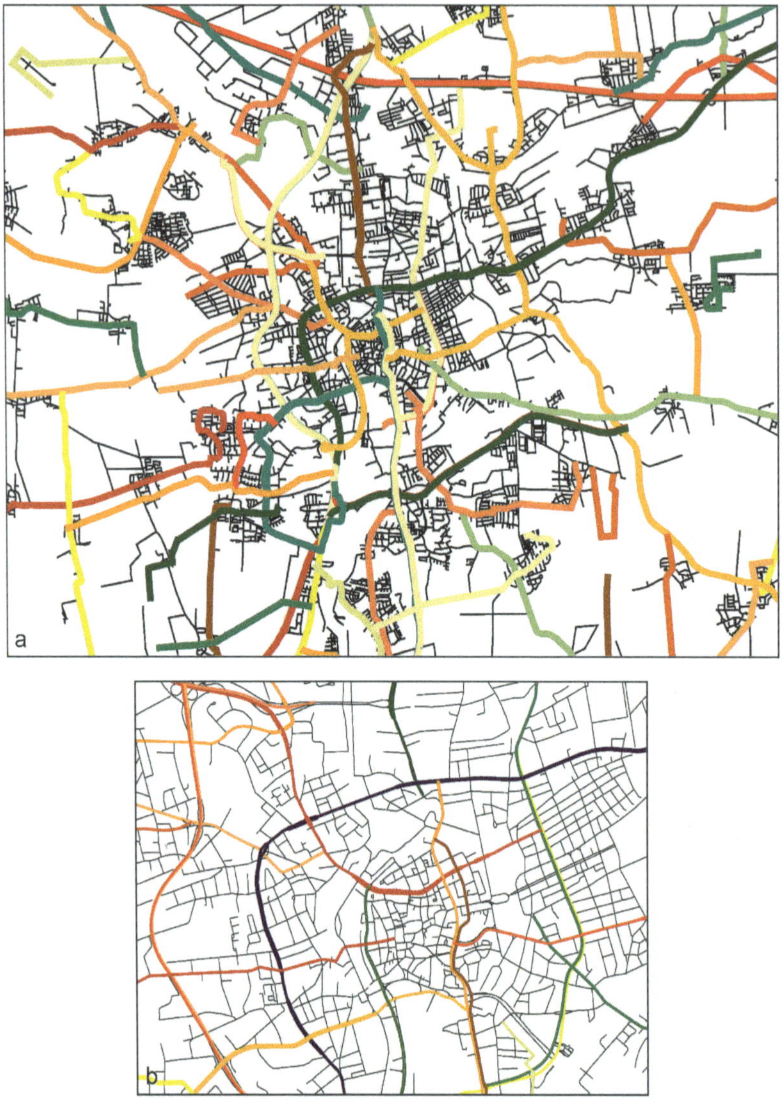

Fig. 12.11. (a) Detected strokes in a road network. (b) Different strokes are visualised by different colours.

internal rectangles and their convexity. By allowing for some variation in these qualities, we can develop techniques that detect these discriminating characteristics, and so detect the grid structures overall, as well as discrete sub-grid patterns that differentiate one part of a grid from another. In making explicit these various characteristics, we begin by identifying points in the grid where four edges intersect – so-called "CRS" nodes. If a direct straight line connection

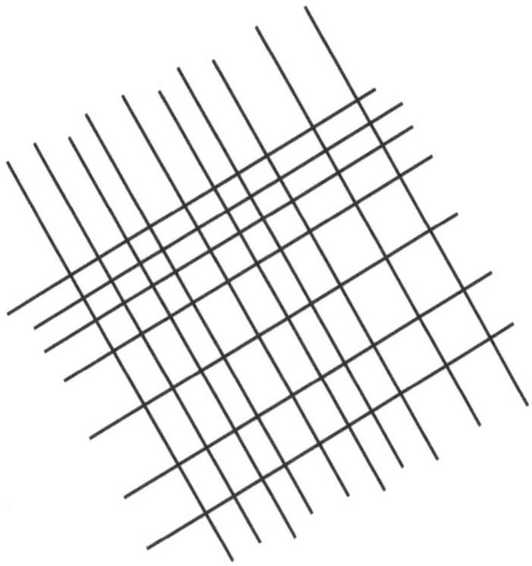

Fig. 12.12. The perfect grid.

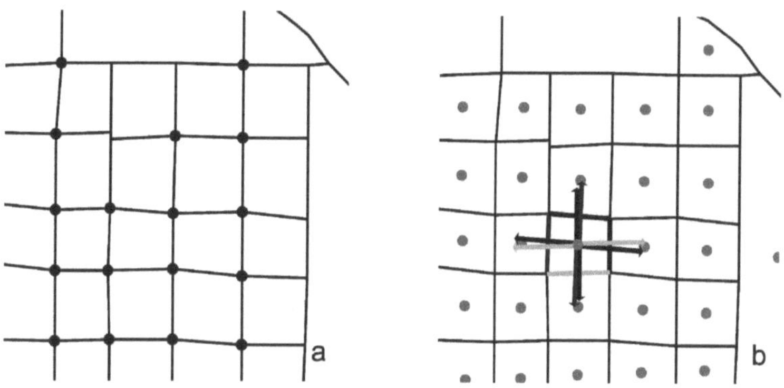

Fig. 12.13. (a) "CRS" nodes as starting points. (b) Determining the alignment of the centroid location among neighbouring polygons.

exists between two or more "CRS" nodes, then they are regarded as a possible basis for pattern detection (Figure 12.13a).

In a regular grid, the shape of the grid faces should be similar. Similarly its centroid should be in rough alignment with centroids in the adjoining grid faces. This idea is summarised in Figure 12.13b. The tolerances in that alignment depend on the size of the grid faces immediately surrounding a particular grid face. As a final stage in the analysis we can examine the distribution in the grid face areas, as well as the convexity of each grid face. We also note that where parallel sets of lines do not intersect at right angles, a more "flattened" or sheared "X" describes the "CRS" node. Examples of such nodes are highlighted in Figure 12.14a. Therefore the program for detecting "CRS" nodes was modified in order to detect such node shapes in addition to those described in 12.3.1.

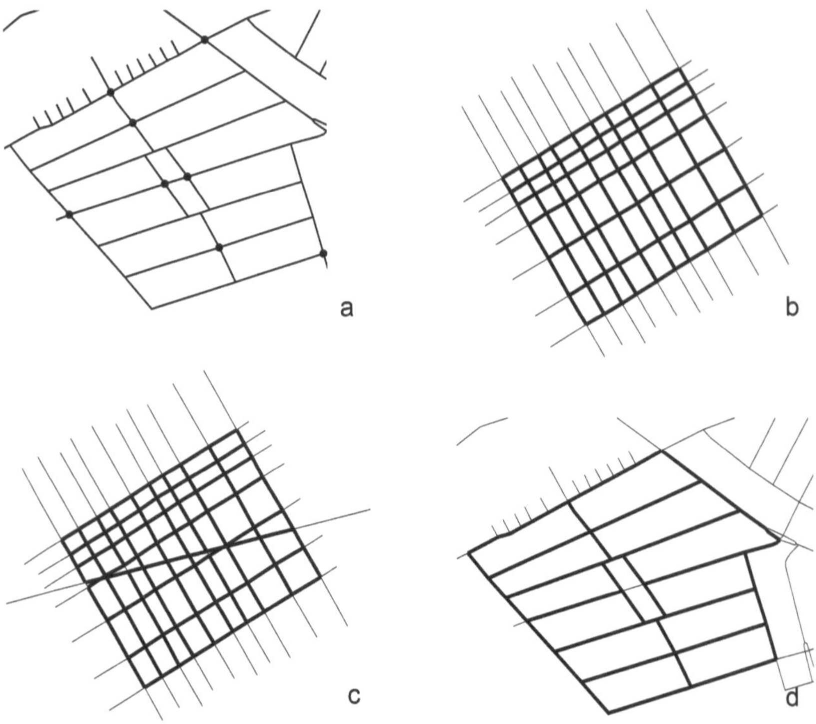

Fig. 12.14. (a) "X" nodes as starting points. (b) Detected grid. (c) Detected grid with another crossing road. (d) Detected grid patterns beginning first with identification of "X" nodes.

In assessing the success of the algorithm, we begin with an idealised grid as the simplest case. Here all grid elements were successfully detected (Figure 12.14b). Interruptions can occur in the grid, for example where another road crosses the grid structure (Figure 12.14c). In such cases, a majority of the rectangles "cut" by this road are successfully assigned to the grid. Only the smaller remaining triangles are expelled from the grid structure.

Figure 12.14d utilises real data, in this case a small subset of a road network from a city in northern Germany. The grid is slightly distorted and the edges of the grid are not completely parallel. The edge directions change continuously from top left down to the bottom right. The handling of the internal regions is also problematic, where two small road meshes are present. In this case they are not detected, because where a grid consisted of only two small polygons it was deemed not to be sufficiently significant. Figure 12.15 is a sample from the city of Zurich in Switzerland. The variously coloured regions depict grid structures detected by the algorithm.

One of the best known examples of a city with clear grid structure is in Manhattan, USA. At a course scale we see one overall grid. On closer inspection, it is possible to identify separate homogeneous regions, where the characteristics of the grid are slightly different. The algorithm was applied to the grid, and able to detect the various homogeneous regions that lie either side of Broadway, a street that crosses these uniform structures, interrupting the

Fig. 12.15. Grids detected by the algorithm (area covers part of the city of Zurich).

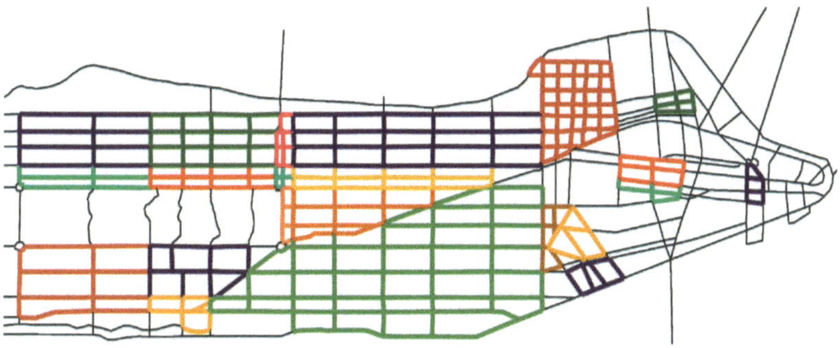

Fig. 12.16. Detected grids in Manhattan (different grids are variously coloured).

uniformity of the grid. The various regions (highlighted in different colours in Figure 12.16) can be used as a basis for generalisation. Broadway, was identified by its high stroke value.

Another useful approach is the application of a Hough transform (Hough, 1959) to the road segments of the grids. The advantage of this application is that it can detect the orientation of the straight lines defining the grid. Therefore all the separate grids in Figure 12.16 can be grouped according to their shared orientation. Again this provides a useful basis for generalisation, preserving the essential characteristics of the road pattern. Figure 12.17 shows those

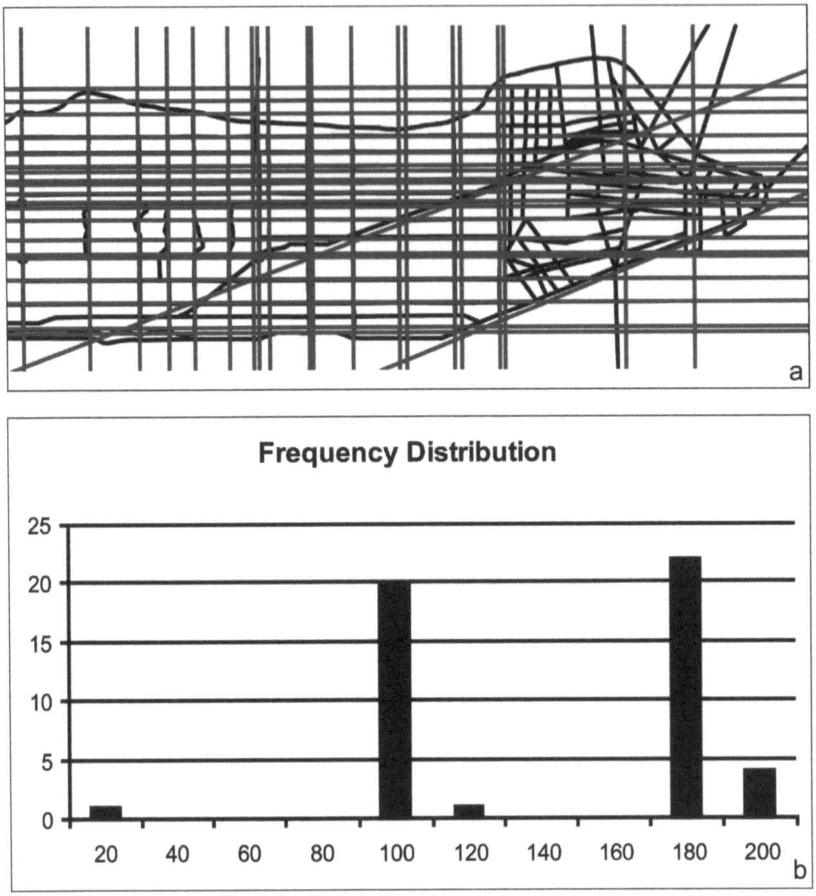

Fig. 12.17. (a) Main directions of the Hough transform and (b) the distribution of all existing line directions.

dominant straight lines in the grid, and a frequency graph of the various orientations from 0 to 180 degrees.

From these examples it can be seen that structures implicit in the grid can be detected by the algorithm. Where grids are well defined, there is little problem in identifying both the subgrids and dominant orientations. Where the structures are more irregular, it has shown that determination of the centroid alignment is not sufficient to avoid misdetections.

12.3.4. *Star pattern*

Star-shaped arrangements of road networks are typically found in dense urban areas. We associate the existence of a star pattern with the centre of a city – reflecting the functional designs of urban planners and the importance of access to the city centre. Thus we can identify the city centre even where no explicit information on the city is provided. The properties of a

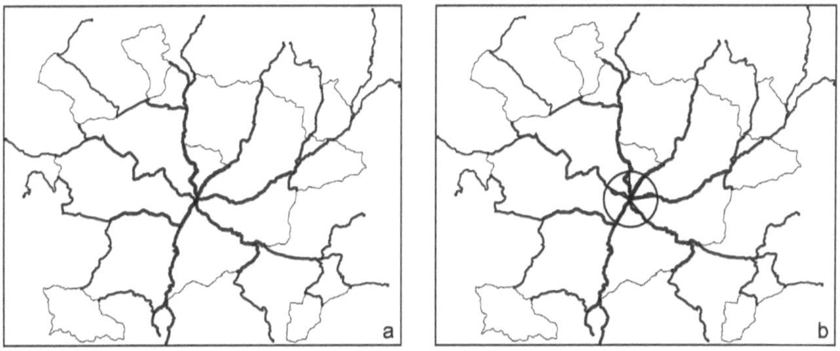

Fig. 12.18. (a) Single-source shortest paths (Dijkstra algorithm). (b) Circular search region in the Dijkstra graph with radial connections.

star-shaped structure can be described as a point – ideally a unique point – from which rays – ideally equally spaced and of equal length – radiate outwards. In road networks, it is rarely the case that the centre of the star is represented by a single point, but we can accommodate this fuzziness by defining a central minimal circle, and then assessing which rays fall within that circle. The problem is made more complex by the fact that roads may not be "ray" like, but contain shallow curves and bends in the road. Furthermore the rays may not be evenly distributed around the central area, nor be of equal length. Some radial roads can end at crossings or intersect with circular roads that ring the city, or are of a length such that they connect to neighbouring cities. Star structures can vary in size – from the scale of pedestrian precincts (with a radius of 100 m) to star structures the size of capital cities. The size of star that should be searched for very much depends on the intended level of generalisation (i.e. the final scale or level of detail required).

In developing a pattern recognition algorithm it was assumed that a set of roads come together at some central point of the star (but without requiring it to be a single Cartesian point). For this reason it is not sufficient to inspect the degree of the node. As a first approach we can take any point in the data set as a potential candidate for being the centre of a star. Using the Dijkstra algorithm (Dijkstra, 1959) a single-source shortest path is computed from that node to all other possible nodes (Figure 12.18a). For sparse graphs ($e \ll n^2$) one can solve this problem in $\Theta(e + n \log n)$ time, where e is the number of edges and n is the number of nodes.

After computing all the shortest paths, we identify those paths that intersect a circle of specified radius, circumscribed around the given point (Figure 12.18b). The length of all the shortest paths intersecting the circle are computed. The assumption is made that all rays intersecting the circle radiate in a relatively straight-lined manner. That is why the computed length of the shortest path would be approximately the same as the search radius, if the path is following approximately a ray. The number and angles of distribution around the circle are used as a basis for classifying the star structure. The challenge in utilising this algorithm is in defining an appropriate search radius. Since there is no a priori knowledge of the size of such a pattern, we must repeat this process for different search radii. The algorithm permits a certain fuzziness in the meeting place of the rays. Utilising information from the shortest path

Fig. 12.19. (a) Detection of star pattern without knowledge about expansion. (b) Selected ring pattern. (c) The intersections of the star rays and the ring determine the correct range.

algorithm the marginal connections between the individual ray intersections can be found and the centre can be fixed in the best location. In tests using geographical data, the algorithm worked well (where the centre is not perfectly defined and the rays are not straight).

A further problem in extracting star shaped structures was the deficiency of knowledge about the range of each configuration. Figure 12.19a shows an example, where star rays were detected correctly, however the detected lengths of the rays of the star were not well determined by the algorithm. By attempting to find ring-shaped structures first, we can partly address this issue (Figure 12.19c). The notion of ring-shaped structure will be described in §12.3.5 but still we do not have a working algorithm to detect these ring structures. Therefore the assumed ring in Figure 12.19b was manually selected. It is not always the case that each star is surrounded by a ring. So far we have assumed a relatively uniform distribution and length of ray. If this is not the case, other measurements for detecting the length of the ray are necessary.

Figure 12.20 shows an example of a city in northern Germany. No knowledge of the road categories, or the relative importance of a street were used in the application of the algorithm. From glancing at the image in the top left, no star is apparent. The central image shows the star pattern automatically detected by the algorithm. The range of examples illustrates the potential of the algorithm and its limits as the shapes become less clear, more irregular and "overlapping". The star structure is very important from the perspective of road network generalisation though further extension to the algorithm is required in order to detect more subtle patterns. One requirement is a method by which search radii can be set – since to try different radii is computationally intensive.

12.3.5. *Ring-shaped structures*

Ring-shaped patterns occur in road networks in a variety of sizes and shapes. There are common ring roads, small circle-shaped pedestrian areas, streets along old town walls, roundabouts inside the city as well as in the countryside and big traffic circles of motorways in the vicinity of important towns. The expansion of such ring-shaped structures depends primarily on their functionality; they vary from very small entities to larger elements. Often they exist in connection with star-like patterns, but this is not always the case. Frequently they present some kind of centrality because most of the time they surround and connect together important parts of the road network. A ring should approximate a circle with all the characteristic properties of a circle. In road networks the appearance of ring shaped patterns adapt to natural

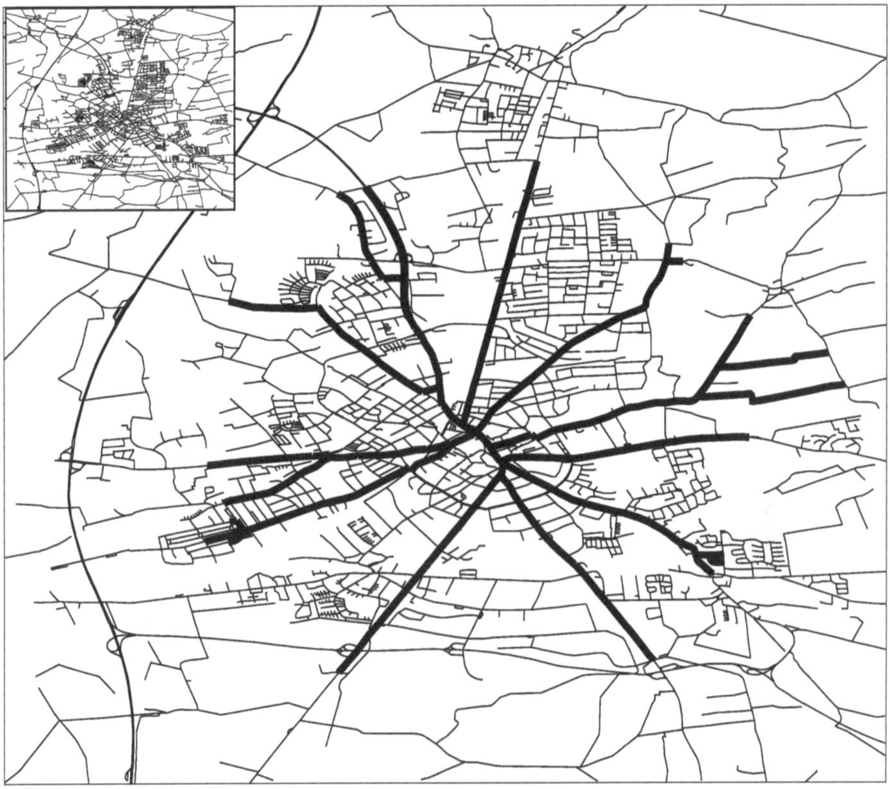

Fig. 12.20. Detected star in a data set of a northern town in Germany.

or human made conditions such as town walls, courses of rivers, differences in elevation or existing building areas (Figure 12.21a). Perfect circles are rarely found. Usually they are deformed, stretched and include irregularities. Correspondingly properties such as centre points, radius or periphery also change.

Figure 12.21b shows a city with two nested rings. Both rings are very well defined, however the outer one is interrupted in the lower part of the picture. The nearly perpendicular conjunction of parts of the outer ring in the upper right part of the picture constitutes another problem that applies to the inner ring too. This makes the development of an algorithm to automatically detect such structures very problematic.

The typical characteristics of a circle-shaped ring can be described by the compactness and the convexity of the ring structure. The compactness C is a measurement of the jaggedness of the object contour. If the compactness refers to a circle then it is defined as

$$C = \frac{P^2}{4\pi A},$$

Fig. 12.21. (a) An approximate ellipse builds a ring around the town centre. (b) Ring-shaped structures with deformations and interruptions.

where P is the perimeter and A the area of the object. In the case of a circle the compactness is optimal and has the value one. C is a measure of the roundness of an object. The convexity of an object is determined by the ratio of the area of the convex hull to the area of the object. In the optimal case of a circle it has the value one. In order to find ring-shaped contours in a data set it is necessary to detect objects with a compactness and a convexity close to one. Additional constraints could be the size of the objects or the minimal number of edges belonging to a ring. Simple rings such as small rectangles should be filtered out. A first approach to detect ring-shaped structures might be to use a Hough transform based on a circle or an ellipse equation (Duda and Hart, 1972). The results of this approach were found to be poor because the shape of rings in real road networks differs too much from circles or ellipses. Another approach was an exhaustive graph search for all existing simple cycles, which then have to be analysed to determine which of the characteristics are present. A cycle in a graph consists of a sequence of successively incident edges and their end vertices, where the terminating vertices are identical. The constraint of a simple cycle is that every vertex is incident to exactly two edges. Because of the exhaustive search this brute force approach is also impractical for large road networks. The automatic detection of ring structures therefore remains an area ripe for research.

12.3.6. *The structure of a city centre*

The centre of a city can be interpreted in different ways. There are various definitions for city centres such as historical city centres, economic, business city centres, centres of town development, politically based or traffic related definitions. A standard (consistent) picture of the "real" centre of a city probably does not exist. Often it is a combination of these – various forces at work in moulding the shape of the city, and its many functions. A goal of our work was to find broad regularities in the structure of a city and its city centre. We would like to call such structures "meta patterns". We acknowledge the existence of overlapping, palimpsest patterns, driven by a range of anthropogenic processes, interacting with a set of natural constraints, such as the morphology and hydrology of the land. This interaction is further complicated by the scale dependency in the shapes that we discern. At small scales the search for the city centre will be limited to the search of a central point of the graph (Freeman, 1979; Jiang and Claramunt, 2004a). A more detailed data set is a substantially more detailed picture of the city centre. This may lead to the idea that a polygon would be a better representation of the city centre rather than an individual point (e.g. the market place or an area within the city wall). However, the road network of a city already contains a great deal of shape semantic information that can be used to draw conclusions about the nature and function of a city centre. This can be used to identify meta patterns.

Since the city centre is a meta pattern it is not sufficient just to examine a few parallel lines or a centre of a circle since a lot of these structures "interact". In summary we note that the following structures may be variously found in the city: small blocks and denser road network towards the city centre; major roads radiating outwards from the city centre; ring shaped structures at various points emanating from the city centre; that "subordinate", previously "satellite" suburbs become captured in the city's expansion; radial major roads connect these suburban satellites with the city centre. The ability to see these patterns very much depends on the size of the city and the scale at which they are observed (Heinzle and Sester, 2004). One approach to identifying localised centres within a city, is to use cluster analysis based on the centroids of the road blocks. For the clustering the Hierarchical Parameter-free Graph CLustering (HPGCL) described in Anders (2003, 2004) was used. The HPGCL algorithm can find clusters of arbitrary shape and needs neither parameters like thresholds nor an assumption about the distribution of the data or number of clusters. The novelty of the HPGCL algorithm lies on the one hand in the application of the hierarchy of neighbourhood graphs (also called proximity graphs) to define the neighbourhood of a single object and object clusters in a natural and common way. On the other hand, in the definition of a median based, threshold free decision criteria for the similarity of clusters. In the HPGCL algorithm the Nearest-Neighbour-Graph, the Minimum-Spanning-Tree, the Relative-Neighbourhood-Graph, the Gabriel-Graph, and the Delaunay-Triangulation are used. In the HPGCL algorithm model, every cluster contains a set of inner edges and a set of outer edges. The inner edges connect nodes which belong to the same cluster and the outer edges connect nodes which belong to different clusters. Every cluster is characterised by the median of the inner edge size (cluster density) and the cluster variance. The cluster variance is the absolute median deviation of all inner and outer edge sizes from the cluster density, which introduce an uncertainty model (tolerance interval) to this clustering approach. Two neighbour clusters X and Y (connected by at least one edge) are merged if they are density-, distance-, and variance-compatible. For instance distance-compatible means that the median distance between X and Y belongs to the tolerance intervals of X and Y. The clustering algorithm starts at the edges of the Nearest-Neighbour-Graph and proceeds up to the edges of the Delaunay-Triangulation.

The results are shown in Figure 12.22, colour coded to show regions of equal centroid density. Though the results show regions of equal density, they do not reveal the identity of the city centre. Regions of high density are potential candidates for the city centre, but measures regarding the centrality are additionally required (Figure 12.23). For this reason the centrality of each node in the road network is calculated in a second step. The centrality measure of a node is calculated by the sum of all shortest paths to all other nodes. The more central a node is, the smaller the centrality sum will be. The combination of both measures (density and centrality) is used to filter potential areas that might be candidates for the city centre. The result is a bounding polygon, which encloses the inner city.

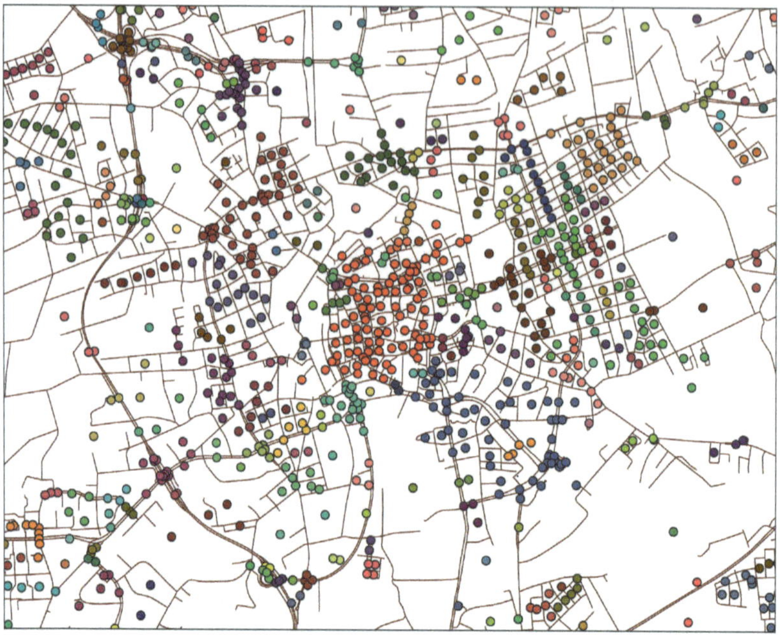

Fig. 12.22. Clustered centroids give a first impression of the density distribution.

Fig. 12.23. The red coloured points represent centroids with the best sum of all path lengths in a Dijkstra graph.

12.4. Summary and Outlook

In this work we have described the importance of patterns in road network generalisation. Structures such as strokes, grids, stars, rings and city centres were explained by characterising their properties and appearance, outlining ways for their detection and giving examples. There are many ways by which we might detect and measure such patterns, though much work remains to be done to integrate and parameterise these algorithms. While it is straightforward to detect isolated patterns, the reality is that the city is a palimpsest of patterns, recognisable at different scales. The discovery of additional structures and rules that model their interaction is our long-term objective. The ability to detect and characterise these patterns is fundamental to the generalisation process – helping to ensure that cities are recognisable whatever the scale at which they are observed. In essence we are trying to make explicit the shape semantics and incorporate this as part of the contextual generalisation process. The application of this type of information extends beyond ideas of generalisation though, and is relevant to applications such as car navigation, visualisation on small displays, internet mapping based on automatic derivation of maps from multi scale databases, and automatic enrichment of geographic data.

Acknowledgement

This work was supported by the EU in the IST-programme Number 2001-35047. We thank the National Mapping Agency of the state of Lower Saxony in Germany (LGN) for providing the ATKIS data, the National Mapping Agency of France (IGN) and TeleAtlas Germany for providing several topographic data sets.

W.A. Mackaness, A. Ruas and L.T. Sarjakoski (Editors)

Chapter 13

Generalisation of Geographical Networks

Robert Thomson[a], Rupert Brooks[b]

[a]*School of Computing, The Robert Gordon University, Aberdeen AB25 1HG, UK*
[b]*Atlas of Canada, Geomatics Canada, Ottawa, Ontario K1A 0E9, Canada*
now at Centre for Intelligent Machines, McGill University, Montreal, Quebec H3A 2A7, Canada

Abstract

This chapter reviews approaches to the model generalisation of geographical networks. Automating such generalisation can be particularly challenging due to the range of possible contextual data and constraints that have to be taken into account. At its simplest this task can be viewed as controlled network attenuation in a predictable and repeatable fashion, which preserves important network characteristics. The basic tree-like structure of river networks supports a natural generalisation based on stream orders. Road networks, in contrast, have more complex structures and semantics, and their generalisation has proved more difficult. The use of graph theory in network modelling and analysis to support generalisation is first considered. Example applications to road network generalisation are described where functional importance of road segments is inferred from their occurrence in optimal paths. Attention then focuses on generalisation techniques inspired by perceptual grouping. Visual grouping by good continuation is particularly important in human analysis of networks. The idea of a network "stroke" was introduced in an attempt to model this process. The practical implementation, at the Atlas of Canada, of network generalisation based on strokes is described. The results of their experience are used to illustrate the theoretical discussion. Generalisation techniques for road networks are found to be equally applicable to rivers (via the use stream order). Elaborations of the basic "stroke-based" generalisation technique are described.

Keywords: network generalisation, network attenuation, road generalisation, river generalisation, stream order, graph theory, good continuation, strokes, stroke-based generalisation, perceptual grouping, Atlas of Canada

13.1. Introduction

This chapter considers the design and implementation of the model generalisation of geographical networks, and so complements reports focussing on cartographic generalisation (for example Mustière and Lecordix, 2002). The principal objectives of model generalisation can be summarised as data reduction in a controlled, predictable and repeatable manner, whilst at the same time preserving the integrity of the data (Weibel, 1995). The reduction of network data should preserve its important characteristics as far as possible (Heinzle and Anders this volume (Chapter 12)). The definition of importance will be context dependent and may involve possibly conflicting factors. In practice, generalisation effort has focused on river and road networks, but has also been applied to generalisation of fault patterns (Downs and Mackaness, 2002), and could just as well be applied to thalwegs, pipelines and watersheds.

Ideally, objects are not generalised in isolation but rather with respect to other objects (Ruas and Mackaness, 1997). This means that network generalisation may have to take into account a wide range of contextual information. A road network, for example, may share geographic space with rail and river networks: each may share important point locations, reflecting an overall multi-modal transportation network. Generalisation of one network needs to take into account interdependence and linkages between the networks, and with other, non-network contextual data which may be as varied as political boundaries or the location of hospitals. A road network need not be uniform in character: it can encompass city streets, rural roads and major highways, and different forms of generalisation may be appropriate for each category. It is not therefore surprising that network generalisation remains challenging. Nevertheless, this chapter will show a development of concepts and techniques that have proved valuable in advancing the technology. Particular emphasis will be given to techniques inspired by perceptual grouping, which have proved useful in supporting the efficient implementation of effective generalisation.

13.2. Early Approaches to Network Generalisation

The literature on network generalisation shows an understandable focus on river and road networks, and that the linear representation of networks has dominated over the complementary areal view. River networks, being the simpler, proved more tractable to automated processing. At its simplest, their generalisation may be viewed as an ordering process: linear elements are identified and assigned an ordinal value. Reducing the level of detail then translates into the progressive removal from the network of the elements ranked lowest according to these derived weights.

Stream orders developed by physical geographers and hydrologists, such as the Strahler (1952) and Horton (1945) orders, can be used to quantify the relative importance of stream segments in a network and provide an ordering that does not disconnect the network during data reduction. Rusak Mazur and Castner (1990) showed that Horton ordering, using segment length as a secondary criterion, supports a simple and effective generalisation that closely approximates the generalisation decisions made by a human cartographer. Complications are introduced into the ordering scheme by stream braiding and delta structures, and when lakes within the network are subject to removal by generalisation. However, these difficulties can be overcome and a reliable generalisation scheme for hydrologic networks has been established on this foundation (Richardson 1993, 1994).

Road networks, in contrast, lack an equivalent unidirectional "flow" and contain abundant loop structures that foil attempts to derive equivalents to stream orders, and whose calculation requires an underlying tree structure. Furthermore, road segments commonly have associated semantic and geometric data that can influence generalisation. Graph theory (Diestel, 2000) has provided tools for handling and analysing such data (Mackaness and Mackechnie, 1999).

13.2.1. *Graph theory in network generalisation*

A graph can be represented diagrammatically as a set of points and edges (curved or straight lines, each connecting a pair of points): each point represents an object and each edge links a pair of points whose corresponding objects are in a relation (Diestel, 2000). This provides an obvious means of modelling the topology of a network: each graph node represents a road junction, terminal (i.e. a dead-end), or location of special interest, and a graph edge links pairs of nodes whose corresponding network locations are connected by a road segment that does not pass through any other node. This model can be improved by associating values with each arc to indicate, for example, the length of the road segment it represents. This so-called weighted graph then carries enough data to support, for example, the calculation of the shortest path between given points on the network.

A single geographical network can be represented by many different weighted graphs, since arc weights can be generated on the basis of many different properties. In practice it is convenient to model a network as a collection of node and arc objects with associated thematic and geometric information and arc-node incidence data. Figure 13.1 shows a typical representation. Possible thematic attributes associated with road segments include: road width or number of lanes, street name, speed limit, road class, surface type, one-way travel/flow restriction. The associated segment geometry could be represented by a polyline structure. These data implicitly contain a representation of the linear graph of the network for which different arc (and node) weights can be generated as required. The various derived arc weights for use in generalisation can also be stored in this database. Similarly, a road junction may have associated data describing delays in taking turns or passing traffic lights. A network structure can be represented using an arc-node topology. The network is treated as a graph structure, where each linear feature is an edge, and junctions between linear features are nodes. For analysis, an intersection table or turntable is often generated. Figure 13.1 shows an example of a road network where the arc-node incidence data has been explicitly modelled along with the turning possibilities at each road junction. Multiple arcs intersect at each node, and as a result there are many possible turns or movements from one arc to another. This example shows a simple intersection of a freeway ramp entering a residential street. Observe that this consists of three arcs, one node, and 6 turns. Impossible turns are indicated with an infinite turn weight.

Mackaness and Beard (1993) showed how graph theory can provide the information required to make intelligent decisions in map generalisation, and in the generalisation of networks in particular. It permits the detection and thus preservation of topological characteristics such as connectivity and adjacency: minimum spanning trees (MST), for example, can be used to preserve connectivity among a set of important network locations. Graph-based clustering techniques can detect special network regions, such as complex junctions or cities (Mackaness and Mackechnie, 1999). Graph-theoretic measures can also quantify the importance of individual arcs and nodes according to different graph structural properties; such information can be exploited for controlling generalisation (e.g. Mackaness, 1995b; Li and Choi, 2002; Jiang and Claramunt, 2004a). More formal, abstract analyses of the generalisation of graph structures for spatial data have also been developed (e.g. Puppo and Dettori, 1995; Stell and Worboys, 1999).

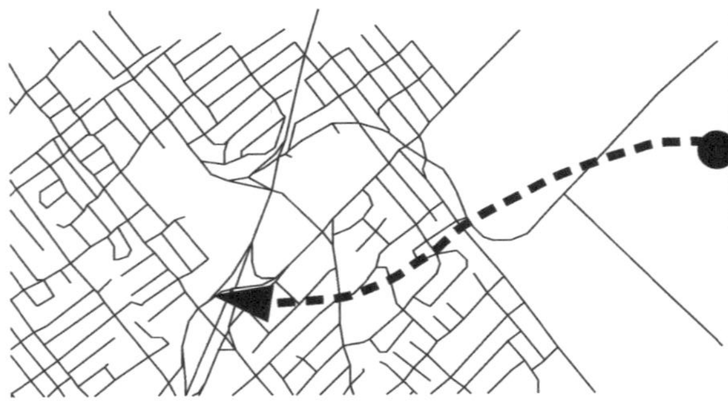

ARC TABLE

ARC#	FROMNODE	TONODE	NAME	LENGTH	CLASS
31	156	148	Exit 51	375.2	Ramp
32	148	150	Carling	125.3	Street
33	147	148	Carling	220.2	Street
34	154	159	Beaulieu	105.4	Street

NODE TABLE

NODE#	CLASS
148	Junction
149	Junction
150	Junction
151	Junction

INTERSECTION (TURN) TABLE

NODE#	FROMARC	TOARC	ANGLE	TURNING	CLASS
148	31	32	294	Inf	Illegal
148	32	31	-294	Inf	Illegal
148	31	33	34	0.5	Yield
148	33	31	-34	Inf	Illegal
148	32	33	0	0	Connected
148	33	32	0	0	Connected
149	33	34	92	1.5	Stoplight
149	34	33	-92	1.5	Stoplight

Fig. 13.1. Typical Arc-Node topological representation of geographic data.

13.2.2. *Road network generalisation*

Given a set of point locations in the network, Thomson and Richardson (1995; Richardson and Thomson, 1996) used graph-theoretic measures of the underlying linear graph to derive measures of the relative importance of the network arcs. They then used this information to support data reduction. The method was based on the assumption that the segments most probably used in travel between the nodes would lie on minimum cost paths. By tracing each optimal path between pairs of points of interest and keeping a tally of the number of times each network arc was used, a set of arc weights could be built up to indicate the arc's functional importance in linking the given node set, so providing direct measures for context-based generalisation. For the generalisation to be valid, arc removal must not disconnect nodes which previously were connected. To achieve this, weight values for the arcs representing one spanning tree of the node were set to the maximum value present in the network. Since it is also possible that some road segments outside the spanning tree could become disconnected during network attenuation a further checking of connectivity, and adjustment of arc weights, was required.

In the absence of relevant thematic information, the costs used in the optimal path calculation would be the length of the arcs. However, travel time through a network is a more important determinant of transportation behaviour than distance (Clark, 1977). Thus, if suitable thematic information such as road-surface type or number of lanes were available, this could be used to derive new costs more closely corresponding to relative travel times. This can be achieved by deriving road "friction factors" from the thematic attributes (Richardson and Thomson, 1996).

On completion of all arc weight calculation and adjustments each segment in the network has an associated value that indicates its relative importance in the given context. Given a percentage reduction of the network, the appropriate number of network segments can be selected by a simple thresholding of these values. The data can be represented to the cartographer in an interactive display with slider control of the threshold percentage. Thus, percentage reductions based on criteria such as Töpfer's radical law (Töpfer and Pillewizer, 1966) can be applied to the data, or by user interaction if greater flexibility is required.

Reynes (1997) and Morisset and Ruas (1997) also used a linear graph representation of a road network in techniques for road generalisation that focussed on the relative use of road segments. Morriset and Ruas estimated relative frequency of each road segment's use by simulating the movement of traffic through the use of "driver agents": the day-to-day movement of vehicles from place to place. The calculation of each path between pairs of locations dynamically selected by the system finds the quickest path on a classified road network. As with the above method, at the end of the simulation a network is obtained where each road segment has a count of its utilisation. Again, additional criteria must be applied to this raw data to avoid the discontinuities that could be generated during subsequent network attenuation.

13.3. Exploiting Perceptual Grouping

Although direct analogies have long been noted between what is required for successful map generalisation and the principles of perceptual grouping (DeLucia and Black, 1987), the principles were first applied to the analysis of geographical networks relatively recently (Thomson and Richardson, 1999) in an effort to find more computationally efficient methods of road network generalisation, and to tap into the mechanism whereby a human cartographer can generalise road networks purely on the basis of their geometric, topological, and thematic properties. Indeed, it is usually possible to infer the relative importance of road segments in a network in the absence of all other thematic information.

13.3.1. *Perceptual grouping*

Perceptual grouping describes the Gestalt phenomena by which the human visual system spontaneously organises elements of the visual field: Some arrangements of picture elements tend to be seen as forming natural groups that often appear to stand out from the surrounding elements, i.e. as "figures" against "grounds". Many perceptual grouping principles have been identified, such as proximity, similarity, symmetry, closure, parallelism, and continuity (Wertheimer, 1938; King and Wertheimer, 2005).

Perceptual grouping is now recognised as the basis for parsing the visual world into surfaces and objects according to relatively simple visual characteristics (Palmer, 1983). These grouping principles play a vital role in understanding two-dimensional images, whether of three-dimensional scenes (Witkin and Tenenbaum, 1983; Lowe, 1985; Biederman et al., 1991), line drawings (Saund et al., 2002), or maps (MacEachren, 1995). Perceptual grouping supports a basic level of image understanding that does not require semantic knowledge yet is sufficiently powerful to identify significant features and infer information about their relations and relative importance.

13.3.2. *"Strokes"*

The perceptual grouping principle that most clearly contributes to the analysis and generalisation of networks is the principle of good continuation: "that elements that appear to follow in the same direction … tend to be grouped together" (Coren and Ward, 1989). Thus a series of smoothly connected elements will be naturally perceived as a single graphical object, perhaps intersected by others.

When the abstraction process used to represent a geographical network in a map preserves enough structural and geometric detail, then the perceived groups make sense as "natural" continuous road or river segments. Moreover, as a rule, the perceptual significance of these elements in the map reflects their functional importance in the network: roads or rivers that look relatively important generally *are* more important. The term "strokes" was coined for these network linear "figures", prompted by the idea of a curvilinear segment drawn in one smooth movement (Thomson and Richardson, 1999).

In the traditional Arc-Node topology, for example, the network shown in Figure 13.2 would have eight arcs and nine nodes. These elements form four intersecting strokes. Each stroke is a set of arcs representing a path through the network from one terminal node to another (possibly the same) which uses all the arcs in the set, without repetition. In some cases the path may self-intersect in which case nodes may repeat. Once a network has been analysed into its constituent strokes, these strokes can be ranked according to relative importance, providing an ordering that can support data reduction. The ranking based on individual stroke length and

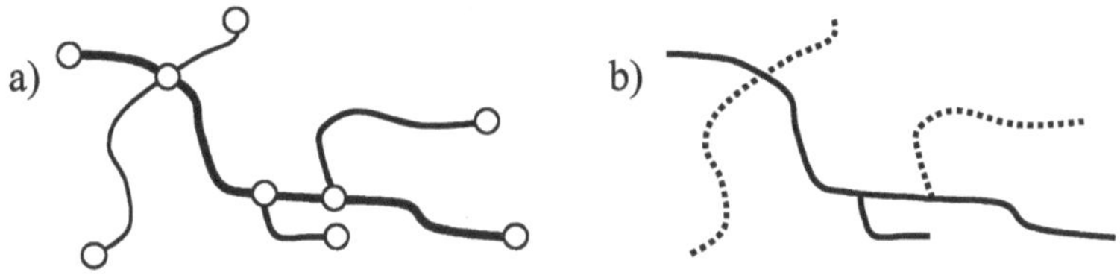

Fig. 13.2. (a) A simple network with 8 arcs and 9 nodes and (b) resolved into 4 strokes.

thematic attributes can be adjusted by further processing to additionally reflect structural or functional importance in the network.

13.3.2.1. *Stroke extraction*

For each node, the stroke building process determines which arcs incident on the node should be connected together. A number of criteria may be applied at this point. The simplest is the angle of deflection that the join would imply: concatenate the pair of arcs with the best continuity of direction, but only if that continuity is good enough. The decisions on arc concatenation at nodes (junctions) are controlled by a small set of rules. The rule sets can be varied according to context, and so different rules may be used for urban and non-urban junctions, for example. Thus stroke building in this manner is a local process, considering each node in turn and dependent completely on the properties of the network at that node neighbourhood. This implies that strokes can be locally adjusted, with the update requiring only small processing time. Furthermore, the stroke-building method can be made consistent and symmetric – the orders in which the nodes and their incident arcs are considered are irrelevant to the final set of strokes produced.

Where other relevant arc thematic data are available this may be allowed to dominate over continuity of direction in the concatenation – preventing, for example, the concatenation of two perfectly aligned arcs if their categories are too divergent (Thomson and Richardson, 1999). Other rules can be easily inserted: it has also been found useful, for example, in both road and river networks, to constrain the concatenation process so that arcs with the same name attribute (e.g. Bank Street or Mackenzie River) are connected into a stroke. Once all decisions have been made at each node the strokes are assembled by accumulating sets of arcs that connect to each other. When all arcs have been assigned to some such set, the stroke construction process is complete. The resulting strokes can now be used as a basis for generalising the network. Figure 13.3 shows strokes derived for the road network of central Ottawa, Canada.

13.3.3. *Generalisation using strokes: experience at the Atlas of Canada*

An automated generalisation system was constructed for production use at the Atlas of Canada (Brooks, 2000) which implemented a stroke-based approach. The system was based on previous work by Richardson and Thomson (Richardson, 1993; Thomson and Richardson, 1995, 1999; Richardson and Thomson, 1996). The Atlas software is available under an open source license (Brooks, 2003).

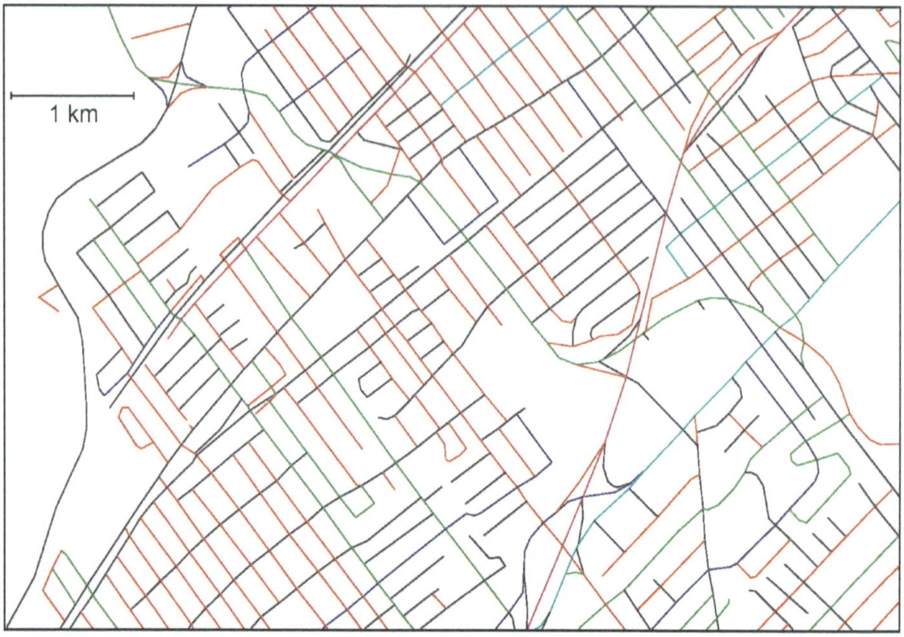

Fig. 13.3. Strokes in the road network for downtown Ottawa, Canada.

13.3.3.1. *Road networks*

The strokes must be ordered by some measure of their importance. To devise an ordering scheme for a road network, three simple principles were applied:

- All else being equal, a longer stroke is more important than a shorter one.
- All else being equal, a stroke made up of higher quality roads is more important than one made up of lower quality roads (e.g. tracks *vs* highways).
- No connected components of the original network would be split into two disconnected parts.

It is relatively simple to define a workable salience value for each stroke based on length and weighted by a factor derived from thematic attributes and reflecting road quality. The ordered list based on the raw salience may then be adjusted to avoid disconnecting the network. This is achieved by a relatively simple limited search procedure (Thomson and Brooks, 2000); the result is an ordered list from which strokes can be removed in sequence without disconnecting the network.

Figure 13.4 shows the effect of this generalisation on the road network data for western Ontario (Thunder Bay and surrounding area). The darker lines in Figure 13.4 show the resulting selection, after generalisation using the method described above. Road segments not selected by the generalisation process are shown lighter, so showing the complexity of the original dataset. No post-processing has been applied to this result so that the effect of the algorithm is not obscured. The importance of length in computing the salience of a stroke is evident in these figures.

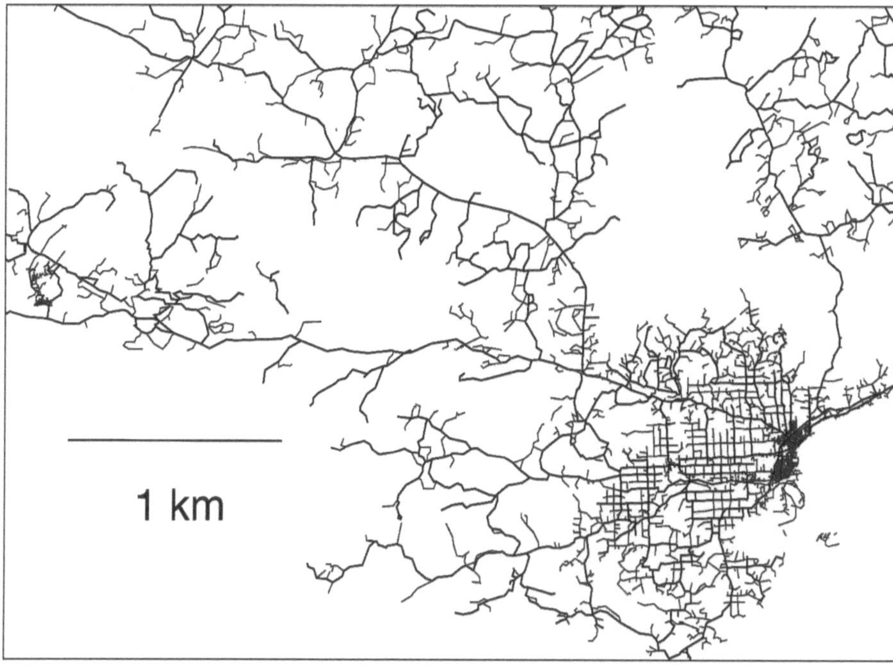

Fig. 13.4. Generalised road network for western Ontario (Thunder Bay and surrounding area).

13.3.3.2. *River networks*

A stroke based approach can also be used to attenuate river networks in combination with ordering based on Horton's method. The strokes are used to define the main paths needed for the second of two processing stages (Thomson and Brooks, 2000). Horton's original criteria for what constitutes the "main stream" were that the longest and straightest path should be used to define the main stream. Other determinations of the main stream have included the straightest path (Richardson, 1993) or the largest drained area (Pfafstetter, 1989). The use of the stroke data model does not enforce any particular choice of main stream; the rules for stroke-building can use these or more sophisticated approaches for selecting the main stream. Braided streams add an additional level of complexity to the ordering process, but a simple extension of the construction process (Thomson and Brooks, 2000) can prevent disconnections.

The system has been used to generalise the hydrology for a published 1 : 4 M scale map of Canada's three northern territories from source material in the GeoBase Level 0 hydrology dataset that had been cleaned and for which attributes had been added, connectivity corrected and directionality computed (Brooks, 2000).

Figure 13.5 shows the generalisation process using strokes as applied to a river network (from Baffin Island). The original data and two levels of generalisation are shown, with no post processing applied to the resulting selections.

The stroke-building algorithm itself is very efficient. Measured computational cost in practice appears approximately linear to the number of nodes. The algorithm will operate on erroneous datasets. Although the results will be progressively degraded as source data degrades,

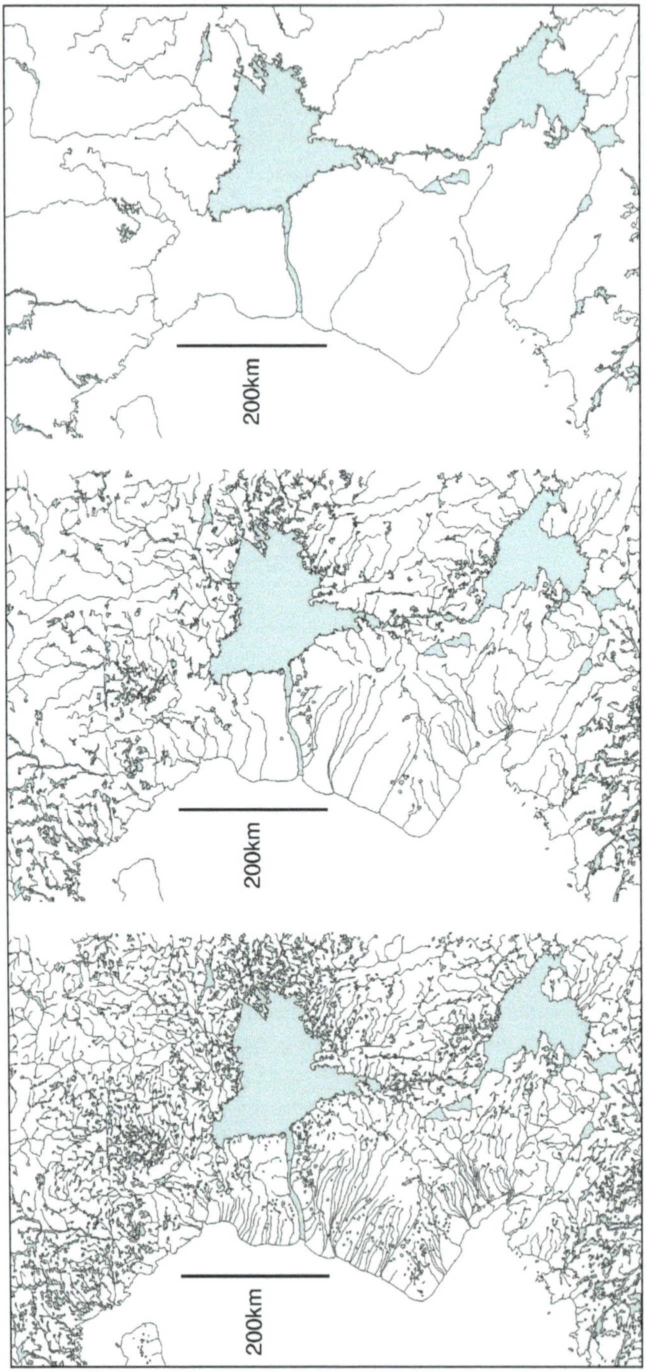

Fig. 13.5. Generalised river network at three levels of generalisation (Baffin Island).

this robustness allows operation in a production environment on large datasets, where detecting and correcting all errors in a dataset may not be an option. For production generalisation, a number of constraints were applied to the stroke building process to increase the cartographic accuracy:

- Strokes could never connect two arcs with incompatible direction of flow;
- Strokes would always follow a named path, if one existed;
- Strokes would never cross from a braid to a non-braid stream;
- Strokes could never jump more than two Strahler stream orders;
- The user could enter constraints to force the stroke to be built along a certain path at a node. These constraints overruled all other considerations.

With these constraints satisfied, the stroke building algorithm would find the "longest and straightest" paths. The final selection that resulted from the generalisation technique described here was processed using cartographic generalisation techniques, including line simplification, displacement and exaggeration.

The automated generalisation was imperfect but acceptable, and remaining problems could be fixed by an experienced cartographer in a reasonable amount of time. The resultant map was considered on a par with any other produced by traditional means at the Atlas (Brooks, 2000).

13.3.4. *Evaluation and further developments*

The perceptual strokes technique allows the delineation of meaningful units of a network in a manner that echoes human perception. Relatively simple properties of these units may then be used to determine an order of importance. This method is robust enough to use on faulty data with reasonable results. Furthermore, local constraints could be added to the stroke building method without affecting its overall performance. This allowed additional information, such as names, where available, to be used to adjust the system's performance.

The stroke-based generalisation applied to hydrographic networks has proven useful and effective in practice, and has been successfully used in commercial map production. Tests on road networks at the Atlas of Canada, and more recent investigations elsewhere, have produced good results, and showed the feasibility and potential of the approach. Chaudhry and Mackaness (2005), for example, applied their implementation of the approach to the generalisation of mixed road networks over a large scale change (directly from 1 : 1250 or 1 : 2500 to 1 : 250 000) and reported results comparable with published paper maps. However, limitations of the method for road networks have been recognised by its authors and others, and extensions to the basic method proposed. As the method stands, for example, additional special handling may be needed for roundabouts and complex junctions. Mackaness and Mackechnie (1999), however, present techniques for the automated recognition and simplification of road junctions that could be integrated into the system to address this problem.

The method may produce strokes that are very close and form structures with unwanted visual impact. It has also been observed that since, for example, the road networks of cities, small villages and countryside require different levels (and perhaps different styles) of attenuation to be applied in generalising, this leads to a problem when handling road networks that encompass urban and rural areas: an issue researched by Edwardes and Mackaness (2000) as part of the AGENT project (Lamy et al., 1999). Edwardes and Mackaness address a general problem – arising particularly in the generalisation of urban networks – that the stroke-based approach is concerned only with the set of linear road or street objects and so may not provide sufficient consideration of the network's areal properties. Their solution use strokes as one tool for characterising the network in order to provide global information to the generalisation process. Their method adopts the areal dual of the network as another structure for characterisation, using the two structures simultaneously to perform the generalisation.

The approach also addresses the problem of generalising networks that comprise both urban and non-urban regions. Urban areas are automatically delimited and intersected with the road network to segment it into intra-urban and inter-urban sections, which can be treated separately. Roads that intersect the urban boundaries are preserved so that the sections re-connect after generalisation.

The urban spaces are partitioned into city blocks using cycles of streets, with block adjacency data stored in a graph. This represents the dual of the linear representation, and also contains explicit perceptual properties relevant to generalisation. These two representations are integrated and used to control the rate and sequence of generalisation. A weighted graph is constructed where nodes represent partitions, the links represent adjacency relationships and the weights are used to integrate stroke information.

Generalisation proceeds by sequential fusion of the partitions where they are below a minimum size constraint. A partition always aggregates with its neighbour across the weakest boundary, hence the weakest strokes are removed from the network. This proved more robust than a similar, rule-based merging scheme of Peng and Müller (1996). Network connectivity is handled implicitly, since block aggregation cannot disconnect the network, although some special cases need additional processing.

The algorithm produced good results, identifying and retaining the essential areal, linear, semantic and density patterns of the network and its constituent roads (Edwardes and Mackaness, 2000).

13.4. Concluding Remarks

One finding from the hydrography generalisation project at the Atlas of Canada was that generalisation requires fairly sophisticated knowledge to be embedded in the data. The investment required to achieve this is questionable for generalisation alone: it may only make sense when adding that additional level of sophistication allows other uses of the data. The atlas drainage basins/hydrology dataset is a good example where the stroke model was also used for other non-generalisation applications. For example it was used in matching name attributes to river tributaries, greatly increasing the efficiency of that process, and in defining upstream drainage basins (Thomson and Brooks, 2002).

It has been observed elsewhere (Ruas and Mackaness, 1997) that successful generalisation requires enrichment of the data through cartometric analysis and information gathering, and this will surely hold true for the generalisation of all geographical networks. Given the success of perceptual strokes further analyses may be inspired by perceptual organisation and visual pattern recognition.

Historically, the view of a network as a graph with attributes has dominated generalisation. Increasingly, however, properties of the dual areal representation are being exploited and incorporated, which may reflect the growing awareness that generalisation takes place in the context of other objects and not in isolation. This growing use of contextual information can be expected to lead to increasing use of constraint-based and agent-based approaches in implementation.

Acknowledgements

Much of the work described in §13.2 and §13.3 built on methodologies and techniques developed at Geomatics Canada by, or under the direction of, Dr Dianne Richardson. Statistics Canada provided road network data from their database which was used for the road network experiments.

Generalisation of Geographic Information: Cartographic Modelling and Applications
W.A. Mackaness, A. Ruas and L.T. Sarjakoski (Editors)

Chapter 14

A Prototype Generalisation System Based on the Multi-Agent System Paradigm

Anne Ruas, Cécile Duchêne

Institut Géographique National (IGN), Laboratoire COGIT, 2-4 av. Pasteur, F-94160 Saint-Mandé, France
e-mail: {anne.ruas,cecile.duchene}@ign.fr

Abstract

Automation of the generalisation process is one of the main research subjects of the COGIT laboratory. The automation of this process is a real challenge for a NMA that in the short term wishes to reduce costs and time taken to produce series mapping, and in the longer term wants to be able to deliver maps over the internet, providing generalisation on demand. Research in this area, undertaken at the COGIT laboratory has resulted in the award of nine PhDs and a further two that are ongoing. This chapter presents various results with a particular focus on two generalisation engines conceived at the COGIT laboratory and based on the Multi-Agent System paradigm. The first one is based on the concept of constraints, ideas of autonomy and levels of details. It models the micro- and meso-generalisation of roads and urban areas. This model has been reused during the Agent project and has been commercialised in the form of "Clarity" – a Laser-Scan product. The other engine is based on interactions between micro-agents and has been optimised for the generalisation of rural areas. This recent development has produced some promising results. Work is ongoing to develop linkages between the two models.

Keywords: agent methodologies, multi-agent systems, constraint based modelling, level of detail, meso-generalisation, incremental generalisation, strategies, interactions, CartACom

14.1. Introduction

This paper presents two pieces of research undertaken at the French National Geographic Institute's (Institut Géographique National, IGN) COGIT laboratory that used Multi-Agent Systems (MAS) to automate the generalisation process. These two complementary pieces of work (described in §14.2 and §14.3) were linked together to build AGIT, which is a generalisation research platform based on MAS technology. The work reflects efforts to apply multi-agent systems to map generalisation, a core activity within the COGIT laboratory. The research reflects work that commenced in 1992. The work has links with the AGENT project, and previous research developed on the PlaGe and Stratège platforms.

14.1.1. *Multi-Agent System (MAS)*

The concepts of agent and multi-agent systems stem from the field of artificial intelligence (AI). The agent paradigm is increasingly being used to solve complex problems in a broad set of domains, from robotic agents to software agents. The adopted definition was taken from Russell and Norvig (2003): "An agent is anything that can be viewed as perceiving its environment through sensors and acting upon that environment through effectors (...) We will make the assumption that every agent can perceive its own actions; A rational agent is one that does the right thing (...) As a first approximation, we will say that the right action is the one that will cause the agent to be most successful" (Russell and Norvig, 2003, pp. 32–35).

A multi-agent system (MAS) is one that is composed of several interacting agents. The definition of the agents depends on their capacity to choose, perceive and communicate. The category of agents that is closest to the one we used in our research work is the goal-based agent: A goal-based agent is one that uses "some sort of goal information that describes situations that are desirable" (Russell and Norvig, 2003, p. 49). An agent can be thought of as "an object that has a goal and acts autonomously in order to reach that goal due to the capacities of perception, deliberation, action, and possibly communication with other agents" (Weiss, 1999, p. 32).

Cartographic agents were developed in this research, defined as a set of cartographic objects (such as a building object, a road object) that have the goal to generalise themselves individually and all together in the most effective and efficient manner.

14.1.2. *An approach to automate generalisation*

Our research takes place in a localised manner, as a step by step approach to generalisation (Brassel and Weibel, 1988; McMaster and Shea, 1988). In 1997, a test of interactive generalisation performed by the OEEPE[1] working group on generalisation revealed in a detailed study that the choice of algorithm on a given object is guided by the characteristics of the object and the nature and violation state of the cartographic constraints applying to it (Ruas, 2001). Thus the idea is to model geographical objects as agents that are able to perceive and evaluate their current state, and to choose and apply to themselves generalisation algorithms that improve that state (Ruas, 1998b, 1999). The proposal for modelling geographic objects as agents which generalise themselves follows on from previous attempts to use expert systems in map generalisation (Mackaness et al., 1986) and to use constraints as a way to represent user needs (Beard, 1991a; Weibel and Dutton, 1998). The work presented here is original in that it links the choice regarding the order, algorithm or parameter values according to the general map specification and the characteristics of each map object.

14.1.3. *AGIT: An agent based research platform*

AGIT is the name of the COGIT research platform used to perform generalisation that integrates MAS methodologies presented in §14.2 and §14.3. It inherits from the AGENT prototype that was conceived during the AGENT project funded by European Union (ESPRIT

[1] Organisation Européenne d'Etudes Photogrammétriques Expérimentales (now EuroSDR).

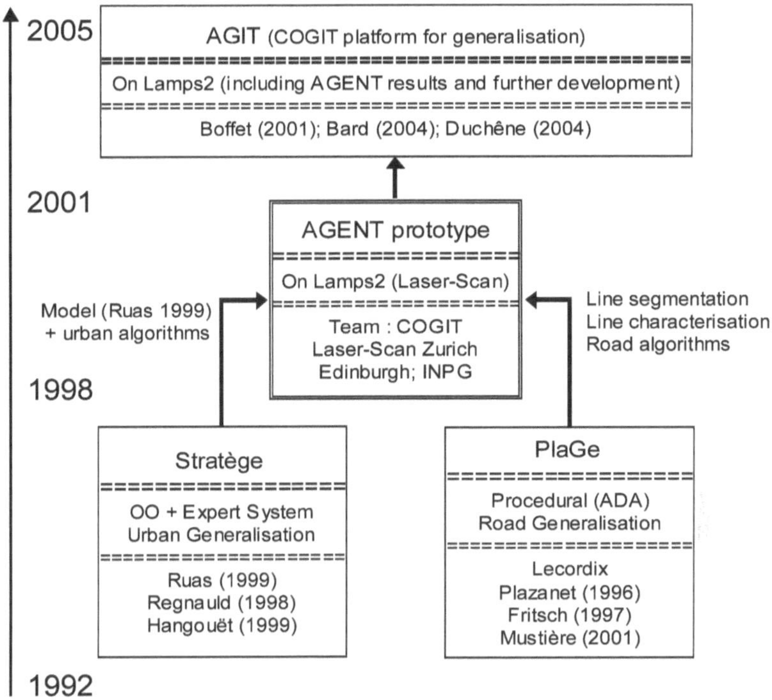

Fig. 14.1. COGIT Generalisation platforms and PhD research work during 1992–2004.

24939) (AGENT, 2000; Lamy et al., 1999). The partners of the project were the Institut Géographique National (France), Laser-Scan Ltd (UK), the Institut National Polytechnique de Grenoble (France), the University of Edinburgh (UK) and the University of Zurich (Switzerland) (see AGENT, 2000). The model used in the AGENT project inherited some algorithms from research performed at the COGIT laboratory on two platforms named "PlaGe" (for road generalisation) and Stratège (for urban generalisation). AGIT also integrates recent research work performed at the laboratory (Boffet, 2001; Bard, 2004a; Duchêne, 2004). AGIT inherits from a large pool of PhD research (COGIT, 2005) that have been demonstrated on the AGENT project platform (Figure 14.1). This collaborative research work has also as its seed point the research vision of Jean-Philippe Lagrange from back in the early 1990's.

14.1.4. *Structure of this chapter*

This chapter presents the MAS principles used to generalise data. Section 14.2 focuses on the AGENT model which originates from Ruas's PhD (1999), and enriched during the AGENT project by the Hill Climbing strategy proposed by Nicolas Regnauld as well as the Object Decomposition principles proposed by Duchêne. Section 14.3 presents Duchêne's PhD results (2004) used to enrich the agent's capabilities. Agents can exchange information to improve their generalisation: each micro-agent perceives its neighbourhood and integrates it into the decision making process. Section 14.4 critiques the results and presents current work and future ambitions of the COGIT laboratory.

14.2. Principles of the Generalisation Model

In this chapter we use the following terminology:

- *agent:* a geographical object such as a road, a house, an urban block or a town that is able to generalise itself;
- *constraint:* a function that describes the required value of the characteristic of an object (e.g. the size), two objects (e.g. the distance) or a set of objects (e.g. the density);
- *conflict:* the fact that one object does not satisfy a specific constraint;
- *operator:* a generalisation action (implements the algorithm together with a parameter value), (Sarjakoski, this volume (Chapter 2); McMaster and Regnauld, this volume (Chapter 3)).

14.2.1. *A generalisation engine based on dynamic choices*

The first principle of the generalisation model is that each geographical object is autonomous. Each object dynamically chooses its own generalisation operations according to its own set of conflicts. This capacity of choice makes each object an agent. As a consequence, instead of using predefined processes such as: Sequence 1: Apply (Douglas–Peucker simplification, $\lambda = 15$ m) to ALL Roads – Sequence 2: Apply (Gauss smoothing, $\sigma = 15$ points) to ALL Roads, etc., each agent (each road, each building) uses rules to dynamically find its own generalisation solution. Thus the agent may have a set of predefined algorithmic solutions (one could be (Gauss smoothing, $\sigma = 15$ points)), and chooses one at a time by means of its rules. In essence the actual generalisation process is not predefined.

At each step of the process, an agent chooses an operator to reduce its own conflicts and it triggers it. Then this agent checks if it succeeded in reducing the quantity of conflicts. If yes, it validates its new state and it continues. If not, it does not validate the last operator and chooses an alternate operator to try. This cycle continues until an acceptable state is reached. This is a state that minimises the conflicts or is a point where the agent is not able to improve itself given its choice of operator and its knowledge.

In order to generalise itself, each agent follows a template of predefined actions named the "generalisation engine" which is inspired from Simon's work on decision making and process design (Simon, 1977). He defined the classic steps of decision making as: intelligence, design, choice and review. The review by Pomerol and Adam (2004) describes each agent as a decision-maker that accepts sub-optimal or "satisficing" decisions. The generalisation engine is composed of the following decision making steps:

1. "Intelligence and Design": the agent computes its own conflicts, and finds (according to its procedural knowledge base – its intelligence) which algorithms might decrease its own conflicts;
2. "Choice": the agent chooses one operation amongst the set of selected ones and applies it;
3. "Review": the agent checks if its state is improved. If it is, it carries on (step 1 in Figure 14.2), if not, it chooses another solution (step 3 in Figure 14.2).

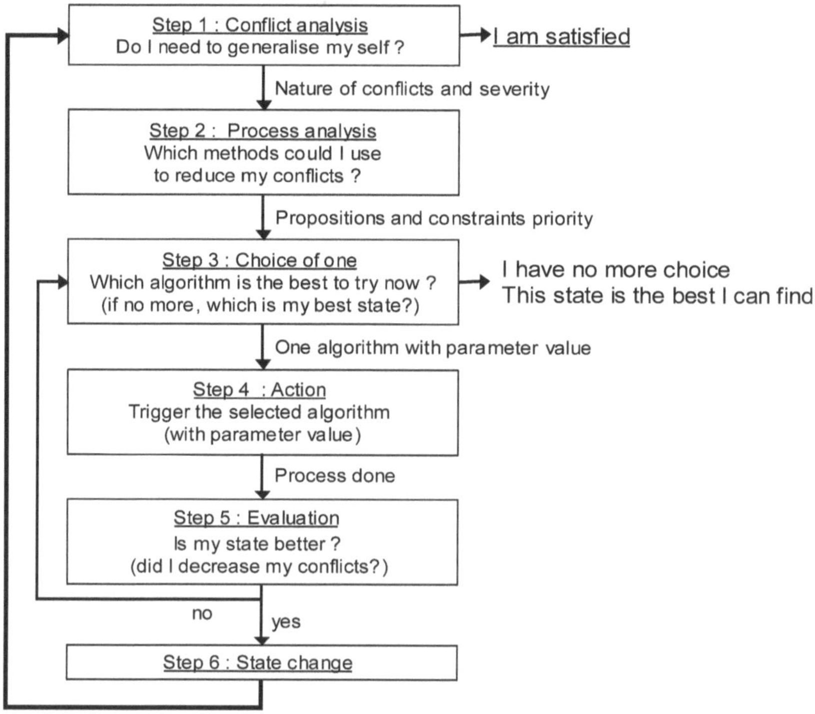

Fig. 14.2. A schematic figure of the AGENT generalisation engine.

For example, a building – which is initially too small and too detailed – improves its own state if after applying an algorithm, it is less small or less detailed.

As a consequence:

- This "*generalisation engine*" is common to all agents, irrespective of their nature;
- The *knowledge* used by the agents to detect their conflicts (constraints modelling) and to seek solutions (procedural knowledge) depends on the nature of the agent. Thus the set of constraints for buildings is different from the set of constraints for roads; the procedural knowledge for buildings is different from the procedural knowledge for roads;
- The *choices* of an agent are governed by an assessment of itself: each agent makes its own choices according to its nature, its own characteristics and its conflicts.

The generalisation engine is the core of the AGENT model. It is often termed "the agent life – cycle" (summarised in Figure 14.2). The engine allows a progressive generalisation of each agent by means of a succession of operators chosen and validated by the agent itself.

Some more functionalities added to improve convergence were proposed by Regnauld and involved the use of a hill climbing mechanism (Russell and Norvig, 2003). This consists of saving some intermediate solutions (at step 5 in Figure 14.2) – even if they are not ideal, and comparing them to other solutions found. In this manner, a solution that is identified as being sub optimal – but better than previous states – can be identified as the best solution found. Hill climbing strategy also prevents from testing all possible states in case of non-perfect results.

14.2.2. *Modelling agent behaviour with constraints and procedural knowledge*

To execute any action, an agent first needs to detect and to characterise its own conflicts and choose an action to reduce those conflicts. For the first step it uses constraints, for the second, it uses procedural knowledge.

Constraints

In the AGENT model, an agent generalises itself only if it does not comply with some constraints. The constraints correspond to the user needs and are translated into *functions* such as:

- Size (building) > 300 m^2 or
- $|$Shape (generalised-building) − Shape (initial-building)$| < \varepsilon$.

Each agent has the specific goal of satisfying its own constraints. Thus these agents can be seen as problem-solving agents. "Problem-solving agents decide what to do by finding sequences of actions that lead to desirable states" (Russell and Norvig, 2003, p. 59). In map generalisation, most of the time it is not possible to satisfy all constraints (Mackaness, 1991). Where a problem is deemed to be *over constrained*, it is necessary to relax a constraint (i.e., to ignore a less important constraint). However, in cartography it is generally better to improve each cartographic criterion instead of abandoning one of them. An alternative approach is to minimise the severity of a set of conflicts. The notion of severity is defined as a distance between an ideal state and the current state (Ruas, 1998b). Thus the goal of each agent is to reduce the severity value for each of its constraints (i.e., to improve its happiness step by step). One of the problems with this approach is that constraints are difficult to compare. It is necessary to normalise the severity values before evaluating the impact of an action on the severity of various heterogeneous constraints. In the context of this research, severity (the "quantity" of constraint violation) varied between the values of 1 and 5 (irrespective of the nature of the associated constraints). A novel aspect of the AGENT model was this qualitative and normalised representation of conflicts that facilitated the progressive (step by step) convergence of the agents towards a state where the conflicts were solved.

In terms of modelling, each constraint is a function (e.g. size (building) > 300 m^2) related to a feature class (the building) and a characteristic (the size). The threshold (the "goal-value" 300 m^2) can be set either at the beginning of the generalisation process or during the process according to specific analysis (for more detail see (Ruas, 1999)). Each agent computes its conflicts (step 1 Figure 14.2) and represents each conflict as a severity value associated with each characteristic. To model this, each agent is described by a set of "constraint objects". One constraint-object holds the "initial value" of a characteristic (the size), the "current value", and the "severity" computed from the current-value, the initial-value and the goal-value.

$$\text{severity} = \text{compute\_severity}_{\text{feature-type,characteristic}}(\text{initial-value, current-value, goal-value}).$$

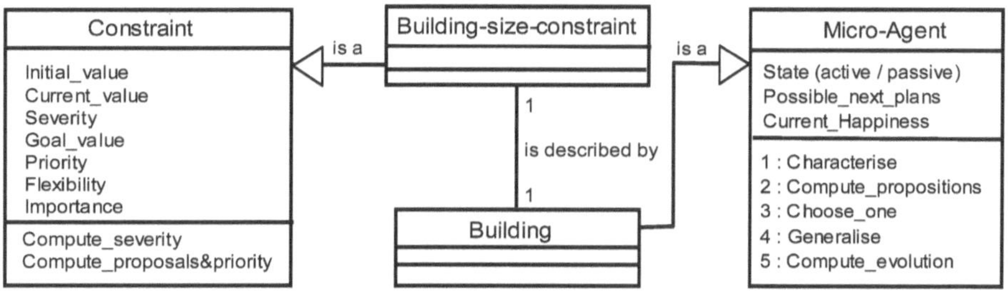

Fig. 14.3. Constraint and Agent Classes.

As an example, for a building – build.45, with an initial-size of 89 m^2 and a current-size of 270 m^2, the function would be:

$$\text{severity}(\text{build.45}) = \text{compute\_severity}_{\text{building,size}}(89, 270, 300) = 4\ (\geqslant \text{nearly good}).$$

Figure 14.3 shows a simplified version of the data model in UML (Booch et al., 1999) and includes the *constraint class* and the *micro-agent class:*

- A building *is-a* Agent so it inherits attributes and methods. These methods are the ones presented in Figure 14.2: they are the "generalisation engine" methods;
- A building-size-constraint *is-a* Constraint. It inherits attributes and methods.

Each building-size-constraint object describes the initial-value, the current-value, the goal-value and the severity-value of the size of one specific building agent (i.e. cardinality 1 : 1). It also computes the procedural proposals and the priority used by the building-agent to reduce its conflicts (i.e., to improve its current happiness).

Procedural knowledge

Procedural knowledge is the knowledge related to the use of procedures: the operators, the algorithms and the parameter values (Beard, 1991a). The procedural knowledge represents heuristics to solve conflicts. Examples of procedural knowledge (how to generalise) are given by McMaster and Shea (1988). It contains rules such as: for a given situation, you can try to use this algorithm with this parameter value to solve this conflict.

The method "Compute_proposals_and_priority" (step 2 in Figure 14.2 and the second method of the micro-agent class in Figure 14.3) proposes a list of possible operations that are supposed to improve a situation. The procedural rules are based on the severity value and each proposed algorithm is weighted (w):

if $\text{severity}_{\text{building-size}} = \text{high}$,
then proposals $= \{(\text{building-emphasize}, \lambda = \text{goal-value}, w = 0.9), (\text{building-removal}, w = 0.1)\}$.

The rule can also integrate some other criteria such as:

if $\text{severity}_{\text{granularity}} = \text{high}$ and if shape = not-complex,
then proposals $= \{(\text{Gauss}, \lambda = 10\ \text{pts}, w = 0,9), (\text{Douglas}, \lambda = 3\ \text{m}, w = 0,1)\}$.

or it can contain advice such as:

if $\text{severity}_{\text{granularity}} = \text{high}$, and shape = complex, then do not use Douglas–Peucker.

Wherever possible, the list contains more than one choice of algorithm, so that an alternative solution can be explored (Figure 14.2, steps 3 and 5). Often an agent has several conflicts at one time and ideally a set of condition rules could be used to prioritise those conflicts (Burghardt and Mathur, 2004). Unfortunately these rules are difficult to build, at least empirically. Moreover this solution is difficult to update as each introduction of a new constraint or a new algorithm in the system may require modification of all the rules. The alternative is to try to solve one conflict at a time and to use the generalisation engine to converge step by step towards an acceptable solution. In such a case, the agent needs to choose which conflict should be solved first. The process analysis step (step 2 in Figure 14.2) seeks to determine the best choice according to advice given by the constraints by means of the procedural knowledge. The attribute "priority" is used to help the agent select the best order. The priority value is a priority of treatment: if "priority = high" then this signals that the conflict should be solved as soon as possible.

The priority value changes according to the severity value. For example, if within an urban block the severity of the density is very high, and the severity of the proximity is also very high, then it is better to remove objects before displacing them. In the case where severity of the density is medium, it might be better to displace the buildings first. These heuristics are procedural knowledge that can be built either experimentally or by means of learning techniques. The first version of Stratège (Figure 14.1) contained knowledge derived from experiments and from the OEEPE test (Ruas, 2001). The distributed versions of the AGENT prototype and Clarity^{TM} package (Laser-Scan, 2005) contain their own procedural knowledge. The modelling of procedural knowledge remains an area of research in further need of study. Section 14.4 proposes further work on improving the quality of procedural knowledge.

14.2.3. *The specific case of meso-generalisation*

Generalisation is often described as a holistic process because a "good" generalisation requires us to take account of *all* information for *each* decision. Some authors have proposed the use of "information theory" to model the process (Bjørke, 1997), while others have used optimisation techniques to iteratively solve conflicts (Harrie and Sarjakoski, 2002). Some generalisation operators are performed on single objects (emphasising, simplification), others concern two objects (local displacement, aggregation), others concern a set of objects (displacement, removal) and the last set concern all objects of a type (semantic simplification of all buildings, removal of all small vegetation areas) (Regnauld and McMaster, this volume (Chapter 3)). One approach to automating this process is to explicitly represent these different levels as different kinds of agents. We use the following terminology (Ruas, 1998b):

- *micro-agents:* the agents that are only responsible for their own generalisation (e.g. building, road);
- *meso-agents:* the agents that govern the generalisation of a set of agents (e.g. a building block *composed of* buildings, a street network);
- *macro-agents:* the agents that govern or control the generalisation of a population of agents (such as "all buildings" or "all building-blocks").

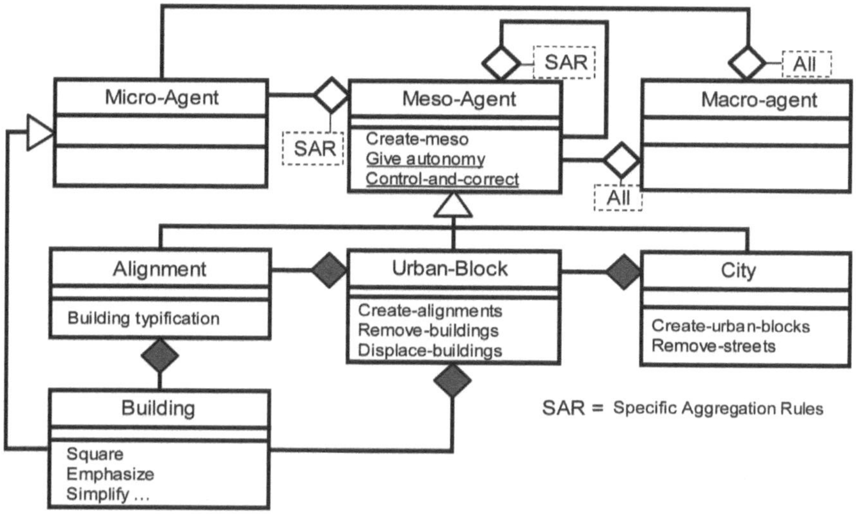

Fig. 14.4. Examples of Meso-Agents classes and their methods.

A *micro-agent* (such as a complex road) may temporarily become a *meso-agent* if it has to be segmented. In such a way the meso-agent road manages the generalisation of its parts by giving them their own autonomy. This idea of agent decomposition–recomposition was proposed by Duchêne during the Agent project.

An organisation based on levels of detail is very convenient and facilitates decision making. The challenge then becomes one of finding "appropriate and reasonable" groups of objects on which to perform the contextual operations. Some of the meso-agents have a well defined geographical meaning (such as a town, a street network or an urban block). They are created at the beginning of the generalisation process. Specific research has been undertaken to automatically detect and characterise such objects (Boffet, 2001). Other groups can be computed on the fly based on their spatial pattern (such as building clusters and building alignments (Ruas and Holzapfel, 2003)) or according to contextual and complex conflicts.

The meso-agents necessary to perform contextual generalisation are grouped into new geographical classes (City, Urban block, Alignment) which inherit from a *meso-agent* class (Figure 14.4). By means of inheritance, the geographical classes (such as the urban-block class) hold both the generic methods for the generalisation engine (Figures 14.2 and 14.3), as well as methods to activate their parts (give-autonomy), methods to control and correct the generalisation of their parts (control-and-correct) and specific generalisation methods related to their nature (see Figure 14.4).

14.2.4. *Benefits and limitations of the Agent approach*

The principles presented in §14.2 were first implemented on Stratège (to undertake urban generalisation), and then on the AGENT prototype including urban and road network generalisation, and later on different versions of this prototype. At the end of the AGENT project in 2000, the AGENT prototype contained about 25 generalisation algorithms, 30 measures

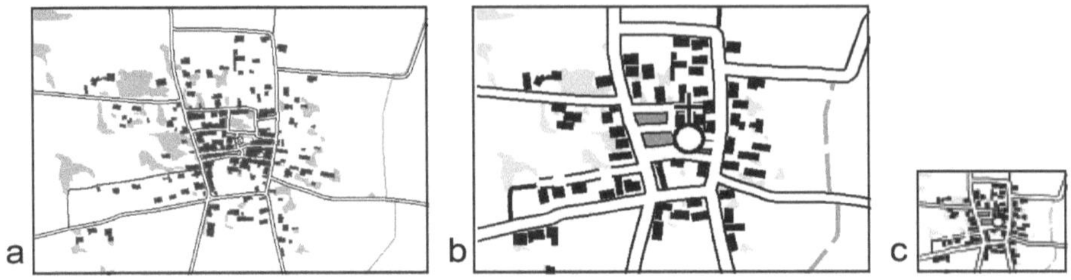

Fig. 14.5. Generalisation with AGENT (a: input, b: generalised for 1 : 50 000, c: b mapped at 1 : 50 000).

to compute the constraints and eight algorithms to control the process and to create meso-agents. Since 2000 the COGIT laboratory has continued to develop the prototype with new constraints, new algorithms and new meso-agents, such as those proposed by Boffet (2001). Improvements have also come from other IGN projects seeking to produce a 1 : 100 000 scale map from the IGN BDCarto© (10 m resolution DLM data base). Chapter 15, this volume, presents some of the results from this project.

In 2004, a new IGN project was launched to produce the 1 : 25 000 and 1 : 50 000 scale maps from the IGN BDTopo© (1 m resolution DLM data base). Studies are on-going to tune the system (to formulate the constraints and goal values) and to enrich the procedural knowledge. Figure 14.5 illustrates the first results based on the AGENT prototype, including urban block and building clusters. Current studies of this IGN project are being made on the Clarity™ Package and seek to take into account building alignments in order to obtain better generalisation results.

The AGENT model was developed to generalise urban areas and road networks (Figure 14.1). When the system was extended to manage the generalisation of rural areas, it became apparent that it was necessary to enrich the model in order to manage more efficiently the constraints between agents. This work is now described in more detail (§14.3).

14.3. Making the Agents Communicate

14.3.1. *Motivations*

In the AGENT model, every constraint involving two or more agents is handled by the meso-agent containing those agents. For instance, overlapping conflicts between buildings are handled by the urban block that contains those buildings. During generalisation agents interact hierarchically. In other words a meso-agent can give an order to one of the agents it contains. No interaction occurs between agents at the same level. This presupposes that no agent belongs to two meso-agents of the upper level – i.e., meso-agents at the same level are disjoint.

The AGENT model has shown to give good results, especially in the generalisation of urban areas and roads, where a hierarchical organisation can easily be designed. However the fact that agents cannot have transversal interactions has been identified as a limitation. This is especially the case in rural areas, where no clear hierarchical organisation of the space can be

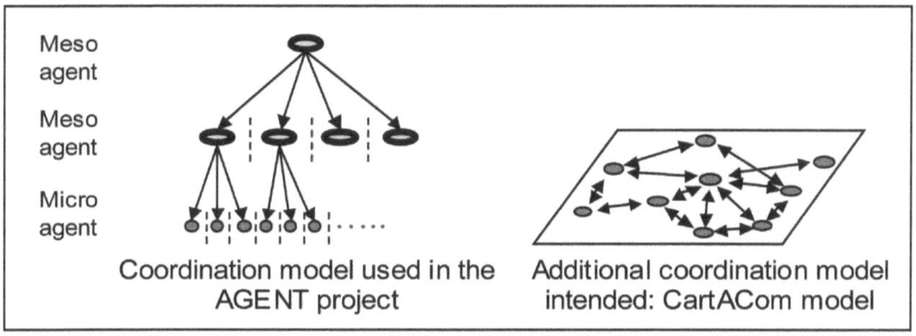

Fig. 14.6. The CartACom model enables transversal interactions between agents.

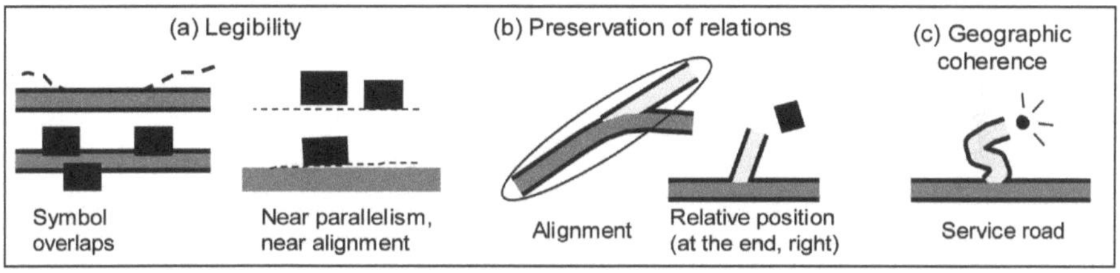

Fig. 14.7. Examples of relational constraints.

found. Thus a complementary model of coordination was designed, based on transversal communications between agents at the micro-level (Figure 14.6). This model is called CartACom (Duchêne, 2004). It was designed independently of the AGENT model, with the intention of combining them afterwards. Some plug-ins between them have already been designed and implemented.

14.3.2. *A model based on relational constraints*

The first principle of the CartACom model relates to the kind of cartographic constraints considered. The CartACom model is intended for geographic spaces that are not structured by obvious disjoint groups of objects, but where cartographic constraints are more local. Thus, in the CartACom model, we consider *relational constraints*. A relational constraint is a constraint on a relation between two agents. We identify three kinds of relational constraints (Figure 14.7):

(a) *legibility constraints*. These include the constraint that prevents symbols from overlapping, and constraints that prevent a relation from being almost present;
(b) *constraints of preservation*. These seek to preserve certain relations between objects;
(c) *constraints of geographic coherence*. These concern particular relations that make sense geographically speaking – where there is a meaningful association between features. For example the instance where a road leads to a viewpoint, the road should be kept unless the viewpoint is removed (Figure 14.7).

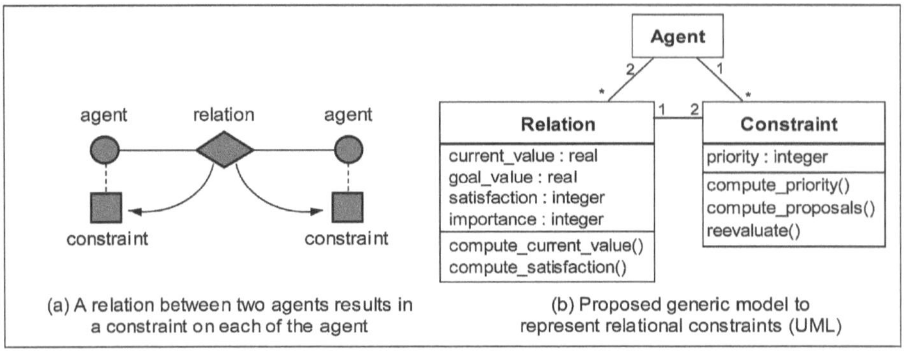

Fig. 14.8. Modelling a relational constraint.

The main difference between a constraint considered in AGENT and a relational constraint, is that the former concerns only one agent (be it *micro* or *meso*), whereas the latter is shared by two agents. CartACom provides a model for the representation of the relational constraints, based on the following idea: when a relation between two agents is constrained, it constrains the behaviour of both agents (Figure 14.8a). Thus we have adapted the model used in AGENT to represent constraints. In CartACom the representation of a relational constraint is split into two parts (Figure 14.8b):

- the first part is relative to the objective description of the *state* of the relational constraint, which is identical from the point of view of both agents and can thus be shared by them. This description is carried by a *Relation* object linked to both agents;
- the second part is relative to the *analysis and management* of the constraint, which is different for each agent and should thus be described separately. This part is described by two *Constraint* objects, each of them linked to the *Relation* object and to one of the agents.

14.3.3. *Communication model based on Speech Acts*

In order to identify and to assess their relational constraints, the agents are provided with the capacity to perceive their spatial environment, i.e. the surrounding space and the neighbouring agents (Figure 14.9a). However this kind of interaction – perception – is not sufficient to enable an agent to choose the right generalisation algorithms to apply to itself: an agent often needs to know not only about its own relational constraints, but also the constraints of its neighbours. This is why in CartACom the agents are also provided with capacities to communicate (Figure 14.9b).

The model used by agents for communication, based on the Speech Acts theory (Austin, 1962; Searle, 1969), has been described by Duchêne (2003). In CartACom, the agents can have conversations of two types:

- *request for action:* an agent asks another one to perform an action; this one either accepts and performs the action, or refuses;
- *information transmission:* an agent informs another one of a fact (e.g., it has just eliminated itself).

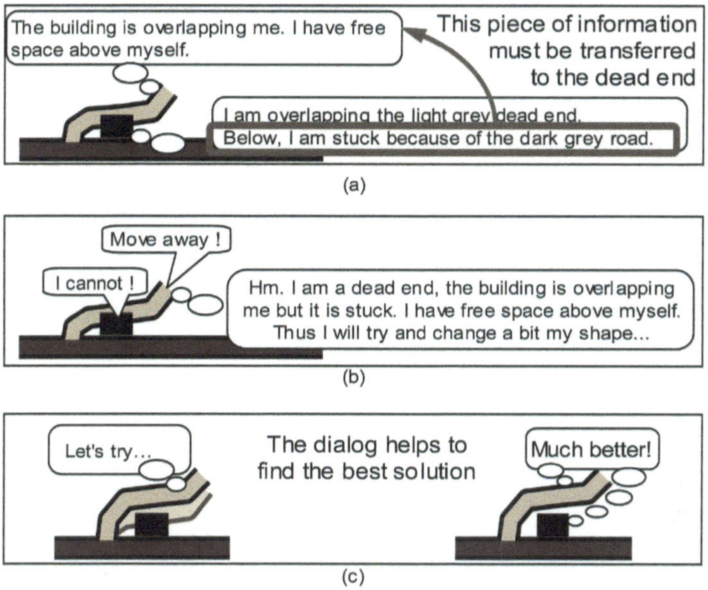

Fig. 14.9. The CartACom agents interact in two ways: by seeing their environment and talking to each other.

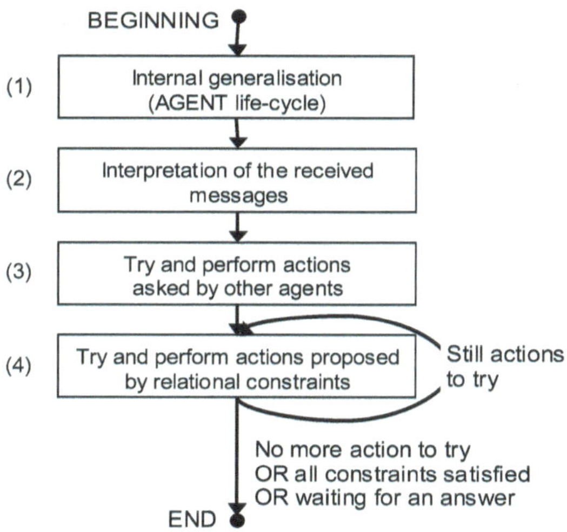

Fig. 14.10. Generic behaviour of a CartACom agent.

14.3.4. *Generic behaviour of a CartACom agent*

Figure 14.10 shows the generic behaviour of a CartACom agent. The aim of a CartACom agent is to try and satisfy its relational constraints. The behaviour (summarised in Figure 14.10), has similarities with the behaviour of an AGENT agent, the main difference being that an action is not just in response to the state of the constraints, but also to a request for action received from another agent. First, the agent performs its internal generalisation (1). This is achieved via a plug-in from the AGENT life-cycle (Figure 14.2), which is considered as a black box from the CartACom point of view. Then the agent analyses its received messages (2). This can lead it to modify some of its knowledge (e.g., if another agent informs it that it has just eliminated itself), or it can result in proposed actions. The agent then tries to perform the actions requested by other agents (3) and answers to its interlocutors following the success of those attempts. Then the agent tries to perform actions aimed at solving its relational constraints, according to the possible actions computed by these constraints (4). This stage is very similar to the AGENT life-cycle, except that an action can consist of asking a neighbour to undertake an action, in which case the agent has to stop and wait until it receives an answer.

14.3.5. *Results*

The CartACom model has been implemented on the GIS LAMPS2 and applied to the generalisation of topographical data stemming from BD TOPO – the 1 meter resolution database of the IGN. The geographic themes are roads, rivers, railways, buildings and land use. It is assumed that the buildings can move whilst respecting constraints of non-overlapping and relative orientation with the roads, rivers and railways. Buildings and roads perform their internal generalisation via a plug-in of AGENT, and land use follows the deformations of the roads to which they are topologically connected.

Figures 14.11–14.13 show results obtained with CartACom at three different scales (1 : 25 000, 1 : 35 000 and 1 : 50 000). In general the solutions appear to be good. However, in dense regions and at smaller scales, we start to see the limits of the CartACom model. In these cases it would be more appropriate to locally create AGENT meso-agents in order to use a group generalisation strategy.

14.4. New Research Challenges

The AGENT prototype (§14.2) and CartAcom (§14.3) coupled with various IGN developments, such as research on urban analysis (Boffet, 2001) and research on evaluation (Bard, 2004a) are being brought together on a new research platform called AGIT. In 2005, we migrated the AGIT agent engine from the AGENT prototype to the Clarity^{TM} package (Laser-Scan, 2005) in order to use a better and easier interface and programing environment.

Some very encouraging results have been realised since the commencement of this work in 1992. New theoretical solutions have been tested and validated at least on topographic data to produce medium scale maps (between 1 : 20 000 to 1 : 100 000). However, our objective is to provide theoretical solutions for *generalisation on demand*. The current prototype has demonstrated a significant reduction in production costs, and future plans will further this work in order to develop a system that is more flexible, adaptive and robust. In particular this research has helped to identify future research directions:

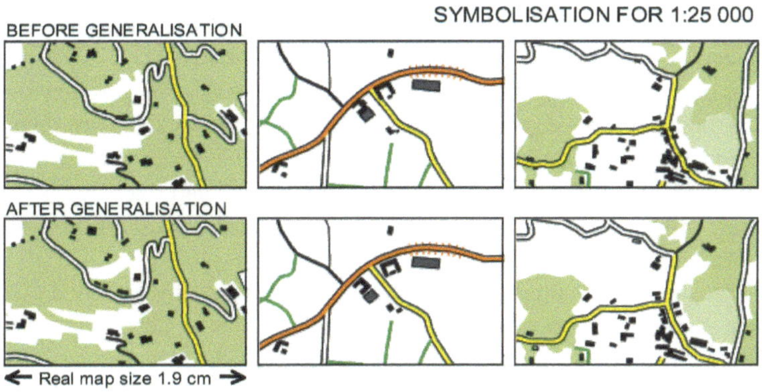

Fig. 14.11. Results obtained with the CartACom model for a display scale of 1 : 25 000 (display enlarged to 1 : 10 000).

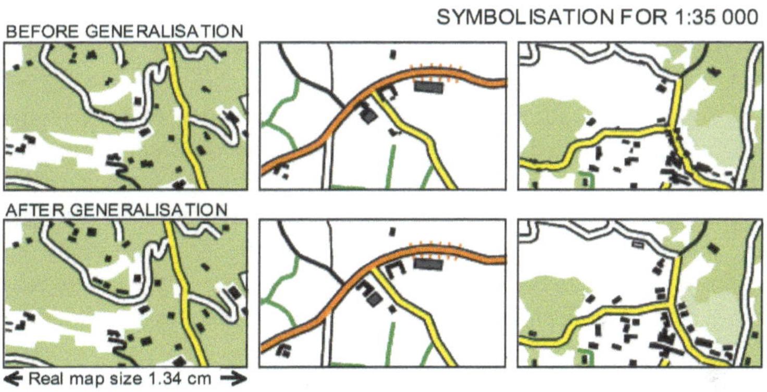

Fig. 14.12. Results obtained with the CartACom model for a display scale of 1 : 35 000 (display enlarged to 1 : 10 000).

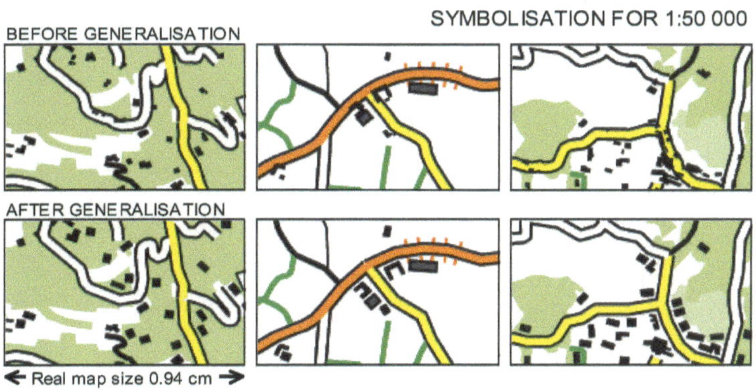

Fig. 14.13. Results obtained with the CartACom model for a display scale of 1 : 50 000 (display enlarged to 1 : 10 000).

1. A study examining closer linkage between the AGENT and the CartACom engines should be undertaken:
   - For example, when a set of agents cannot reach a good generalisation state in a rural area (via the CartACom engine), meso-agents should automatically be created as part of the continuing generalisation process;
   - Conversely, the CartACom engine could be used where an urban area is not very dense and where conflicts occur between two agents. Such interactions raise interesting research challenges;
   - The backtracking mechanism (such as the hill climbing technique) used to improve convergence should be equally applicable in both the Agent or CartACom engines.
2. More convenient methods should be found to improve the completeness and quality of knowledge contained within the system:
   - Early experiments made in 2004 proposed an interface (named Maacol) that supported the acquisition of expert knowledge: some samples were marked by an expert and linked to existing measures. This mechanism improved the quality of functions used to compute conflicts and their evaluation;
   - Further research is required to improve the knowledge contained in the procedural knowledge (§14.2.2). An early solution may consist of recording the activities of each agent to detect systematic errors, and thus allow revision of underlying knowledge. A new COGIT PhD started in late 2005 is focusing on knowledge revision.
3. Finally there is a need to model "secondary objects" in the generalisation process. These are "background" objects such as relief, or the vegetation areas on a topographic map. Secondary objects tend to be generalised late in the process or not at all. The challenge here is to find a computational way of modelling their role in the generalisation process. Here again a new COGIT phD started in late 2004 is focusing on mixing together solid and field object generalisations.

In order to allow an optimal integration of the different generalisation approaches including stretching operators such as the Beams proposed by Bader (2001), it is planned to revise the constraints and agent modelling proposed above (Figures 14.3, 14.4 and 14.8). Then, hopefully, we would be able to start working on the linkage between such systems together with a user friendly Internet based interface to support high quality on demand map generalisation.

Generalisation of Geographic Information: Cartographic Modelling and Applications
W.A. Mackaness, A. Ruas and L.T. Sarjakoski (Editors)

Chapter 15

Managing Generalisation Updates in IGN Map Production

François Lecordix, Cécile Lemarié

Institut Géographique National (IGN), 2-4 av. pasteur, F-94160 Saint-Mandé, France
e-mail: {Francois.lecordix,Cecile.lemarie}@ign.fr

Abstract

This chapter reviews practical experience gained from the application of generalisation technologies in the production environment of the French National Mapping Agency (IGN). Two main stages in production that require generalisation solutions are presented. The first step concerns the process of creating from a geographic database a first version of a cartographic database in which generalisation conflicts have been solved. This allows us to print a first map edition. This complex derivation step is made only once for a map which will be published over many years after updating. The second step focuses on the repetitive process to propagate updating in a cartographic database in order to provide new map editions with a less expensive solution than via a new derivation process. In this chapter, generalisation solutions to produce a 1 : 100 000 topographic map series are presented in detail and illustrated with a number of examples illustrating road generalisation. These solutions come from different research projects on generalisation. The benefits of such new solutions are discussed.

Keywords: generalisation, updating, label placement, map, Top100

15.1. Introduction

15.1.1. *Top100 maps from the IGN BD Carto® database*

The French National Mapping Agency (IGN) has decided to launch a new 1 : 100 000 scaled map series called Top100. These maps will be derived from the reference database BDCarto® which covers the whole French territory at a resolution of 10 meters. The "Carto2001, Cartographic Space Odyssey" project has consequently been entrusted with designing a process leading to the derivation and updating outlines of the future 1 : 100 000 IGN topographic map series using the IGN BDCarto® database as a unique data source. The main challenges identified were (1) automated generalisation, (2) label placement and (3) an updating process.

The Carto2001 project is now complete. All solutions presented in this article are results from that project. A first map prototype has been made using the methods described in this

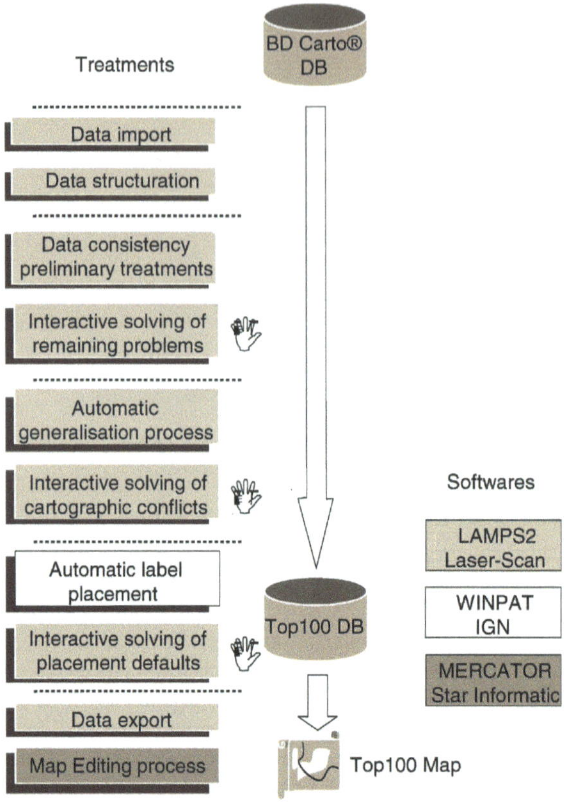

Fig. 15.1. Key stages in the production of the first map series.

chapter and the production of the Top100 map series has begun at the IGN. The first derivation outline consists of different treatments used to transform the geographic database BDCarto® into a cartographic database Top100 using GIS LAMPS2 (Laser-Scan, 2005). These include data import and the creation of a topologically correct database, data consistency tools to merge data in different layers (for example roads and administrative limits), and generalisation to modify the geometry of elements to solve legibility constraints. Label placement is another important step in creating a cartographic database from a geographic one. This is done in part automatically with the IGN software WINPAT. The cartographic database Top100 allows the easy exporting of data into MERCATOR (Star Informatic, 2005) in order to edit Top100 paper maps (Figure 15.1).

15.1.2. *Modelling updating constraints as part of the derivation process*

A fully automatic process for derivation is not yet achievable. To update the map, it is not worth deriving again the full cartographic database from the geographic database. Therefore an incremental updating process was needed to propagate the geographic database updates in the cartographic database and to obtain the next map editions (Figure 15.2). A major economic

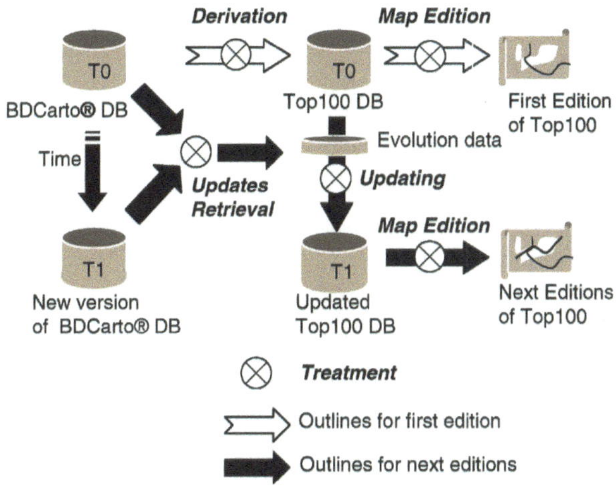

Fig. 15.2. Derivation and updating outlines.

objective of the Carto2001 project was to drastically decrease production costs by automating the incremental updating process by using source database evolutions. The evolutions describe the source database modifications between two moments: creation, deletion, geometric or semantic modifications, fusion and splitting. It is important to emphasise that, in this chapter, the update process concerns only the propagation of updating from the geographic database BDCarto® to the cartographic database Top100. The process of collecting and updating information from the field and to digitise it in the geographic database is quite a separate matter, and not discussed here.

To automate the incremental updating process, a unique numerical key is generated and assigned to each object in the BDCarto® geographic database. This numerical key is preserved during the derivation process. At the point of updating (creation, deletion or modification) the numerical key is the basis by which modifications are managed. A second requirement of the automatic updating process is to be able to automatically generalise updated data. For this reason the Carto2001 project developed generalisation tools specifically for the derivation process that were as automated as possible. A description of these tools will be detailed in the first part of this chapter, and the second part will explain how these generalisation tools were used again in the updating process.

15.1.3. *An industrial solution*

The Carto2001 project was launched after ten years of research in generalisation, label placement and updating methodologies carried out at the IGN COGIT research laboratory. It can be considered as one of the first implementations that brought together these three research areas. The Carto2001 implementations used mainly COGIT laboratory research. The emphasis was on producing industrial solutions that were robust (thoroughly tested), predictable, and efficient. The Carto2001 project improved upon solutions proposed by the COGIT laboratory and provided a framework in which a first Top100 map prototype was produced. Given the applied nature of this work, the Carto2001 project did not attempt to examine and compare all research solutions but instead focused on research output from the COGIT laboratory. But other references pertinent to this work include: the difference between geographic and cartographic databases together with the concept of Digital Landscape Model (DLM) and Digital Cartographic Model (DCM) (Brassel and Weibel, 1988); and incremental updating (Kilpeläinen, 1995; Hampe et al., 2003). To implement generalisation and updating solutions, the Carto2001 project used the LAMPS2 GIS. This choice was mainly guided by the automatic generalisation solutions already implemented on LAMPS2 during the AGENT project (Ruas, 1999; Lamy et al., 1999; Barrault et al., 2001). This GIS also provides an environment in which to model the updating process:

- LAMPS2 manages updates by creating versions, so during an updating process it is possible to query the database not only in its current version but also in its older states. It is then very easy for example to cancel an update and retrieve the initial state of an object;
- LAMPS2 relies on a robust and efficient Database Management System (DBMS) that is able to store and handle the whole French territory in a single dataset (thus avoiding the need to make multiple updates across a number of datasets);
- LAMPS2 implements Object Oriented concepts which allow users to easily define the behaviours of any cartographic object. This aspect has especially been enhanced in LAMPS2 by means of a particular type of method (called a reflex method), that are automatically triggered by the system when objects are involved in a process. This facilitates the handling of interactions between objects (especially useful in the updating process);
- LAMPS2 manages topology;
- LAMPS2 can be used in a multi-user environment which makes the simultaneous updating of different areas possible.

15.2. Map Generalisation in the Creation of the First Edition

15.2.1. *Generalisation requirements*

The Top100 cartographic database is derived from the BDCarto® database – a 10-meter resolution database originally designed for 1 : 50 000 scale map production. The BDCarto® contains basic topographic themes: roads, rivers, railways, administrative limits, land cover, relief details as well as tourist information and labels, which will all be used in the Top100 map (a sample is presented in Figure 15.3a). It is important to note that buildings, as elementary objects, are not present in the BDCarto®; and the related generalisation problems have not been examined by the Carto2001 project.

Two distinct issues need to be considered in the generalisation process (Jahard et al., 2003):

- **Independent generalisation**: Independent generalisation resolves internal conflicts arising from coalescence and “noise” within an object. This addresses challenges in the representation of mountain roads for example.
- **Contextual generalisation**: Contextual generalisation is where the object is considered in its environment and in response to symbol widths and local densities. It is sometimes necessary to displace objects to preserve a minimal distance between them or to ensure the

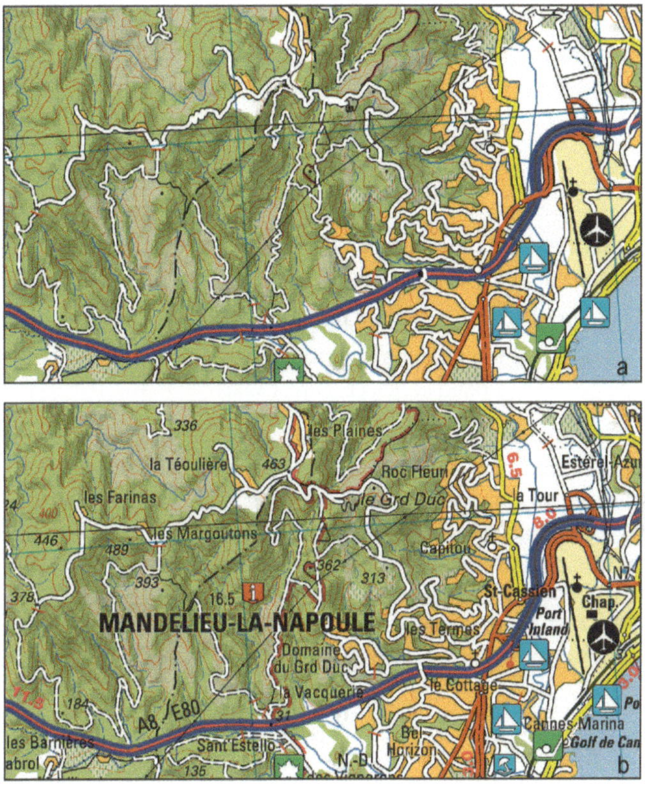

Fig. 15.3. (a) BDCarto database (at a scale of 1 : 100 000). Bend coalescence and overlapping symbols are visible. Text labels are attributes of the objects and have no absolute position. (b) Top100 map after generalisation and label placement.

legibility of roads junctions. Contextual generalisation is used to separate rivers which flow close to road symbols and to process road and railway networks.

In designing the derivation process it was important both to develop generalisation tools that could support a high level of automation but also easily detect the specific cases in which generalisation tools were not sufficiently efficient. Figures 15.3a and 15.3b are examples from the first prototype map produced by the Carto2001 project and illustrate the difference between the initial data, BDCarto® and the derived data, Top100 map.

15.2.2. *Conflict detection*

Automatic detection of cartographic conflicts is an important prerequisite to generalisation automation or to guiding interactive generalisation and editing. The European project AGENT used LaserScan's LAMPS2 system as a platform for implementing a whole range of measures to detect cartographic conflicts such as bend coalescence, overlapping symbols, and oblique junctions. These measures have been re-used by the Carto2001 project. These tools are essential to reducing the interactive component and to optimise the level of automation. The experience gained with the first Top100 map prototype produced by the project Carto2001 highlighted the efficiencies of the AGENT system in detecting cartographic generalisation conflicts.

15.2.3. *Independent generalisation of roads using agent based methodologies*

The derivation of BDCarto data into map objects at the 1 : 100 000 scale required solutions to intra-object conflicts to be found:

- To remove tiny details which would be viewed as noise to the naked eye;
- To "open up" some bends to make loops more legible;
- To remove some bends in a bend series (Figure 15.4a)

In AGENT (Ruas and Duchêne this volume (Chapter 14)), each road is modelled as an agent. If the degree of coalescence is unacceptable then the road will be divided up into several parts, each one with a homogeneous coalescence value. For each section in turn (in the form of an agent), a variety of solutions are explored, and the one that delivers the best transformation is chosen. The resulting road is the best that can be drawn with the available measures and algorithms (Figure 15.4b). AGENT is used for independent generalisation on each object separately. The AGENT solution for resolving bend coalescence conflicts was excellent with a success rate of 99%. The unresolved conflicts were detected and passed to the user for interactive correction.

15.2.4. *Contextual generalisation using elastic beams and flexibility graphs*

Initial tests using AGENT revealed that the agent technique was not able to solve overlapping symbols. The cartographic results were not always good and too few conflicts were solved. Moreover, the processing time was not acceptable, even for small sets of objects. Each time an object was displaced, the other objects re-evaluated themselves to check whether the displacement was acceptable. The problem came not from the agent technology but from the displacement algorithms used by AGENT. Bader (2001), developed work based on "*elastic beams*" – another type of displacement algorithm which follows an optimisation approach

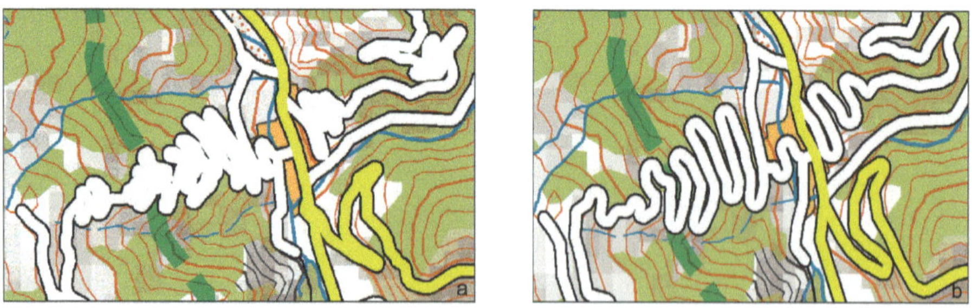

Fig. 15.4. (a) BDCarto data (initial state) and (b) result after application of the AGENT process.

and differed from the sequential techniques used in the displacement algorithm developed in AGENT.

The beams approach consists of defining internal and external forces on each of the objects' vertices (Figure 15.5a), and in finding the minimum energy for this system. Internal forces result from bend coalescence and external forces from symbols overlapping. Objects will move according to the resultant force with varying degrees of resistance, depending on their ability to be distorted, compressed or extended. This technique which brought a noticeable improvement for road displacement (Figure 15.5b) was developed by LaserScan at IGN's request and the "moveability" of objects was set up by the Carto2001 project. The algorithm produced very pleasing results, both in highly crowded areas and in specific configurations such as interchanges (Figures 15.6). The advantage of beams is that it models compromise by taking into account all the various "forces" acting to improve the legibility of the symbology.

Once the cartographic quality of beams had been recognised, the next problem was an operational one since the technique is extremely computationally intensive. It is not possible to launch beams on a large area. Two approaches were developed to overcome this problem:

- The first one was to segment data from the network of main roads and to process beams per partition. However in this solution, conflicts which are often on the main roads are not processed in the same way in the left-hand and right-hand partitions. The partition limits become visible on the map in the form of a break instead of a nice continuous displacement.
- The second approach was to detect all the conflicts and for each one to define the set of roads which could be displaced to solve the conflicts. This set of roads is called the **flexibility graph**. It represents the extent of the conflict in terms of the displacements needed to resolve it. These graphs are then aggregated: if a road belongs to two graphs, these two graphs are aggregated into one. Beams are then launched on the set of roads of each flexibility graph. Conflicts are detected with the available AGENT measures. This was the chosen solution in the final implementation.

The "flexibility" of a road is a measure of its sinuosity. The more bends the road has, the more flexible it is deemed to be. During displacement, the more flexible a road is, the faster the displacement will be cushioned and "absorbed" into the surrounding region. The first stage of the analysis is to determine from an object in conflict (the source object) the set of objects which are required to be displaced. A geographic object belongs to the flexibility graph of a source object if it is possible to find a path from the source to the geographic object with a flexibility path less than a given threshold. The flexibility path from a source object to a destination object is given by the following formula:

$$\text{Flexibility path} = \min\left(\sum_n 2^{n-1} F_n\right),$$

where F_n is the flexibility of each road and n the number of roads which have already been traversed prior to considering the current road. In Figure 15.7, the flexibility path from the source to the destination object is $F1 + 2 \times F2 + 4 \times F3 + 8 \times F4 + 16 \times F5$. The flexibility graph of a source object can be seen to be the object which can be reached in a given time if

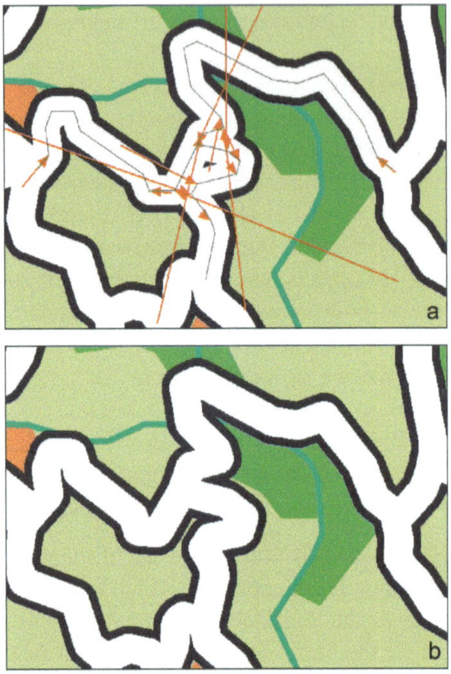

Fig. 15.5. (a) Application of beams forces (in orange), and (b) resulting solution.

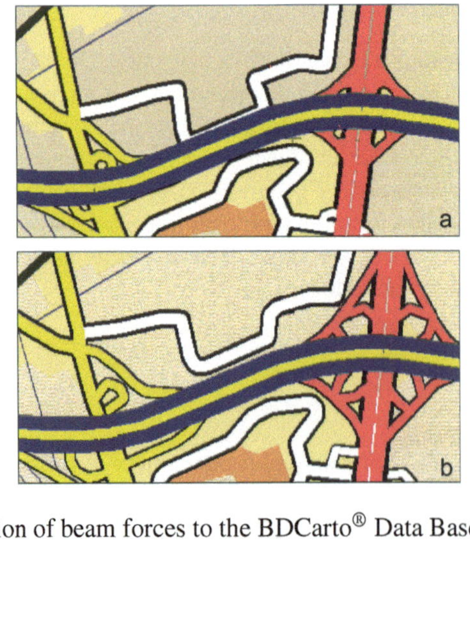

Fig. 15.6. (a) Application of beam forces to the BDCarto® Data Base, and (b) resulting solution.

we consider that the time to reach an object depends on its flexibility and on the number of previous roads.

With this formula of the flexibility path, the farther an object is, the more by which flexibility of this object will be increased. This solution allows for an increasing of the cushioning with distance. The flexibility graph of a source object is then composed of all the objects relying on a flexibility path which has a total flexibility less than a given value (the value of the parameter depends on the type of conflict being solved). Once the flexibility graph has been calculated for each object in the conflict, these graphs are merged so that each object occurs in only one graph. Beams are launched on each merged graph. Facilities have been added by Laser-Scan to freeze some objects, for example "land/water borders", to prevent displacement of roads into seas or rivers. Figure 15.8a is a sample of source data. In Figure 15.8b, the roads have been displaced without encroaching upon the sea.

95% of symbol overlaps were resolved by combining beams with the flexibility graph strategy. The processing time is quite long (30 hours on a Pentium IV for a sheet size of 96 cm by 121 cm), but this was not deemed to be a problem and the cartographic results are very satisfying. This compares favourably with an interactive manual process that typically requires several hundred hours of interactive work.

15.2.5. *Maintenance of data consistency*

To preserve the "geometry sharing" between objects, Laser-Scan developed a "diffusion" function which allows diffuse road displacements on objects that are connected or share the same geometry. This function has been interfaced with the AGENT and beams tools. We have used this tool to maintain relative positions between the road network and isolated features by adding a fictitious line object which links an isolated object to the road network. In Figure 15.9a the isolated object (circle) is linked to the track by a fictitious line and in Figure 15.9b, after the track generalisation (smoothing), the isolated object can move thanks to the fictitious line. At the end of the generalisation process, data in the LAMPS2 database are still in a topologically coherent state. This topological consistency is integral to the map updating process.

15.2.6. *Benefits from the derivation process*

In 1998, the IGN tried to produce the Top100 series from the BD Carto® database and undertook tests to create a prototype without automatic tools. It was estimated that generalisation and label placement would have taken more than 2000 hours of interactive work for each map and that the process for complete derivation of the first edition would have taken approximately 16 months for each map. With the solutions introduced by the Carto2001 project, the automatic generalisation takes 50 hours of computational time for the whole sheet and the interactive re-working required to finish the generalisation of the whole sheet takes a further 100 hours. The time required to carry out the same generalisation of the whole sheet using solely an interactive system amounts to about 1200 hours.

The Carto2001 project has introduced another important innovation in label placement utilising the WINPAT software developed at IGN (Lecordix et al., 1994; Barrault, 1998). Automatic label placement takes 12 hours of calculation for the whole sheet. Interactive placement

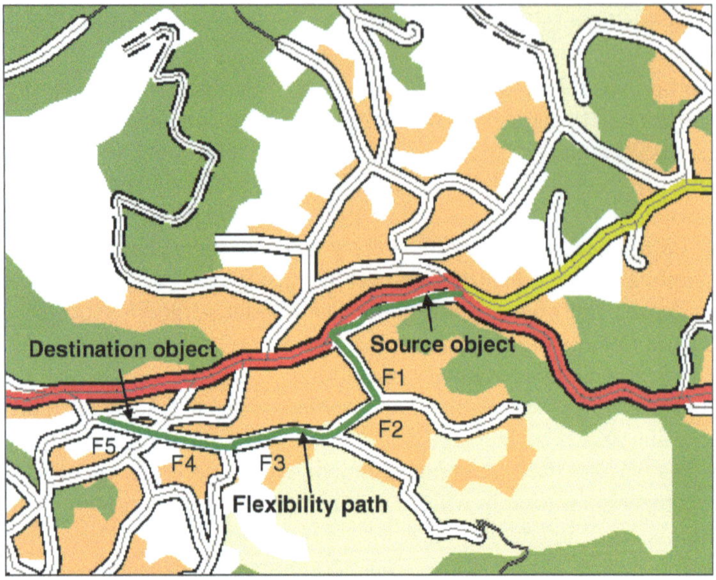

Fig. 15.7. Determination of the flexibility path.

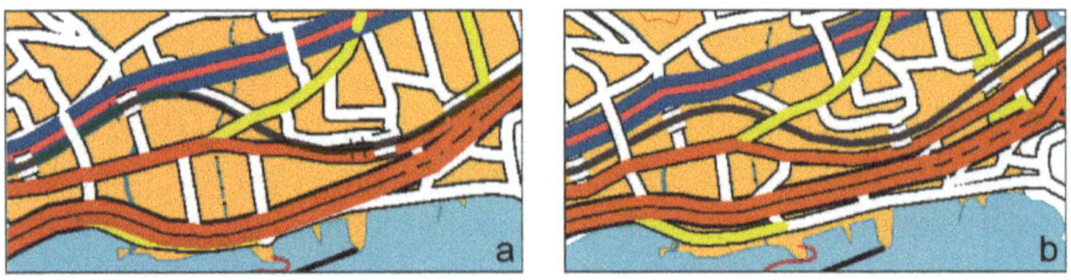

Fig. 15.8. (a) BDCarto data (initial state) and (b) result of applying "beams" without displacement of roads towards the sea.

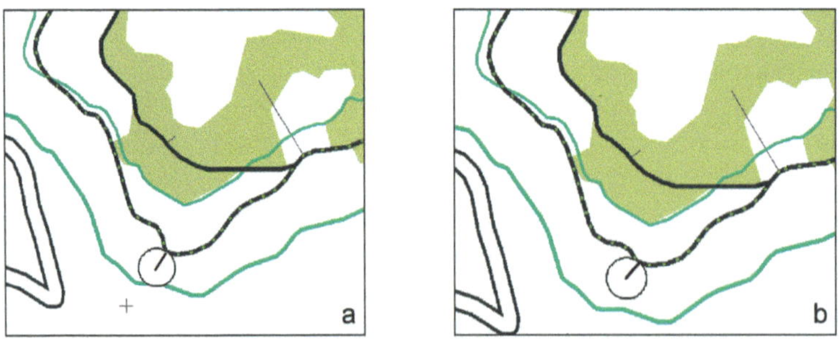

Fig. 15.9. (a) BDCarto data (initial state) and (b) Top100 Data Base with the preservation of consistency between the road network and isolated points.

of text that was missing from the BD Carto® database or that could not be placed automatically takes 160 hours. The time required to carry out all the placement of all labels using a purely interactive system amounts to about 800 hours. At the moment, only one map prototype has been drawn by the project, but the first estimation of derivation time for any sheet in production is 6 months. Despite these efficient innovations on generalisation and label placement described here, the derivation process is still a time-consuming process (6 months for one sheet). The need for further efficiencies highlights the need for development in updating processes to create revised editions and the need for incremental generalisation techniques (Kilpeläinen, 1995).

15.3. Incorporating Generalisation in the Updating Process

15.3.1. *Updating*

Automatic map updating is another highly innovative area in which the Carto2001 project is at the forefront of technology. It requires two specific mechanisms when the initial database is set up:

- **Digital signature of items**. This technique is used to associate a unique identification (a numerical key) to each item in the geographic database, which is deduced by the MD5 encoding algorithm from the attributes and shape of the item. The digital signatures were developed at IGN by the "Distribution of digital data through networks" project and were applied to the BD Carto® database.
- **Updating retrieval**. Also developed by the "Distribution of digital data through networks" project based on research carried out at IGN COGIT laboratory (Badard, 1999, 2000; Badard and Richard, 2001), this technique is used to recover data that are modified in the archived database between two dates, using matching algorithms. This technique provides detailed updating information for each updated object. Each object is defined by its updating operation: creation, destruction, fusion, aggregation, splitting, geometric modification, semantic modification and relation modification.

Updated data are identified and extracted for all map themes such as roads, rivers, and administrative boundaries. They are extracted from the BDCarto® database and must then be integrated in the Top100 database taking generalisation into account.

15.3.2. *Updating in the data integration process*

Digital signatures on updated data and updated retrieval facilitate the introduction of updated data into the generalised database where the digital signature of each object is registered. The updating process has been developed in LAMPS2 and can be decomposed into two software components which interact:

- **The updating engine** implements a set of functions called in a predetermined order. This part is not specific to the reference database and to the cartographic database, and can be used for the updating of other maps series;

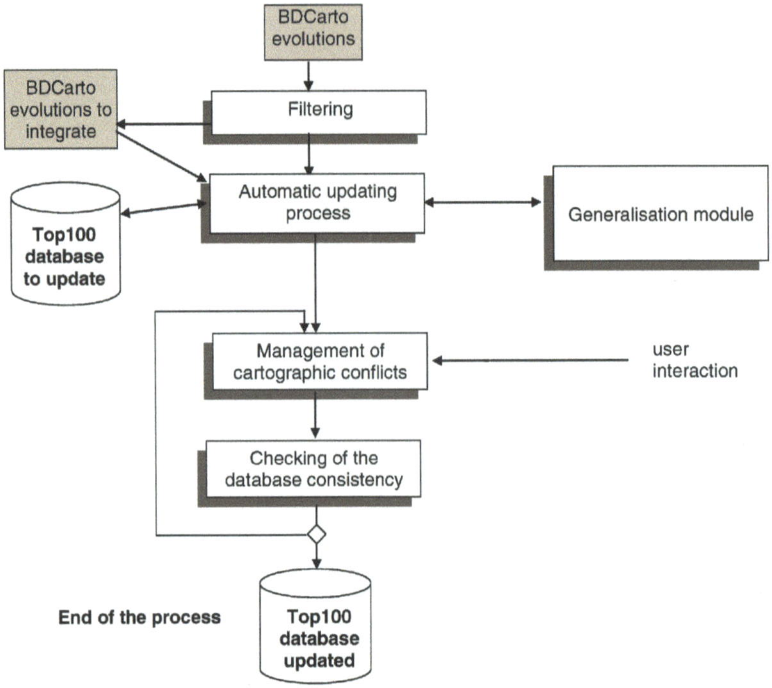

Fig. 15.10. Summary of the key stages in the updating process.

- **The updating rule base** allows for the definition of behaviours to execute when a particular operation is performed on a specific object. For example symbol modification if the update is a semantic modification, or label modification (font, size, text modification, etc.). These rules express the cartographic consequences of updating. The updating rule base is taking advantage of the object oriented concepts implemented in LAMPS2 and is coded in the form of LAMPS2 reflex methods. This part is specific to the reference and cartographic databases.

The updating engine enables the integration of updates and assumes that objects are modified without creating any conflicts in the database and without altering any of the topological constraints. The updating rule base deals with the cartographic aspect of the database, by controlling the effects of the updates on the cartographic objects. There are various interactions between these two components, but they are mainly managed by the LAMPS2 engine. Each time an object is modified by the updating engine or by users, LAMPS2 automatically refers to the reflex method (call-back mechanism) that implements the updating rules. The updating engine is comprised of four main steps (Figure 15.10): filtering, automatic updating, interactive conflict solving and consistency checking (Lemarie and Badard, 2001).

Filtering step

The cartographic database is a generalised image of the reference database, and some modifications performed in the reference database are not always legible on the map. In the filtering step, BDCarto® evolutions are analysed: updates which are not "cartographically" legible are ignored while the other updates are imported in the database (in particular classes). Criteria involved in this filtering step include: minor modification of the geometry (i.e. distance of positions less than some threshold), and updating of attributes which are not used for the cartographic database. At this step in the updating process, about 30 to 50% of the geometrical updates can be ignored. Even if a geometric modification is defined as not being relevant for the Top100 database, the numerical key is nevertheless updated to preserve the link between the BDCarto® database and the Top100 database.

Automatic updating step
Automatic updating consists of the integration of updates and the propagation of their effects in the cartographic database. Updates are processed in a defined order to avoid incoherent states: for example, deletions are integrated before any other updates. For each update, the geographic objects involved in the modification and the methods for maintaining (at each step) the consistency of the database are identified. Updating rules are then automatically called by LAMPS2 to define the effects of the update on the cartographic aspect of the object (symbol change, symbol orientation) and on the surrounding objects (road number, bridges, road lengths, etc.). Propagation of updates on surrounding objects cannot always be defined during the updating of an object since the context surrounding the object may not be up-to-date. Therefore for specific treatments (road numbering for example where the number is placed according to a set of road sections) the effects of updates are calculated at the end of the process (when data directly corresponding to the reference database are at last up-to-date). The two main characteristics of the process are as follows:

- The consistency of the database is always maintained – the BDCarto® geometry on each object of the Top100 database is stored. This geometry is used to maintain the connectivity of the network and to ensure that an updated object is correctly connected to an existing part of the network. An update can thus be integrated in the cartographic database only if one of the objects connected to the updated object is already in the cartographic database and up-to-date. This constraint is the easiest way to preserve a correct network topology.
- The previous generalisation process is used for the integration of updates. To avoid local distortion and topological errors, it is necessary to replicate the existing displacement in the BDCarto®/Top100 on the updated object.

Conflict solving steps
The third step consists of solving conflicts. All the propagation effects cannot be fully defined through the updating rule base, so during the automatic step, objects which cannot be automatically processed are "marked" with a description of the problem. The main point in this interactive step is that all the updating rules are still automatically triggered by the LAMPS2 engine: the cartographic processing is indifferent as to whether the objects are updated by the automatic process or by the operator.

Consistency checking step
The last step is consistency checking. This step ensures that the updates are correctly integrated and propagated: all the evolutions have been taken into account, and topology is in

accordance with BDCarto®. The versioning mechanism of LAMPS2 is then used to check the consistency of updates. For example avoiding deletion of an object which has not been removed in the BDCarto®, or checking that the modification of a numerical key has taken place in order to preserve future updating.

15.3.3. *Cartographic conflict detection in the updating process*

Once updates have been integrated in the dataset, cartographic conflicts are evaluated on the objects which have been created or moved. Cartographic conflicts remaining on objects which have not been affected geometrically by the updating are then no longer taken into account: they are supposed to have been validated by the operator in the previous map version. This process uses the same conflict detection tools as the derivation process but is launched only on geometrically updated objects.

15.3.4. *Independent generalisation with updating constraints*

The AGENT tool is used to resolve the internal conflicts of coalescence and "noise" in road objects, but taking care not to move the end points of the road. Independent generalisation methods are used to calculate the generalised geometry of the object to avoid coalescence conflicts and noise. This solution is not applied wholesale, but is readjusted so that the extremities of objects are not moved. This approach avoids large diffusions that could create conflicts in areas which have not been updated. It also prevents objects from overlapping with labels or indeed moving too far away from their labels.

15.3.5. *Contextual generalisation with updating constraints*

A requirement of the contextual generalisation is that it avoids moving too many objects. Once conflicts have been calculated on the updated data, a buffer search is used to identify other objects nearby. Flexibility graphs are then calculated and beams launched on each merged graph. This solution is very efficient because updates are usually localised and it is not necessary to take into account lots of objects in order to solve the conflicts.

15.3.6. *Maintenance of consistency*

To maintain consistency, updates use the same diffusion tool as in the derivation process but some post treatments are required once automatic updating has been performed, for example in order to detect objects which used to share their geometry and no longer do so. Furthermore the updating process has to take into account problems that could not be resolved by the derivation process because these objects were created afterwards. The updating process has to take into account consistency of label placement, and check that the distance between the geographic object and its placement is not too great, and that labels are not overlapping with important cartographic details.

15.3.7. *Benefits from the updating process*

The updating process described here has been applied to the first edition of the whole map produced by the derivation process. The map reflects changes in the BD Carto® database over

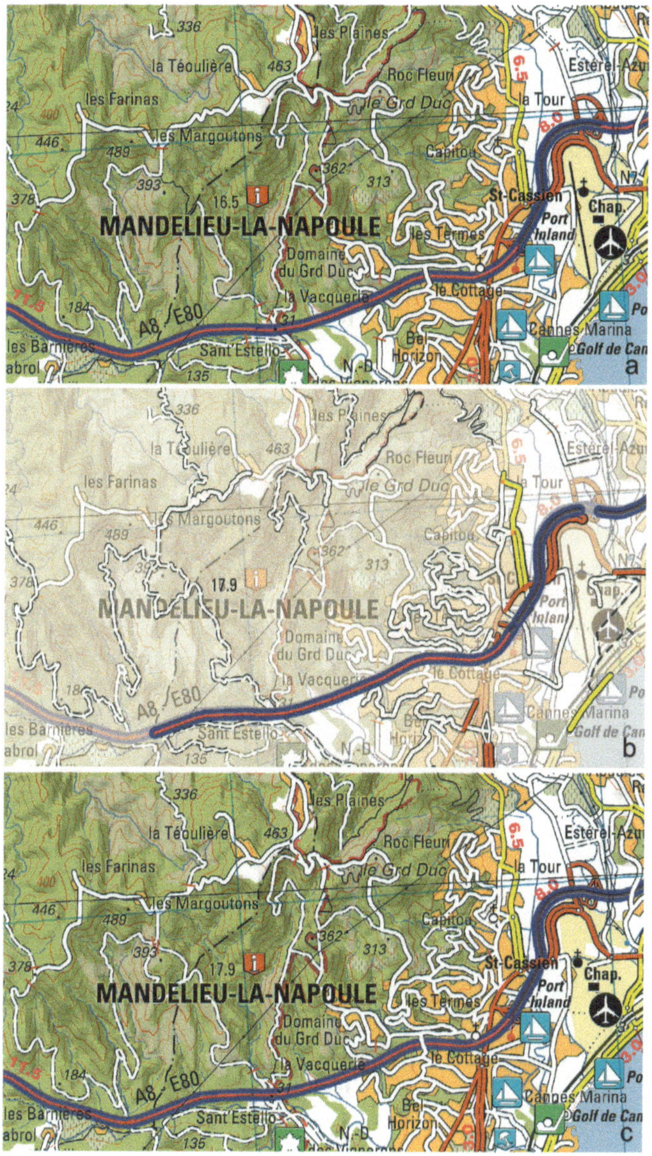

Fig. 15.11. (a) Top100 first edition (scale 1 : 100 000), (b) Updated data (drawn without the town drawn) prior to integration in the first edition, (c) Top100 database after automatic updating. Updates are integrated with generalisation into the map whilst taking into account the generalised elements.

a two and a half year period. Figure 15.11a shows the result obtained after the first derivation. Figure 15.11b shows the BD Carto® difference data (without generalisation) overlaid on a lightened copy of the previous edition of the basic map. Figure 15.11c shows the new map after automatic updating, and automatic generalisation but before any interactive retouching.

The new edition of the whole Top100 map was obtained after 10 hours of automatic processing including generalisation of updated data and 50 hours of guided interactive editing. This is to be compared with the current production process that IGN uses to update a similar map at the same scale and with the same extent: namely 300 hours for each sheet. Thus it can be seen that the Carto2001 project has provided an updating process that decreases costs drastically.

15.4. Conclusion

The Carto2001 project has provided new cartographic techniques for generalisation, label placement and map updating. These techniques have migrated from research and been applied in a production environment. A first map prototype has been produced with significant cost savings, even where interactive editing is still required to solve difficult cases (particularly for label placement during derivation and updating). These new solutions are so efficient that IGN decided immediately to use the incremental updating process described here for another series at IGN (the Departmental Map: 1 : 140 000 scaled roads map series issued from BDCarto®). In the future, these solutions and new research on the generalisation of buildings will be used to produce simultaneously new 1 : 25 000 and 1 : 50 000 scale maps issued from IGN topographic database BDTopo®.

W.A. Mackaness, A. Ruas and L.T. Sarjakoski (Editors)

Chapter 16

Automated Generalisation in a Map Production Environment – the KMS Experience

Peter West-Nielsen, Marlene Meyer

Cartographic Department, Kort & Matrikelstyrelsen, Rentemestervej 8, DK-2400 Copenhagen NV, Denmark
e-mail: {pw,mlm}@kms.dk

Abstract

This chapter gives an overview of the development of a small-scale map production system at the Danish National Survey and Cadastre (KMS). The current production system is based on automatic generalisation combined with manual editing. This chapter describes the development of methods applied and the various considerations relating to the creation of the derived databases. In the set up of the production line it was first necessary to find solutions to a broad range of technical problems and to take a pragmatic view of resources, technology, and cartographic traditions. Despite all of these challenges, this chapter demonstrates, that it is possible for KMS to employ automated generalisation techniques in both the production of printed maps and creation of digital multi-scale databases.

Keywords: map production, national mapping agency, GeoDB

16.1. Introduction

KMS is the Danish governmental organisation responsible for mapping and charting, for cadastral registrations and the authorisation of licensed surveyors. KMS also coordinates the public mapping, charting and registration of spatial information. Fields of activity include the countries of Denmark, Greenland, the Faroe Islands and their surrounding waters. KMS seeks to supply citizens, government, private business and the public sector with advice on how to invest and use map and geographical data in an efficient way – both in the short and long term.

The digital topographic database – TOP10DK, is the national digital topographic base map recorded at a scale of 1 : 10 000. It contains basic information regarding landscape, municipal boundaries, names, and elevation. Similar to the digital cadastral map, TOP10DK is built using vector mapping, which makes it suitable for use in advanced GIS. TOP10DK has an update cycle of 5 years. KMS also produces topographic maps at the scale of 1 : 25 000, 1 : 50 000, 1 : 100 000, 1 : 200 000, 1 : 500 000 and 1 : 1 000 000. The current production of printed maps at 1 : 50 000 is based on an automatic generalisation of the TOP10DK database. It is planned that the future production of paper maps as well as various digital products will be based on a generic multi-scale database derived from TOP10DK through the process of automatic generalisation.

16.2. Key Considerations in Automating Map Production Lines

KMS, like many other national mapping agencies, has traditionally produced paper maps at different scales, each individual scale having its own individual specifications, update cycles, production methods and traditions. In recent years, focus has changed from paper map production to the production of digital geodata and digital infrastructure enabling the production of topologically well-structured datasets. At the same time, dissemination of geographic data has changed from being mainly topographic maps to a much broader range of thematic maps. Geographic data now serve a wide range of purposes ranging from a simple backdrop to thematic information to the fully integrated geospatial systems associated with geographic information systems (Ormsby and Mackaness, 1999). These new demands from users combined with reduced government grants and increasing demands for currency have required KMS to remodel their production environment, including developing new approaches to data handling, storing and distribution of geographical information.

KMS has adopted the "single-stringed data collecting" approach to maintaining specialized geographic datasets. "Single-stringed data collecting" means maintaining a detailed database based on unified data sources, and from this deriving smaller-scale databases. The derived multi-scale database forms the basis for creation of products, and service delivery. The change towards the "single-stringed data collecting" approach started 10 years ago, when KMS began the production of TOP10DK. From this base scale, it is the intention to derive smaller-scale multi purpose datasets (**GeoDB10**, **GeoDB25**, **GeoDB50**, **GeoDB100**, **GeoDB250**, **GeoDB1000**) from a multi-scale database, reflecting the demand for geospatial information services as well as the demand for the more traditional printed map products.

To initiate this process in KMS, the **GeoDB** project was started. The project had four sub-projects, each deals with specific challenges associated with the derivation process. Key deliverables from these projects were:

- specifications for the derived datasets;
- a data model for the derived database;
- conceptual models of derivation;
- modernisation of the visualisation methodology for printed map.

16.2.1. *Dataset specifications (feature catalogue)*

One important goal of the first sub-project was to guarantee a high level of consistency in the semantic generalisation of the feature classification system, thus ensuring streamlined application of data based on GeoDB specifications in multi-scale geospatial services. A set of separate specifications for each of the individual scales had to be written, as well as a specification for elements stretching across the full-scale ranges (Figure 16.1). The figure ilustrates the GeoDB family of specifications and their interconnected relationships, emphasising the fact that the individual specifications cannot be seen in isolation.

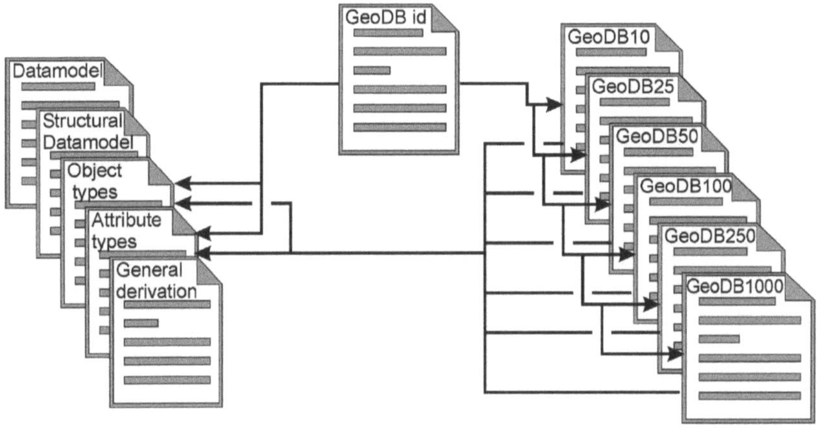

Fig. 16.1. The figure shows the specifications in the GeoDB family and how their content are dependent on each other. On the left side are the general specifications, and on the right side the specifications, for specific scales.

Because of the interconnected nature of the project, it was hard to define a start point in specification writing. The question was whether the specification should be designed for one scale at a time – thereby maintaining the consistencies within that scale (horizontal), or should the different object types be defined one by one for each of the scales – thereby maintaining the consistency across the scales (vertical). Eventually it was decided that the specifications would be written for one scale at a time while trying to maintain some degree of vertical consistency. A further issue, when defining the specifications, was implementation of international standards. The question was how much should KMS adopt standards from ISO (ISO, 2005), and should the organisation adopt FACC/DFDD/EDCS (Digest, 2005; Sedris, 2005) as the coding system. Eventually it was decided that ISO standards would be used only in defining metadata. Due to production constraints, and legacy data, it was decided, for the moment, to carry on using the Danish system for feature coding.

16.2.2. *GeoDB data model*

The data model developed in the GeoDB Project is a further development of the data model used for TOP10DK. The GeoDB data model attempts to solve some of the problems when dealing with a single database that includes many related scales, as opposed to a data model where each scale is stored individually. The strength of the model lies in the modelling of links between objects at different scales. It means updates can be made once, but propagated across all scales in which that object is represented.

Figure 16.2 illustrates a simplified version of the **GeoDB** data model. The datamodel is implemented in an Oracle database. At the top is the **GeoDB** database with a dataset instance for each scale. The **GeoDB** data model has two types of objects: "**idObjects**" and "**mapObjects**". A **mapObject** belongs to one **GeoDB** dataset (scale) only, whereas an **idObject** is a part of the entire **GeoDB** stretching across all the scales. An **idObject** has no geometry of its own,

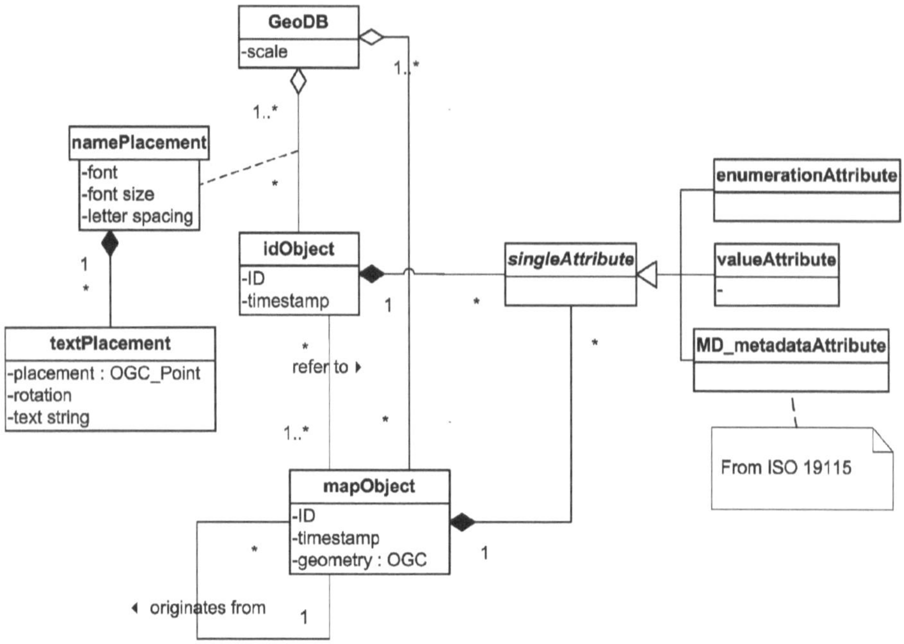

Fig. 16.2. The concept of the GeoDB in UML.

but points to one or more **mapObject**. **MapObject**s can point to one or more **mapObject**s represented at other scales, and can thus point to one or more "parent" **mapObject**s. Both **idObject**s and **mapObject**s can have attributes associated with them.

Text placement and name placement are modelled as objects having various attributes such as font size and rotation. An **idObject** can have a name as an attribute. If the name is placed on a map the **idObject** will point to one or more **namePlacement** objects, varying according to scale. Each **namePlacement** can point to several **textPlacement**s.

Figure 16.3 shows an example of the use of the GeoDB datamodel. An **idObject** of the object type **idBuilding** has two attribute types, "Name" and "Function2". The **idBuilding** is scale-independent, but depending on the scale, the **idBuilding** points to **mapObject**s of the types "Building", "Building_p" and "Silo". At the same time, an **idObject** of the type **idAddress** with the attributes "address" and "number" points to some of the same **mapObject**s.

Each **mapObject** has only one geometry, and all **mapObject** types have their own individual attributes. The **idBuilding** has the name "attribute" which is shown at scales 1 : 10 000 and 1 : 50 000. Therefore the **idBuilding** points to a **namePlacement** in both scales, and the **namePlacement** again points to some **textPlacement**. Although not shown in this example, the "Building" at scale 1 : 50 000, and the "Building_p" at scale 1 : 100 000 could both point back to the "Buildings" at 1 : 10 000, from where they originate.

By adopting this data model, a number of advantages were gained though some new challenges arose! This model provides the KMS with a method to produce spatial data for services, general purpose as well as specialized data products. It is possible to deliver data from a single scale; transferring ID-data attribute information to their associated **mapObject**s or to select

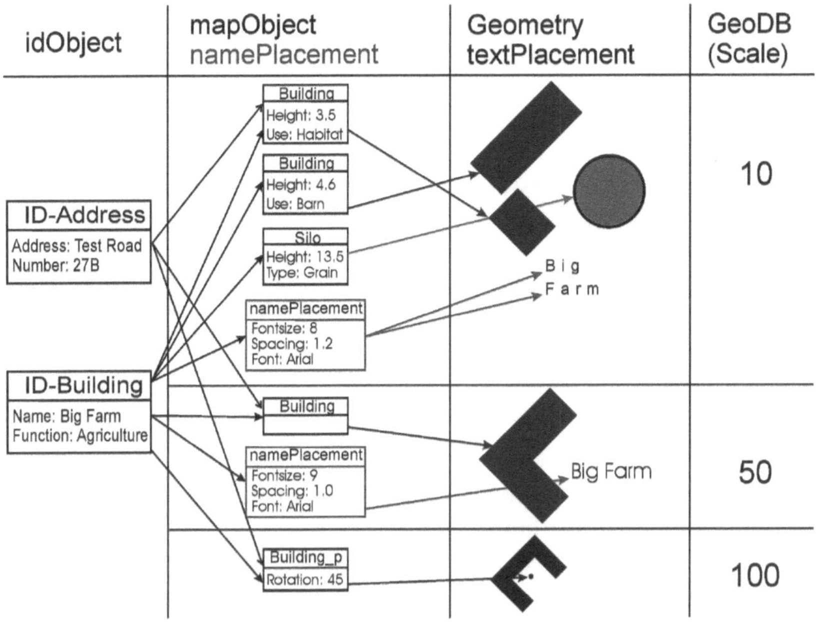

Fig. 16.3. Information flow in the GeoDB.

a subset of objects within the scale. It is also possible with care, to combine data from the different scales. For example, the watercourses at 1 : 10 000 and everything else at 1 : 100 000 as a background can be combined – if the user is only interested in watercourses. In mentioning these possibilities, it is important to note that the topology (geometrical match) between the data can only be guaranteed by using data from one single scale (van Oosterom et al., 2002). In future it is envisaged by KMS that users will have access to all components of the **GeoDB** – not just a subset of the individual scale-dependent components. Therefore, this new structure is advantageous to users not just KMS. Instead of linking information to each scale, it would be sufficient to link the data to an **idObject**, whereby the information would occur at all scales.

This datamodel offers several benefits in the context of updating. Because much attribute information is stored at the **idObject**s level, it is only necessary to update information connected with the base scale. An updating mechanism will then propagate the updates by cascading them into the structures of the **GeoDB**. The links between the derived **mapObject**s and the "parent" **mapObject** will be helpful in this regard. With these links it is possible to identify associated **mapObject**s at the smaller scale only affected by revisions at the large scale.

There are of course, challenges in adopting this model. First of all, the whole process of generalisation is more complicated. Generalising geometry is difficult as it is, but with this model, all information about relationships between **idObject**s and **mapObject**s, as well as relationship between **mapObject**s and "parent" **mapObject** have to be taken into account. In some cases this is straightforward, as when one building is generalised to a new building.

But when several **mapObjects** of different types are aggregated to a new object type during a generalisation, maintaining the connection becomes complicated.

16.2.3. *Conceptual models of derivation*

With a set of unified specifications, as well as a strong data model for the **GeoDB** as background, the GeoDB Project considered four conceptual models for generalisation. Each model described in what order the generalisation should be carried out, and which factors govern the workflow. It is considered that there are two ways of generalising, but can be combined to arrive at four conceptual models. Either generalisation of the base scale to derive each individual scale ("fan shaped"; model 1 in Figure 16.4), or stepwise from one scale to the next ("ladder shaped"; model 2 in Figure 16.4). These can be combined in various ways to create hybrid solutions (models 3 and 4 in Figure 16.4). We consider briefly the pros and cons of each of these four approaches:

- By choosing to derive individually each scale from the base scale (model 1), each scale will be independent from the others. It is not necessary to produce the 1 : 25 000 scale data before starting work on the 1 : 50 000 scale data, thus each scale can have its own individual update cycle. On the other hand, the gap from 1 : 10 000 to 1 : 250 000 may be too great, and when deriving individually, the mutual coherence between the scales may be lost.
- By choosing the ladder shaped approach in model 2, there is a certainty that what is in 1 : 250 000 is present in the larger scales, but the production line is interdependent, which may create delays in production.
- Model 3 shows a combination of the fan shaped approach and the ladder shaped approach, where the three largest scales are derived directly, but where the large step from 1 : 10 000 to 1 : 250 000 has been broken into two steps.
- Finally in model 4, a method is shown where different object types are taken from different steps in the process. For instance at a scale 1 : 100 000, one would derive the areas from scale 1 : 50 000, whereas the buildings might be derived directly from the base scale data. So potentially, one could take the elements that were difficult to generalise and use a step-

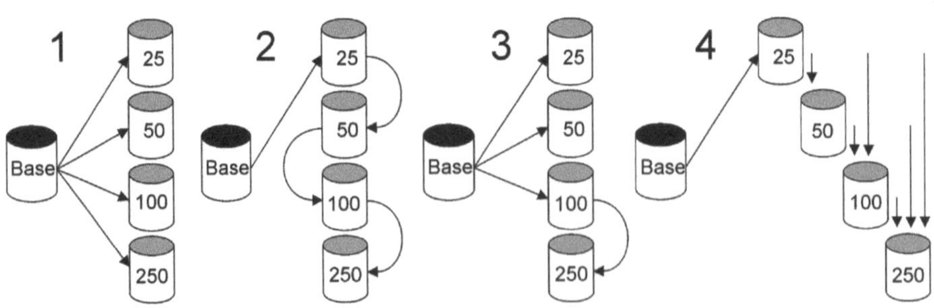

Fig. 16.4. Different models of generalisation. KMS is still evaluating the choice of process and derivation model.

wise approach, and for elements that had a faster update cycle, the direct input approach might be preferable.

It is not only the process of establishing the database that has to be considered. Even more important is the strategy for propagating updates throughout the range of scales. Specific demands on update frequency in small target scale datasets raises the problem of "backwards cascading" such updates to the larger scales. KMS has not yet chosen a generalisation model, but it is likely that the ladder shaped approach will be used for establishing the databases and the fan shaped approach will be chosen for the update of databases.

16.2.4. *Modernisation of printed map production*

The aim of this part of the project was to produce a number of test plots using the data automatically generalised at 1 : 50 000 and to see if the proposed modernisation of the cartographic design of the traditional military topographic maps would facilitate production and minimize the manual editing component.

Figure 16.5 shows two ways of symbolising a viaduct using the same data. The left example is based on the traditional cartographic legend. It takes more space in the map and is difficult to visualise automatically. Thus a manual correction is needed. The right example is generated automatically in an ordinary GIS platform and shows an acceptable map design with sufficient readability. Thus to meet the needs of traditional cartography as well as the requirements of a digital infrastructure, KMS would need two databases with somewhat conflicting requirements – a cartographic database supporting high quality production of traditional map products as well as a cost-effective generic database for digital purposes. Realistically, the resources available only allow KMS to support one, yet the aim of KMS is to produce one database that meets both requirements. This will require a remodelling of cartographic needs, in which map design reflects the opportunities afforded by modern production systems. The **GeoDB** specifications, datamodel, generalisation model and visualisation methodology are thus a compromise between minimizing production cost and obtaining high cartographic standards as well as modern digital data.

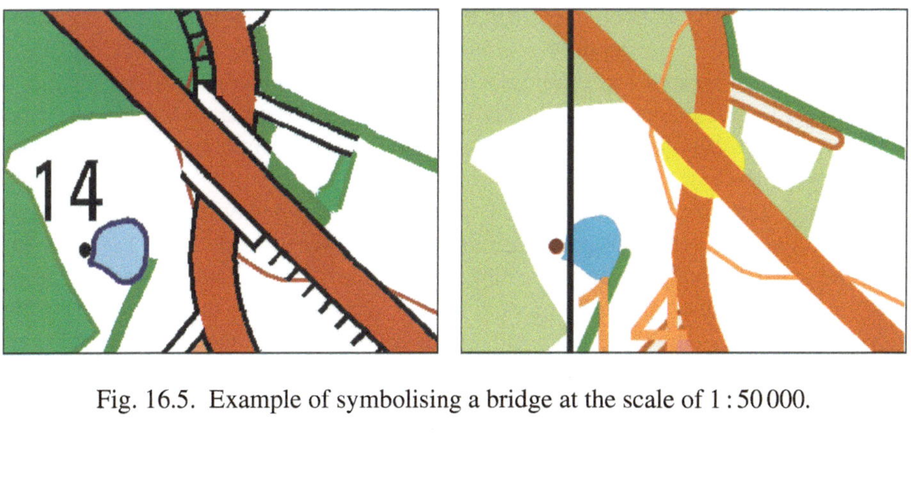

Fig. 16.5. Example of symbolising a bridge at the scale of 1 : 50 000.

16.3. Generalisation Platform

In 1999 and 2000, KMS carried out an investigation into possible generalisation platforms to help in the rationalisation of map production by means of automatic generalisation. Defining the general GIS application requirements was not a problem. Specifying the generalisation was much more problematic and could only be described in broad terms. So instead of defining the requirements for generalisation, a set of twenty generalisation cases were created. A geographic region was chosen, and within this area twenty situations in need of generalisation were marked and the generalisation-desired output of each location was described as a benchmark. The general requirements were:

- the system should include a generalisation package;
- the system should be able to work with 2.5/3D data;
- the data to be generalised would come from and be stored in an Oracle database;
- the system should be able to work with a range of attributes;
- the generalised data should be topologically well structured;
- the system should have a good interface for development and user adjustments since early in the preliminary study it was realised that many functions would have to be developed in-house at KMS.

At that time the AGENT concept from the software company Laser-Scan was considered the most advanced product for generalisation (Sheehan, 2001). Laser-Scan was at that time a member of a European development project together with a number of universities, and the COGIT research laboratory of the French National Mapping Agency focusing on utilising agent based methodologies in generalisation (Lamy et al., 1999; Ruas, 2000a; Ruas and Duchêne, this volume (Chapter 14)). Having taken all this into account, it was eventually decided to choose Laser-Scan as the generalisation platform provider (Hardy et al., 2003).

Participation in research and development projects

The university research in Denmark on automatic generalisation can best be described as modest, and KMS as an institution has no resources to do intensive research on its own. KMS remains very dependant on the research and activities being undertaken by organisation such as the International Cartographic Association (ICA), EuroSDR, and ISPRS. KMS is also very dependent on research and development projects being undertaken at universities and other National Mapping Agencies (NMAs). During the last five years KMS has participated in a NMA and university generalisation network and has gained much knowledge from these activities.

KMS is also a member of the MAGNET Group, which is a newly formed cooperation between NMAs using Laser-Scan generalisation software (Woodsford, 2003; Lecordix et al., 2005). The scope of a cooperation is to exchange experiences as well as to coordinate requirements for the generalisation platform. Additional benefits include the sharing of software development costs. Small NMAs such as KMS do not have the financial resources to launch large scale developments. By joining the group, KMS has greater influence in improvements to the generalisation platform, and in the exchange of knowledge.

16.4. Present Generalisation Workflow

Modernisation in production

In 2000, it was decided to establish a new vector dataset at a scale of 1 : 50 000 for the purposes of producing paper maps, and to derive it from the TOP10DK via automatic generalisation. Initially the end goal was to produce a traditional topographic paper map, but from the start of the project, the emphasis was on topologically consistent and attributed databases – not only on the pure visual/cartographic side of things. Automated generalisation that takes into account the topological rules and attributes of the database is more complex than a simple graphical approach. Increasingly the focus at KMS has changed from one of topographic map production to a geospatial data and digital infrastructure. The additional effort and sophistication of the model are seen as being necessary in order support broader application and services.

Figure 16.6 illustrates the 2005 version of the workflow developed to derive data suitable for making a paper map at 1 : 50 000 from the TOP10DK data. The first step of the workflow is the preparation of the data. As shown in Figure 16.6, TOP10DK data are combined with other data sources such as elevation data. To begin with TOP10DK does not contain all the object types needed to make the KMS 1 : 50 000 scale map, and secondly, various additional information facilitates the decision making associated with the generalisation process. For instance data from the cadastral maps are used to help determine the grouping of buildings.

The different data sources are stored in different formats and with different specifications. In order to bring the data into the generalisation workflow, FME from SafeSoft is used for data translation and restructuring, as well as for the validation of topology. In future it is planned to remodel all sources of data into a common database – GeoDB10.

A series of different data handling and generalisation processes are handled by the generalisation platform. First a series of methods are used to reorganize the data, such as extracting information from some of the additional datasets and combining it with the topographic data, or using add-on data to make overlay analyses, and thereby add more information. Once reorganized, the actual generalisation can begin. The generalisation consists of a broad spectrum of standard generalisation procedures (selection, simplification, aggregation, merging, collapsing, exaggerating, classifying and displacing) – these techniques being combined, one

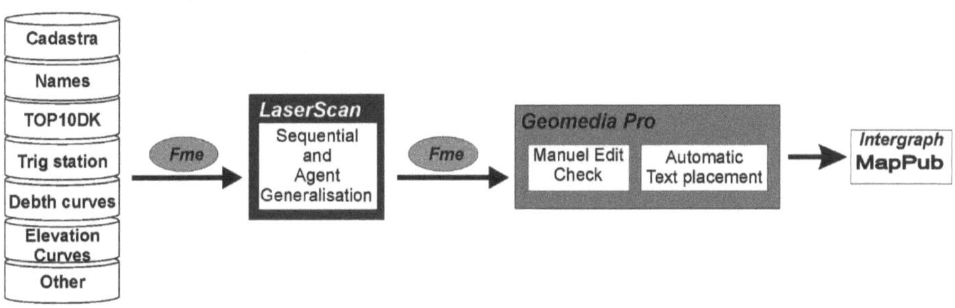

Fig. 16.6. From TOP10DK to Paper map at 1 : 50 000.

after the other, in a long sequence. In all, more than a hundred methods are used sequentially in the first block of generalisation.

Of the hundred plus methods, some originate from the Laser-Scan generalisation package, some have been developed for KMS by Laser-Scan, but by far the largest number of methods has been developed in-house by KMS. The methods include simple as well as more complex methods, from very specific methods used on specific object types at a single scale, to general methods usable on many object types at a variety of scales. Two examples of in-house developed methods are:

- A simple generic method for reclassifying areas depending of their visible size, their shape and their "surroundness" of other areas (Figure 16.7).
- A very specific method for generalising fishing ponds from 1 : 10 000 to 1 : 50 000.

Currently there are two approaches to generalisation:

- The sequential method, where all themes or layers are generalised individually in a sequence, and
- the context driven approach, where all objects are considered in their context and generalised using contextual rules.

KMS is applying a mix of these two concepts; some data are generalised sequential, whilst other generalisation takes into account the context of the surrounding. The contextual generalisation is done either by applying contextual rules into the generalisation methods used in the sequences, or by using the agent based approach which, by its nature, is context driven. The sequencing of different procedures is also important to a successful outcome. Long sequences of procedures in generalisation are seldom ideal. Often the chosen sequence is set up in an order that produces the best overall result, and not necessarily a perfect result in every case.

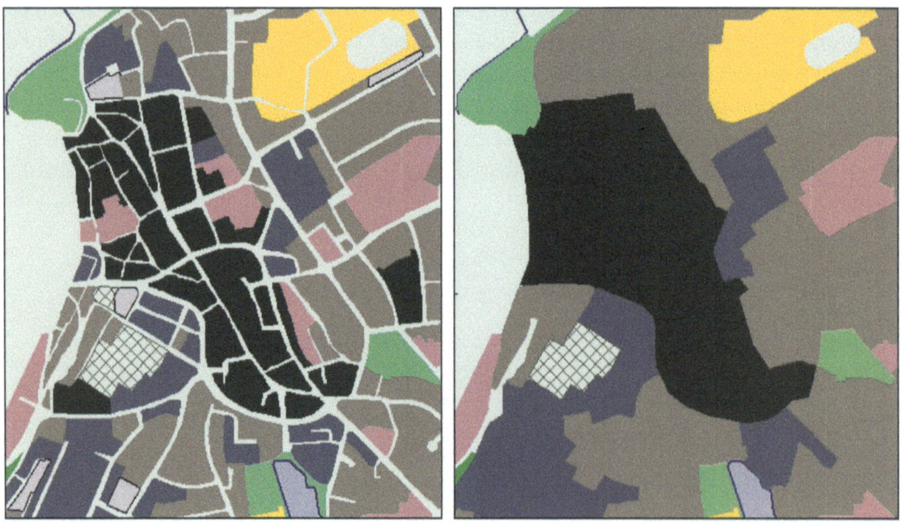

Fig. 16.7. Source data (1 : 10 000) and target data (1 : 50 000) after generalisation.

It is difficult to handle the variety of complexities with such a long generalisation sequence. For example the roads are used to demarcate regions in need of area generalisation. Similarly the roads are selected depending on which areas they lie within. The sequences are chosen based on an understanding of how they interact with one another and partly by trial and error. To overcome some of the complexity, many modular generalisation methods are developed in anticipation of being used in a specific dependent order, and not developed to stand-alone.

The buildings are generalised using the AGENT technology that performs simplification, enlargement, amalgamation and displacement (Figure 16.8). The setting up of the KMS AGENT-building generalisation was originally done for KMS by Laser-Scan, as a prototype to validate agent based technology. This eventually proved to be sufficiently reliable to be incorporated into the production workflow. Figure 16.8 shows the result of applying the AGENT-building generalisation (source buildings are shown in orange outline and the resulting buildings in black). This agent-based generalisation resolved some of the contextual problems that could not be resolved using a sequenced approach. It should be remembered however that considerable effort was required in setting up the AGENT system. Where the problem is deemed to be simple, it may be more cost effective to seek simple application of generalisation algorithms.

When the automatic generalisation is completed, the data are translated to Geomedia readable format for manual editing. At this stage human operators are used either to correct generalisation mistakes or to perform tasks for which no automatic solution exists. Figure 16.9 shows an example of an unresolved conflict whereby the road and railway are too close and need to be displaced apart. Finally cartographic text placement takes place. The text placement is performed automatically using Label-EZ from MapText (Label-EZ, 2005), again followed by manual revision and corrections.

During development of methods and workflows, the focus has been on minimizing but not excluding manual intervention. This means that instead of trying to make every method completely automatic, we often settled for something less. KMS feel that rather than devote effort to developing entirely autonomous solutions, it is more effective to devote resource to developing other generalisation methods. Because of the relatively high cost of manual intervention, resource is targeted at those methods that are most likely to reduce the amount of manual effort required.

Once these vector based techniques are completed, the data are rasterized, and processed by Intergraph Mappub and finally the data is transferred to the printing press. Currently the last stage of the workflow is to produce the actual paper maps (Figure 16.6). In future it is planned to store the data in the GeoDB database at the 1 : 50 000 scale. Before this can be done, changes need to be made to the GeoDB specification. More importantly all methods in the generalisation have to be modified in order to handle idObject identifiers, and IDs from the parent objects.

Data evaluation

The data are evaluated in two ways. First the topology is checked even though the generalisation methods are designed to create topologically well-structured data. During the manual revision the data are topologically validated and visually inspected since some of the generalisation methods are designed to report where the generalisation has failed. Additionally

Fig. 16.8. Buildings generalised using the Agent method (input data shown as orange outlines).

Fig. 16.9. Example of an incomplete solution: the road and railway are too close; one or both should be displaced.

spatial queries are used to highlight where the generalisation has failed, for instant in the case of overlapping features. Currently a general method to automatically evaluate the results from automated generalisation is not available.

16.5. Discussion

Over these last few years of development at KMS, numerous questions have arisen that remain unanswered. A major challenge remains in the development of evaluation techniques that minimize the need for manual editing after automatic generalisation. This can be achieved either through the improvement of existing algorithms or through the revision of the data specification itself – there being an intrinsic link between the generalisation methods, the data model used and the map specifications.

As generalisation becomes more automated, it is thought that the area of automatic update will become increasingly important, and here too many problems remain. For example how do we update a database that has been produced by a variety of automatic and manual processes and how do we handle the fact that different themes in TOP10DK are updated at different frequencies. Another concern is that of quality and the quality testing of data. At present we have neither automatic evaluation methods nor a formal set of evaluation criteria. There is increasing interest in modelling and representing 3D data. Handling and maintaining the third dimension of generalised data is another area that has to be dealt with.

The results presented here are focused towards production and updating of a number of coherent databases at various scale intervals. With the development of faster and more efficient tools, it may be possible to create maps instantaneously, tailored in response to customers' wishes rather than being produced and stored in databases. This again is an area in need of further research. Last but by no means least, is the issue of the generalisation of attributes. Generalisation of geometry is complicated enough – but the desired output is relatively straightforward to specify. Conceptual work remains to be done in the area of attribute generalisation, as well as the generalisation of names.

A question often asked by senior management is what efficiencies have been gained by incorporating these automated techniques. This is a difficult question to answer. In the old raster production the 1 : 50 000 maps had an update cycle of 10 years. With the new production environment we are currently able to produce the same amount of new maps with a third less staff. With continuing development, productivity will increase thus allowing a faster update of our maps. The update cycle is currently ten years, but the aim is to reduce this to five years (the same update cycle as for TOP10DK). Beyond these productivity gains it is important to acknowledge the significant modernisation of KMS. KMS are now producing a well-structured multi-purpose attributed vector dataset derived from our topographic database that will support a broader range of products and services, better tailored for future customer needs.

Generalisation of Geographic Information: Cartographic Modelling and Applications 315
W.A. Mackaness, A. Ruas and L.T. Sarjakoski (Editors)

Chapter 17

Observations and Research Challenges in Map Generalisation and Multiple Representation

William A. Mackaness[a], Anne Ruas[b], L. Tiina Sarjakoski[c]

[a]*Institute of Geography, The University of Edinburgh, Drummond St, Edinburgh, Scotland, EH8 9XP, UK*
e-mail: William.Mackaness@ed.ac.uk
[b]*Institut Géographique National, COGIT Laboratory, 2-4 av. Pasteur, 94165 Saint-Mandé, France*
e-mail: Anne.Ruas@ign.fr
[c]*Department of Geoinformatics and Cartography, Finnish Geodetic Institute, P.O. Box 15, FIN-02431 Masala, Finland*
e-mail: Tiina.Sarjakoski@fgi.fi

Abstract

Whilst technology may fundamentally change how we handle spatial information, the underlying principles of effective visualisation and cartographic design remain the same. Generalisation is but one essential piece of the jigsaw, integrated among a set of techniques and technologies that extend from capture and storage through to visualisation and interaction. There are many challenging issues in generalisation research: from modelling the complex decision making processes inherent in the art and science of cartography, through to interaction models that support easy interaction and exploration of geographic phenomenon across a continuum of scales. Moderately high levels of automation are now achievable but there remains an exciting set of research challenges – challenges that are evolving in response to the disruptive nature of IT. In this chapter we present a set of challenges as a pointer to a research agenda that includes among other things, generalisation of global datasets, better evaluation methodologies, and better integration of reasoning about space in order to support autonomous solutions in map generalisation. The importance of all research connected with generalisation is built around the concept that scale is intrinsic to ideas of geographic meaning and is therefore critical to geographical problem solving.

Keywords: research agenda, modelling cartographic design, geographical modelling, GIS

17.1. Some Observations

Attempts to automate cartography are humbling in the way they reveal the science and art of human cartography. They reveal the powerful combination of the mind, the eye, and the hand in the creation of maps – abstractions of the world in which we live. They reveal cartographic design to be a complex decision making process, in which the solution is one of compromise, working within a multiple set of constraints and opportunities. The design task takes advantage of gestaltic properties inherent among phenomena represented by a set of visual variables, of the user's knowledge and experience of the world, and their prior use of maps. The design task can be limited by constraints in the resolution of the media, the cognitive load associated with the map interpretation process and the need to deliver solutions of consistent and high quality. Technology has made these constraints and opportunities "elastic"; that is to say we are no longer constrained by traditionally high production costs that led to generalised, fixed scale products (typified in the creation of series mapping and topographic mapping). Technology has simultaneously raised user's expectations of the instant, tailored map, delivered through any media and device, and provided the database technology to support the creation, integration and maintenance of data to support such delivery. When combined with remote sensing technology and automatic feature extraction algorithms, it seems possible that high levels of automation can be achieved throughout the data handling process: from automatic capture all the way through to visualisation. Advances in data capture technology (notably LIDAR) means that attention is now extending to include 2.5D and full 3D modelling of geographic space. A picture is emerging of a seamless set of complementary integrated technologies and concepts (of which generalisation is one) in which high-dimensional data can be abstracted, combined, visualised, and analysed.

Generalisation is more than just mimicry of the human cartographer, it is about modelling geographic space. Many of the solutions presented in these chapters are underpinned by the spatial modelling of surfaces, networks, and discrete geographic objects. It is true that at the most fundamental level, generalisation operates on the geometric primitives of lines, points and polygons. But these operations are governed by the properties of the geographic phenomena being represented – both their internal properties, and their intrinsic relationship among the phenomena being represented. Any observation in the real world is a manifestation of a set of interactive processes operating at a range of scales. Pragmatically speaking, a generalisation solution yields an abstraction, but cannot take into account every property and inter-object relationship when deriving a solution. The challenge therefore is one of modelling a minimum set of properties, opportunities and constraints sufficient to generate a solution that is fit for purpose. But what do we mean by "fit for purpose"? In series mapping (paper based mapping intended for public consumption), the quality of that cartographic solution must be very high indeed. In exploratory, interactive environments, where the user might resolve any ambiguity by zooming in and out, the quality of the solution may be more flexible. And in the context of spatial analysis (where no visualisation is required), the graphic quality may be immaterial, (though model generalisation is still required).

Technology has not obviated the need for generalisation. On the contrary, we would argue that generalisation has become central to the process of integrating data captured at different resolution and scale, and it remains central to the effective portrayal and analysis of geographic space. In highly automated environments, it plays a critical role in translating the user's needs into a visualisation specification. In many respects this has been the role of the human cartographer – understanding the requirements of the client – and based on the cartographer's knowledge, "translating" those requirements into a design that leads to a cartographic product. Research continues to highlight the complexity of that design task; the need to model cartographic knowledge, the need to devise methods of abstraction, as well as methods of evaluation.

These ideas very much governed the process of commissioning the chapters in this book. In reviewing those chapters, we distilled five key themes, which we now briefly expand on, before pointing to some areas worthy of deeper research in the context of map generalisation and multiple representation.

17.1.1. *Knowledge engineering*

It has long been known that map generalisation is a decision making process dependent on knowledge about the phenomena being portrayed, and rules of thumb (heuristics) used in selecting appropriate solutions. Early work using artificial intelligence techniques were stymied by an absence of techniques for analysing and evaluating solutions. High numbers of candidate solutions could be produced through the permutations of generalisation operators which meant the solution space was large, and the task of modelling cartographic and geographic knowledge proved to be poorly constrained. The use of expert systems has given way to agent based methodologies, and multi agent systems – MAS (better able to manage a knowledge based approach). MAS are able to take advantage of developments in database modelling, and a much richer repertoire of generalisation techniques, analysis and evaluation techniques. Agent based methodologies have made the knowledge engineering process more manageable, and provided an intuitive way of modelling interactions between phenomena being visualised. Machine learning techniques are being incorporated into research systems in order to make the decision making process more efficient (capturing the heuristics of cartographic design). But the notion of a cartographic syntax by which we can formalise or reason about cartographic knowledge remains elusive. It is not always clear when and how cartographic knowledge should be applied throughout the multifarious process, from model to cartographic generalisation. Furthermore it has proved hard to measure or formalise information content, or model the changing nature of information content over scale and theme. Generally speaking, many would argue that much research remains to be done in this area.

17.1.2. *Data modelling*

We have witnessed an increasing sophistication in the data models that underpin automated cartography. Indeed attempts at knowledge engineering have helped to identify the objects and attributes that need to be retrieved (and created) during the generalisation process. There has been considerable debate about the degree to which attributes must be explicitly stored in the database, or whether certain characteristics and qualities can be inferred or derived during the analysis phase. A relatively simple data model is easier to maintain, but tends to have a greater processing overhead (for example in creating a topological or partanomic model, or applying pattern analysis techniques to identify gestaltic properties across distributions), and requires a more complex knowledge based environment to accommodate the scarcity of knowledge about the phenomena being mapped. And precisely the converse is true. "Enriched" databases may simplify the application of model and cartographic generalisation techniques. Some research has focused on semantic and ontological modelling as a way of supporting a much richer interaction between the underlying model and the generalisation processes that are applied to it.

17.1.3. *Paradigms of use*

The Internet coupled with developments in technology has created fundamentally new ways of using and interacting with (geographical) information. Information has become ubiquitous (any type, in any form, anywhere) – exemplified by delivery of geographical information over mobile devices. These new paradigms of use have raised important challenges in how we interact with geographic information, and precisely how this information becomes incorporated into the decision making process. Carroll et al. (1991) argue that the technology fundamentally changes how you go about undertaking a task – the idea of the "cognitive-task-artefact". Therefore it is not sufficient to merely "slot" new technology into existing ways of doing things; at best this merely ossifies the traditional approach and thereby limits the benefits of technology. We would argue that the effective incorporation of new ways of interacting and utilising geographic information requires careful consideration of the context of use, user requirements and examination of the entire decision making "train". For example, how should model and cartographic generalisation techniques be used in geovisualisation and exploratory environments? What interaction models can best support the map generalisation process?

17.1.4. *Human computer interaction*

The complexity of the design task and importance of deriving meaning through interaction means that it is neither cost effective nor desirable to exclude the human entirely from the generalisation process. The role of the human should not be considered as an after thought; instead their role should be treated as an integral part of the orchestration process. The critical points at which the human must interact with the system are (1) in specifying the product, and (2) resolving problems in the design process irreconcilable by the system itself. In the former case the challenge is in designing intuitive interfaces to complex systems, where the user may have poor cartographic knowledge. In the latter case, we must avoid engaging users in tedious and therefore error prone activities associated with the generalisation process.

These ideas of paradigms of use extend beyond the various end users, to include those interacting with map generalisation systems in the creation of map generalisation solutions. Incorporation of a decision making capacity within the system leads to the questions: how does the system convey to the user, the decisions taken? How does it measure "failed solutions" and how are these communicated to the user? More broadly what is the optimal balance of decision making between the system and the user (the idea of cognitive ergonomics). It is critically important that increasing levels of automated decision making do not lead to a role for the human that is mundane and ordinary, repetitive and tedious.

17.1.5. *Generalisation of high-dimensional data*

Our information hungry society appears to have an insatiable appetite for information at increasing levels of details and currency. Data capture technology (using remote platforms) is advancing to a point where 2.5D and 3D models of the world can be captured (and frequently updated) in a cost-effective manner. Similar to the generalisation of 2D data, there are multiple applications requiring volumetric modelling of the land and city scape at various levels of detail. There is also growing interest in how generalisation techniques can be applied in the temporal domain in the context of geographical modelling (particularly given the links between space and time). Given the detail of such models, and their use in virtual environments, the need for, and challenges of, generalisation are considerable. The long cartographic tradition associated with mapping does not extend to 3D and virtual worlds, though developments and methodologies in automated cartography are certainly informing developments in this area. 3D modelling and generalisation are drawing on expertise in application domains such as town planning, architecture and environmental-visual impact assessment. We can extend this line of thinking to higher-dimensional data (bridging across to ideas of self organising maps and spatialisation – visualisation of the non-spatial), even including the generalisation of "non-cartesian descriptions of space" (for example summarising textual descriptions of space). We would argue that this is an exciting area of research, with many potential applications.

17.2. Research Challenges

Beyond the research agenda implicit in the observations made in §17.1, we additionally list six topics of research which we feel is worthy of greater research.

17.2.1. *Modelling user requirements*

The specialist skills of the cartographer are required precisely because of the complexity of the task. Automated approaches do not obviate the need for an interface that enables users to articulate their requirements in a simple manner. Perhaps something is required that is analogous to a "generalisation wisard" that avoids the user having to comprehend the consequences of parameterising and ordering a large number of generalisation algorithms. Any solution would likely need to include the ability to play back to the user, in an intuitive manner, the consequences of their choice, and allow, if necessary, the modification of those requirements. Given the breadth of potential solutions (varying in thematic content and scale), it seems sensible to incorporate metadata that, if required, can explain the provenance of any given solution. Whilst this information may be too complex for the human to comprehend, it would be important to record this information in appraising "fitness for use" and any subsequent use of the data such as in conflation.

17.2.2. *Evaluation methodologies*

Evaluation methodologies are critical to the development of autonomous systems capable of analysing, synthesising (considering alternate solutions to a single problem) and appraising the final output as being optimal – according to some prespecified criteria. By definition changing levels of detail produce changes in the class of phenomena represented, the topology between those classes, and the emphasis given to various characteristic properties of those phenomena. Evaluation methodologies need to operate at a number of conceptual levels, and be able to prioritise criteria according to changing tasks and scales. Future developments will no doubt continue to draw on a broad range of spatial analysis techniques, in particular the use of pattern analysis techniques able to identify and preserve the most salient characteristics that define a particular phenomenon.

17.2.3. *Generalisation and global datasets*

As recently as 1995, Müller et al. complained of the excessive emphasis given to cartographic generalisation over model generalisation. Evidence suggests that we do now understand the broader relevance of generalisation in modelling space and that current research activities do indeed reflect the critical role of data modelling to the generalisation process (in abstracting the data and in storing "higher order" geographic phenomena). Many of the techniques currently being developed are focused on deriving 1 : 50 000 and 1 : 100 000 scale mapping from source data at 1 : 10 000. Many of these algorithms are starting to be applied at larger scales, across larger sets of object classes, and in categorical mapping. At very large scale geographic phenomenon become subsumed within higher order objects. For example at 1 : 1 million scale, a set of streets, railways, and buildings of every variety might become incorporated within a simple polygon representing the boundary of a city. At these levels of abstraction, it is necessary to model the partonomic structures inherent among geographic phenomena (something that is part of something else), as well as use taxonomies as a basis for reclassifying classes of objects into their more general form. At very large scales, the topology among these higher order objects fundamentally changes, as does the data type (for example the city polygon may simply be replaced by a simple point with the word "London" next to it). It is felt that this type of research is particularly relevant in the capture of geographic information at the global scale, where there is a need to reduce increasingly large datasets (Moore's Law) to sizes that are manageable. Thus generalisation has relevance at the global level (modelling of sustainability and global climate change) in managing and integrating large spatial datasets.

17.2.4. *Generalisation in support of maintenance of multiple representation databases*

Both model and cartographic generalisation techniques can be used to populate multiple representation databases. In the context of maintaining both historical and current databases there is continued interest in methods for automatic update – supporting the notion that we capture once and reuse the data in various derived forms. Automatic update encompasses the idea of change detection, assessing the significance of that change, and propagating that change across the scales. This research extends into the field of version management, digital curation (preservation of historical digital information), and development of systems able to summarise key changes over time. This general shift towards a data centric view offers the flexibility of delivering information via a variety of media (visual, tactile, auditory), and over a range of devices and communication channels (paper, mobile devices, Internet), thus supporting a range of applications.

17.2.5. *Generalisation and interoperability*

In 1997, Peng noted the fragmented nature of research in map generalisation, and argued the need for a common platform that could be used to integrate different solutions. Interoperability extends beyond the sharing of algorithms and includes the development of more flexible data modelling interfaces in order to be able to "chain" different algorithms together. Edwardes et al. (2003b) echo this call arguing that a test bed platform would facilitate the sharing of code thus accelerating developments and opportunities for commercialisation. This idea naturally extends to the idea of Web Generalisation Services (WGS) (Badard and Braun, 2003). Initiatives such as the Open Geospatial Consortium (OGC) make this an achievable goal and such ideas continue to be explored through consortium of stakeholders (for example the MAGNET Consortium). The broader remit of OGC includes the idea of tagging data (metadata) in order to record the pathway by which a dataset has been derived. In the context of generalisation this would include the decisions made, and the types of algorithms applied in arriving at a particular solution. Addressing issue of interoperability and standards will facilitate the sharing of data, enable creation of audit trails and generally create transparency in the reasoning behind any given design.

17.2.6. *Map generalisation and geoinformatics*

Geoinformatics is a science concerned with reasoning about space and making explicit the processes and semantics inherent among geographic phenomena. Analysis and evaluation methodologies provide a basis by which we can create audit trails of how generalised forms are arrived at. Development of autonomous generalisation systems continues to require development of analysis and evaluation techniques. Techniques that make explicit the various characteristics, properties and behaviours within and between phenomena are critical in the context of data mining, automatic feature extraction and pattern analysis. Collectively these techniques will allow us to reason about space at a range of conceptual scales.

Some of the items on our research "to do" list were also on the list proposed by McMaster and Shea in 1992, notably the idea of applying parallel processing technology to handle large detailed databases, enhanced user interfaces, and cartometric analysis and evaluation methodologies able to support decision making as a basis for audit control and to enable transparent reasoning in design. Let us hope that next time this book is written, some of these "to do" items can be struck off!

17.3. Conclusion

Monmonier (1999) observed that academic cartography has proceeded along two lines of inquiry – the theoretical and the applied. We would agree with Monmonier's sentiment that "applied" is not a dirty word; much of the applied work has a strong theoretical basis that guides empirical testing and gives meaning to the notion of robust solutions. Monmonier (1999) also speaks of the "appalling disconnect between cartography's academic and commercial sectors" (Monmonier, 1999, p. 236). This is a point echoed by Fisher (1998) who argues the need for design support in the mapping components of GIS software. But the chapters in this book on practitioner's experiences in utilising the fruits of research in this field present a counter view; we would argue that in some areas the "disconnect" has become a "re-connect". There is no doubt however, that the historical pathways of developments in generalisation have encouraged it to be viewed as a process external to the core functionality of GIS (Peng, 1997) and that a stronger case must be made for making this "communication science" more integral to GIS technology (rather than as an end process in display). As Fisher notes: "In spite of novel methods of display, the design concepts of cartography are as essential now as they have ever been in the creation of effective diagrams of spatial information for examination and exploration" (Fisher, 1998, p. 7).

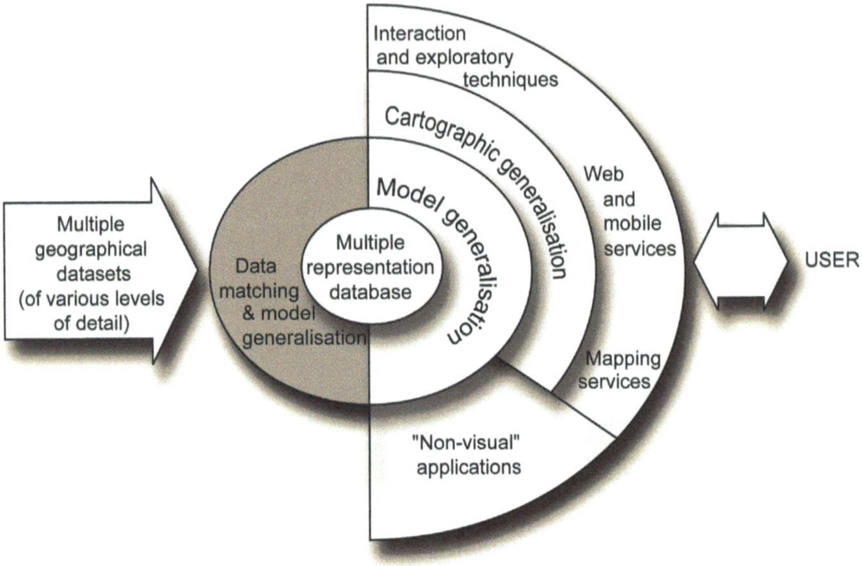

Fig. 17.1. Generalisation as a functional membrane to the integration, analysis and visualisation of geographic information.

We would argue that issues of scale are intrinsic to ideas of meaning and critical to geographical problem solving; that generalisation is related to both data management and visualisation processes, for which there are important implications in terms of broad GIS system architecture, in data modelling, and in human interaction. We go so far as to argue that generalisation methodologies are central to the creation of multiple views, to the process of abstraction and the rendering of that information in a manner capable of supporting multiple tasks. Generalisation is central to the creation of multiple abstracted views of geographic information. It is a key stage in the integration of data, an essential precursor to the revelation of pattern, association and meaning. *Model* generalisation is utilised (in combination with data matching techniques) when populating multiple representation databases (the grey components on the left side of Figure 17.1). In terms of output (the right side of Figure 17.1) model generalisation may be the only requirement in the cases where no visual output is required (for example in response to some sort of spatial query) – demarcated as "non-visual" applications in Figure 17.1. The process of rendering that data in a visual form may be a task in itself (in creating map series), or in support of some human interactive activity or visualisation process. This process of *cartographic* generalisation is achieved in combination with model generalisation in order to render the information in a readily consumable form. Cartographic generalisation supports a range of activities other than conventional mapping. Model and cartographic activities can be used to support exploratory techniques, and a variety of other services (represented by the third ring in Figure 17.1). Overall Figure 17.1 seeks to emphasise the relevance of model and cartographic generalisation as core activities that lie at the heart of geographic data modelling (not merely as peripheral activities associated with making maps).

We feel this is a more useful perspective than the traditional cartographic one. This view attempts to reflect the disruptive nature of technology and how the use of generalisation techniques change – reflecting the breadth of media and mechanisms through which we now share, analyse and interact with data, and thus reason about the world around us. Generalisation as a essential ingredient to the *management* of geographic information is particularly pertinent in the context of the "firehose metaphor" used to describe the ever growing volume of data now being collected at ever finer spatial and temporal scales (Frawley et al., 1992). From this perspective we begin to see generalisation in a new frame of light – concerned with automatically identifying and emphasising salient patterns and associations inherent among disparate and large datasets.

17.4. In Closing

The contributions to this book demonstrate a broadening in the community of cognate disciplines contributing to this field. This includes geoinformatics (developing research in semantic modelling and reasoning about geographic space), and computer science (techniques in exploratory analysis, database design and artificial intelligence techniques). Research has drawn from these and other disciplines, and continues to develop solutions building on new methodologies – one example being the utilisation of agent based techniques in map generalisation. Other fairly recent developments include work in categorical or thematic generalisation – applied to soil and geological mapping. We see growth in the domains to which generalisation is applied – most notably the application of generalisation methodologies to high-dimensional data (beyond 3D and temporal dimensions). Though there are many ways by which we might measure the success of these developments, we note their gradual incorporation in commercial software solutions and map production environments. It is clear that there is huge mutual benefit in continuing the close collaboration between mapping organisations, GIS vendors and research groups. But we would argue that this research is relevant to a broader community that includes internet mapping, those developing location based services, virtual environments and geovisualisation. We argue that research in generalisation and multiple databases is critical to the creation of linkages between data integration, analysis and visualisation. Hopefully the ideas contained in this book will provide a foundation and springboard for further advances in this field.

Consolidated Bibliography

Aasgaard, R., (1992) *Automated Cartographic Generalization, with Emphasis on Real-Time Applications*, Ph.D. Thesis. Department of Geodesy and Photogrammetry, Norwegian Institute of Technology, Trondheim, (1992:56).

Abowd, G. D., Atkeson, C. G., Hong, J., Long, S., Kooper, R., and Pinkerton, M., (1996) "CyberGuide: A mobile context-aware tour guide", *Wireless Networks*, 3(5), 421–433.

AdV, (1989) "Amtliches Topographisch-Kartographisches Informationssystem (ATKIS)", *Das Vorhaben der Landesvermessungsverwaltung zum Aufbau Digitaler Landschaftsmodelle und Digitaler Kartenmodelle*, Bonn: Arbeitsgemeinschaft der Vermessungsverwaltungen der Bundesrepublik Deutschland (AdV).

Affholder, J.-G., (1993) "Road modelling for generalization", *Proceedings of NCGIA Research Initiative 8, Specialist Meeting on Formalizing Cartographic Knowledge*, Buffalo, New York, October 24–27, pp. 23–26.

Agent DC1, (1999) Selection of Basic Measure: Agent technical report DC1. Main authors: University of Zurich. Online: http://agent.ign.fr/ (16/06/06).

Agent DC4, (1999) Measures on MESO level & organisations: Agent technical report DC4. Main authors: University of Edinburgh, COGIT Laboratory, University of Zurich. Online: http://agent.ign.fr/ (16/06/06).

Agent, (2000) Project ESPRIT AGENT: Map Generalisation by Multi-Agent Technology. Online: http://agent.ign.fr (16/06/06).

Agrawala, M., (2002) *Visualizing Route Maps*, Ph.D. Dissertation. Stanford University.

Airault, S., (1996) "De la base de données à la carte: une approche globale pour l'équarissage de bâtiments", *Revue Internationale de Géomatique*, 6(2–3), 203–217.

akademie.de, (2005) Netlexikon. Online: http://www.lexikon-definition.de/Herzlich_Willkommen_im_Net-Lexikon (15/07/06).

Akima, H., (1974) "A method of bivariate interpolation and smooth surface fitting based on local procedures", *Communications of the Association for Computing Machinery*, 17, 18–20.

Anders, K.-H., (2003) "A Hierarchical Graph-Clustering Approach to Find Groups of Objects", *Fifth Workshop on Progress in Automated Map Generalisation*, Paris, April 28–30. Online: http://www.geo.unizh.ch/ICA/docs/paris2003/papers/anders_v0.pdf (16/06/06).

Anders, K.-H., (2004) *Parameterfreies hierarchisches Graph-Clustering-Verfahren zur Interpretation raumbezogener Daten*, Ph.D. Thesis. Institute for Photogrammetry, University of Stuttgart. Online: http://elib.uni-stuttgart.de/opus/volltexte/2004/2024/pdf/AndersDiss.pdf (28/06/06).

Anon, (1965) "Records of the discussion on the papers of Waldo, R. T., Bickmore, D. P., and Boyle, A. R.", *International Yearbook of Cartography*, 5, 30–33.

Apache, (2005) Batik 1.6. Online: http://xmlgraphics.apache.org/batik/ (15/07/06).

Appleyard, D., (1969) "Why Buildings are Known", *Environment and Behavior*, 1(1), 131–156.

Arikawa, M., Kawakita, H., and Kambayashi, Y., (1994) "Dynamic Maps as Composite Views of Varied Geographic Database Servers", In: Litwin, W., and Risch, T., (eds.), *Proceedings, 1st International Conference on Applications of Databases*, Vadstena, Sweden, June 21–23, *Lecture Notes in Computer Science*, (819). Berlin: Springer, pp. 142–157.

Arleth, M., (1999) "Problems in screen map design", In: Keller, C. P. (ed.), *Proceedings, 19th International Cartographic Conference and 11th General Assembly of ICA, Touch the Past, Visualize the Future*, Ottawa, Canada, August 14–21, 1, 849–857.

Armstrong, M. P., (1991) "Knowledge classification and organization", In: Buttenfield, B. P., and McMaster, R. B., (eds.), *Map Generalization: Making Rules for Knowledge Representation.* Harlow: Longman Group, pp. 86–102.

Austin, J. L., (1962) "How to Do Things With Words", Clarendon Press. (Traduction française 1970, "Quand dire, c'est faire". Paris: Le Seuil.)

Avelar, S., (2002) *Schematic Maps On Demand – Design, Modeling and Visualisation*, Ph.D. Dissertation. ETH Zürich, (14700).

Avelar, S., and Müller, M., (2000) "Generating topologically correct schematic maps", *Proceedings 9th International Symposium on Spatial Data Handling*, Beijing, China, August 10–12, pp. 4a.28–4a.35.

Averdung, C., (2004) "Modellierung von 3D-Stadtmodellen mit heterogenen Ausgangsdaten", *KS Band 9 – Der X-Faktor – Mehrwert für Geodaten und Karten*, Bonn, pp. 148–156.

Ayasse, J., and Müller, H., (2001) "Interactive manipulation of voxel volumes with free-formed voxel tools", *Vision, Modeling and Visualization 2001*, Session 8, Stuttgart, November 21–23.

Babaud, J., Witkin, A., Baudin, M., and Duda, R., (1986) "Uniqueness of the Gaussian kernel for scale space filtering", *IEEE Transactions on Pattern Analysis and Machine Intelligence*, 8(1), 26–33.

Badard, T., (1999) "On the automatic retrieval of updates in geographic databases based on geographic data matching tools", In: Keller, C. P., (ed.), *Proceedings, 19th International Cartographic Conference and 11th General Assembly of ICA, Touch the Past, Visualize the Future*, Ottawa, Canada, August 14–21, 2, 1291–1300.

Badard, T., (2000) *Propagation des mises à jour dans les bases de données géographiques multi-représentations par analyse des changements géographiques.* Mémoire de thèse de doctorat en Sciences de l'Information Géographique de l'Université de Marne-la-Vallée, Marne-la-Vallée, France. Online: ftp://ftp.ign.fr/ign/COGIT/THESES/ (28/06/06).

Badard, T., and Braun, A., (2003) "OXYGENE: An open framework for the deployment of geographic web services", *Proceedings of the 21st International Cartographic Conference (ICC), Cartographic Renaissance*, Durban, South Africa, August 10–16, 994–1004, CD-ROM.

Badard, T., and Lemarié, C., (2000) "Propagating updates between geographic databases with different scales", In: Atkinson, P., and Martin D., (eds.), *Innovations in GIS VII: GeoComputation.* London: Taylor & Francis, pp. 135–146.

Badard, T., and Richard, D., (2001) "Using XML for the exchange of updating information between geographical information systems", *Computers, Environment and Urban Systems (CEUS)*, 25(1), 17–31.

Bader, M., (2001) *Energy minimization methods for feature displacement in map generalisation*, Ph.D. Thesis. Department of Geography, University of Zurich. Online: http://www.geo.unizh.ch/gis/phd_publications/mbader (30/04/05).

Bader, M., and Barrault, M., (2000) "Improving Snakes for Linear Feature Displacement in Cartographic Generalization", *Proceedings of the 5th International Conference on GeoComputation*, University of Greenwich, August 23–25. Online: http://www.geocomputation.org/2000/GC034/Gc034.htm (16/06/06).

Bader, M., and Barrault, M., (2001) "Cartographic Displacement in Generalization: Introducing Elastic Beams", *Fourth Workshop on Progress in Automated Map Generalization*, Beijing, China, August 2–4. Online: http://www.geo.unizh.ch/ICA/docs/beijing2001/papers/bader_barraultv1.pdf (16/06/06).

Bader, M., and Weibel, R., (1997) "Detecting and Resolving Size and Proximity Conflicts in the Generalization of Polygonal Maps", *Proceedings of the 18th ICA/ACI International Cartographic Conference*, Stockholm, Sweden, June 23–27, 3, 1525–1532.

Bader, M., Barrault, M., and Weibel, R., (2005) "Building Displacement over a Ductile Truss", *International Journal of Geographical Information Science*, 19(8–9), 915–936.

Balley, S., Parent, C., and Spaccapietra, S., (2004) "Modelling geographic data with multiple representations", *International Journal of Geographical Information Science*, 18(4), 327–352.

Bard, S., (2004a) *Méthode d'évaluation de la qualité de données géographiques généralisées, Application aux données urbaines*, Ph.D. Thesis. University of Paris 6. Online: ftp://ftp.ign.fr/ign/COGIT/THESES (16/06/06).

Bard, S., (2004b) "Quality Assessment of Cartographic Generalisation", *Transactions in GIS*, 8(1), 63–81.

Bard, S., and Ruas, A., (2004) "Why and how evaluating generalised data?", In: Fisher, P. F., (ed.), *Developments in Spatial Data Handling, 11th International Symposium on Spatial Data Handling*, University of Leicester, August 23–25. Berlin: Springer, pp. 327–342.

Barillot, X., (2002) "Mesures et structures d'analyse", In: Ruas, A., (ed.), *Généralisation et Représentation Multiple.* Paris: Hermès Science Publications, pp. 187–201.

Barkowsky, T., and Freksa, C., (1997) "Cognitive requirements on making and interpreting maps", In: Hirtle, S., and Frank, A., (eds.), *Spatial information theory: A theoretical basis for GIS, International Conference COSIT '97, Lecture Notes in Computer Science*, (1329). Springer, pp. 347–361.

Barkowsky, T., Latecki, L. J., and Richter, K.-F., (2000) "Schematizing Maps: Simplification of Geographic Shape by Discrete Curve Evolution", *Spatial Cognition II, Integrating Abstract Theories, Empirical Studies, Formal Methods, and Practical Applications, Lecture Notes of Artificial Intelligence*, (1849). Berlin: Springer, pp. 41–53.

Barrault, M., (1998) *Le placement cartographique des écritures : résolution d'un problème à forte combinatoire et présentant un grand nombre de contraintes varies*, Mémoire de thèse de doctorat en Sciences de l'Information Géographique de l'Université de Marne La Vallée, Marne-la-Vallée, France.

Barrault, M., Regnauld, N., Duchêne, C., Haire, K., Baeijs, C., Demazeau, Y., Hardy, P., Mackaness, W., Ruas, A., and Weibel, R., (2001) "Integrating multi-agent, object-oriented, and algorithmic techniques for improved automated map generalization", *Mapping the 21st Century: The 20th International Cartographic Conference*, Beijing, China, August 6–10, 3, 2110–2116.

Bartie, P., and Mackaness, W. A., (2006) "A Speech Based Augmented Reality System for City Tourists", *Transactions in GIS, special issue on Location Based Services*, 10(1), 63–86.

Bauer, B., Veblen, T., and Winkler, J., (1999) "Old Methodological Sneakers: Fashion and Function in a Cross-training Area", *Annals of the Association of American Geographers*, 89, 679–686.

Baus, J., Kray, C., and Krüger, A., (2001) "Visualization and route descriptions in a resource-adaptive navigation aid", *Cognitive Processing*, 2(2–3), 323–345.

Baus, J., Krüger, A., and Wahlster, W., (2002) "Resource-adaptive mobile navigation system", *Proceedings of the 7th International Conference on Intelligent User Interfaces*, San Francisco, California, USA, pp. 15–22.

Beard, K., (1991a) "Constraints on rule formation", In: Buttenfield, B. P., and McMaster, R. B., (eds.), *Map Generalization: Making Rules for Knowledge Representation.* Harlow: Longman Group, pp. 121–135.

Beard, K., (1991b) "Theory of the cartographic line revisited", *Cartographica*, 4(28), 32–58.

Beard, K., and Mackaness, W. A., (1991) "Generalization Operations and Supporting Structures", *Auto-Carto 10: Technical Papers of the 1991 ACSM-ASPRS Annual Convention.* Baltimore: ACSM-ASPRS, 6, 29–45.

Bédard, Y., (1999a) "Principles of spatial database analysis and design", In: Longley, P. A., Goodchild, M. F., Maguire, D. J., and Rhind, D. W., (eds.), *Geographical Information Systems – Principles, Techniques, Application and Management.* 2nd ed., New York: Wiley & Sons, 1, 413–424.

Bédard, Y., (1999b) "Visual Modelling of Spatial Databases Towards Spatial Extensions and UML", *Geomatica*, 53(2), 169–186.

Bédard, Y., Merrett, T., and Han, J., (2001) "Fundamentals of Spatial Data Warehousing for Geographic Knowledge Discovery", In: Miller, H. J., and Han, J., (eds.), *Geographic Data Mining and Knowledge Discovery.* London: Taylor & Francis, pp. 53–73.

Bédard, Y., Larrivée, S., Proulx, M. J., and Nadeau, M., (2004) "Modeling Geospatial Databases with Plug-Ins for Visual Languages: A Pragmatic Approach and the Impacts of 16 Years of Research and Experimentations on Perceptory", In: Wang, S., et al., (eds.), *COMOGIS Workshop, E/R 2004, Lecture Notes in Computer Science*, (3289). Berlin: Springer, pp. 17–30.

Bédard, Y., Proulx, M.-J., Larrivée, S., and Bernier, E., (2002) "Modeling Multiple Representations into Spatial Datawarehouses: A UML-based Approach", *Proceedings of the Joint International Symposium on "GeoSpatial Theory, Processing and Applications"*, ISPRS/Commission IV, SDH2002, 95th Annual CIG Geomatics Conference, Ottawa, Canada, July 9–12, CD-ROM, WG IV/4, 6 p.

Bernier, E., and Bedard, Y., (2002) "Supporting multiple representations with spatial databases view management and the concept of "Vuel", *Joint ISPRS/ICA workshop on "Multi-Scale Representations of Spatial Data"*, Ottawa, Canada, July 7–8. Online: http://www.ikg.uni-hannover.de/isprs/Program_final.html (29/06/06).

Bernier, E., Bédard, Y., and Lambert, M., (2003) "Cartographie sur demande sur le Web et bases de données multidimensionnelles: De la personnalisation par couches cartographiques à la personnalisation par occurrences", *Revue Internationale de Géomatique, special issue on Web GIS*, 13(3), 339–359.

Bertin, J., (1967) *Sémiologie graphique.* Paris: Gauthier-Villars.

Bertin, J., (1973), *Sémiologie graphique*. 2nd ed., Paris: Gauthier-Villars.

Bertin, J., (1983) *Semiology of Graphics: Diagrams, Networks, Maps.* Madison, WI: University of Wisconsin Press.

Bertolotto, M., and Egenhofer, M., (2001) "Progressive transmission of vector map data over the world wide web", *GeoInformatica*, 5(4), 343–373.

Biederman, I., (1987) "Recognition-by-Components: A Theory of Human Image Understanding", *Psychological Review*, 94, 115–147.

Biederman, I., Hilton, H. J., and Hummel, J. E., (1991) "Pattern goodness and pattern recognition", In: Pomerantz, J. R., and Lockhead, G. R., (eds.), *The Perception of Structure*. Washington DC: APA.

Bjerhammar, A., (1973) *Theory of Errors and Generalized Matrix Inverses*. Amsterdam: Elsevier.

Bjørke, J. T., (1997) "Map Generalisation: An Information Theoretic Approach to Feature Elimination", In: Ottoson, L., (ed.), *18th ICA/ACI International Cartographic Conference*, Stockholm, Sweden, June 23–27, 1, 480–486.

Bjørke, J. T., and Nilsen, S., (2002) "Efficient representation of digital terrain models: Compression and decorrelation techniques", *Computers and GeoSciences*, 28(4), 433–445.

Bodansky, E., Gribov, A., and Pilouk, M., (2002) "Smoothing and Compression of Lines Obtained by Raster-to-Vector Conversion", *Selected papers from the 4th International Workshop on Graphics Recognition, Algorithms and Applications, Lecture Notes in Computer Science*, (2390), Ontario, Canada, pp. 256–265.

Boehler, W., and Marbs, A., (2004) "3D scanning and photogrammetry for heritage recording: A comparison", In: Brandt, S. A., (ed.), *Proceedings of the 12th International Conference on Geoinformatics – Geospatial Information Research: Bridging the Pacific and Atlantic*, University of Gävle, Sweden, June 7–9, pp. 291–298.

Boffet, A., (2000) "Creating urban information for cartographic generalization", *Proceedings of the 9th International Symposium on Spatial Data Handling, SDH'00*, Beijing, China, August 10–12, Section 3b, pp. 4–16.

Boffet, A., (2001) *Méthode de création d'informations multi-niveaux pour la généralisation de l'urbain*, Ph.D. Thesis. University of Marne La Vallée. Online: ftp://ftp.ign.fr/ign/COGIT/THESES, in French (16/06/06).

Bollobás, B., (2001) *Random Graphs*. 2nd ed., Cambridge University Press.

Bollobás, B., and Riordan, O., (2004) "The Diameter of a Scale-Free Random Graph", *Combinatorica*, 24(1), 5–34.

Booch, G., (1994) *Object-Oriented Analysis and Design*. 2nd ed., Redwood City: Benjamin Cummings.

Booch, G. L., Rumbaugh, J., and Jakobson, I., (1999) *The Unified Modeling Language User Guide*. Reading, Mass.: Addison-Wesley.

Brassel, K. E., and Weibel, R., (1988) "A Review and Conceptual Framework of Automated Map Generalization", *International Journal of Geographical Information Systems*, 2(3), 229–244.

Braun, A., (2004) "From the schema matching to the integration of updating information into user geographic databases", In: Brandt, S. A., (ed.), *Proceedings of the 12th International Conference on Geoinformatics – Geospatial Information Research: Bridging the Pacific and Atlantic*, University of Gävle, Sweden, June 7–9, pp. 211–218.

Brazile, F., (1999) "Computational Methods for the Automated Symbolisation of 1:24,000 and 1:100,000 USGS SDTS Data", In: Keller, C. P., (ed.), *Proceedings, 19th International Cartographic Conference and 11th General Assembly of ICA, Touch the Past, Visualize the Future*, Ottawa, Canada, August 14–21, 2, pp. 1155–1164.

Brazile, F., (2000) *Semantic Infrastructure and Methods to Support Quality Evaluation in Cartographic Generalisation*, Ph.D. Thesis. Department of Geography, University of Zurich.

Bregt, A., and Bulens, J., (1996) "Application-Oriented Generalisation of Area Objects", In: Molenaar, M., (ed.), *Methods for the Generalisation of Geo-Databases*. Delft: Nederlandse Commissie voor Geodesie (NCG), pp. 57–64.

Brenner, C., Dold, C., and Jülge, K., (2003) "Fusion, Interpretation and Combination of Geodata for the Extraction of Topographic Objects", In: Maas, H.-G., Vosselman, G., and Streilein, A., (eds.), *Workshop 3-D reconstruction from airborne laser scanner and InSAR data, International Archives of Photogrammetry, Remote Sensing and Spatial Information Sciences*, Dresden, Germany, October 8–10, XXXIV(Part 3/W3), 6 p. Online: http://www.ikg.uni-hannover.de/publikationen/publikationen/index.html (16/06/06).

Brewer, C. A., and McMaster, R. B., (1999) "The state of Academic Cartography", *Cartography and GIS*, 26, 215–234.

Brodersen, L., and Andersen, H., (2001) "Quality of Maps – Measuring Communication", *Mapping the 21st Century: The 20th International Cartographic Conference*, Beijing, China, August 6–10, 5, 3044–3051.

Brooks, R., (2000) "National Atlas of Canada producing first map using automated generalisation of framework data", *Cartouche*, 39.

Brooks, R., (2003) Atlas of Canada Open-Source Generalisation Tools. Online: http://www.cim.mcgill.ca/~rbrook/atlas_gen/ (16/06/06).

Brophy, M., (1973) "An Automated Methodology for Linear Generalization in Thematic Cartography", *Proceedings of the American Congress of Surveying and Mapping*, pp. 300–314.

Bruegger, B. P., (1994) *Spatial theory for the integration of resolution-limited data*, Doctoral Thesis. University of Maine.

Bruegger, B. P., and Frank, A. U., (1989) "Hierarchies over topological data structures", *Proceedings GIS/LIS'89, ASPRS/ACSM Annual Convention*, Orlando, Florida, Baltimore, 4, 137–145.

Bruegger, B. P., and Müller, J.-C., (1992) "Mechanisms of geometric abstraction", In: *Proceedings of 5th International Symposium on Spatial Data Handling*, Charleston, SC, August 3–7, 1, 123–133.

Burghardt, D., and Cecconi, A., (2003) "Mesh Simplification for Building Selection", *5th Workshop on Progress in Automated Map Generalization*, Paris, April 28–30. Online: http://www.geo.unizh.ch/publications/acecconi/pdf/burghardt_cecconi03.pdf (16/06/06).

Burghardt, D., and Mathur, A., (2004) "Derivation of digital vector models – project DRIVE", *ICA Workshop on Generalisation and Multiple Representation*, Leicester, UK, August 20–21. Online: http://ica.ign.fr (16/06/06).

Burghardt, D., and Meier, S., (1997) "Cartographic Displacement Using the Snakes Concept", In: Foerstner, W., and Pluemer, L., (eds.), *Semantic Modeling for the Acquisition of Topographic Information from Images and Maps*. Basel: Birkhäuser Verlag, pp. 59–71.

Burghardt, D., Edwardes, A., and Mannes, J., (2003) "An architecture for automatic generalisation of mobile maps", In: Gartner, G., (ed.), Location Based Services & TeleCartography, Proceedings of the Symposium 2004, January 28–29, *Geowissenschaftliche Mitteilungen*, Technische Universität Wien, (66), 33–38.

Burghardt, D., Neun, M., and Weibel, R., (2005) "Generalization Services on the Web – A Classification and an Initial Prototype Implementation", *Auto-Carto 2005 Research Symposium, Proceedings of American Congress on Survey and Mapping*, Las Vegas, USA, March 18–23.

Burghardt, D., Purves, R., and Edwardes, A., (2004) "Techniques for on-the-fly generalisation of thematic point data using hierarchical data structures", *Proceedings of Geographical Information Systems Research – UK (GISRUK 2004)*, Norwich, UK.

Burnett, G., (2000a) " 'Turn Right at the Traffic Lights': The Requirement for Landmarks in Vehicle Navigation Systems", *The Journal of Navigation*, 53(3), 499–510.

Burnett, G., (2000b) "Usable Vehicle Navigation Systems: Are we there yet?", *Vehicle Electronic Systems 2000, European Conference and Exhibition*, ERA Technology Ltd, June 29–30, pp. 3.1.1–3.1.11.

Burnett, G., Smith, D., and May, A., (2001) "Supporting the Navigation Task: Characteristics of 'Good' Landmarks", In: Hanson, M., (ed.), *Contemporary Ergonomics 2001*. London: Taylor & Francis, pp. 441–446.

Buttenfield, B. P., (1991) "A rule for describing line feature geometry", In: Buttenfield, B. P., and McMaster, R. B., (eds.), *Map generalization: Making rules for knowledge representation*. Harlow, Essex: Longman Group, pp. 150–171.

Buttenfield, B. P., (1993) *Research Initiative 3: Multiple Representations, Closing report*. Buffalo: National Center for Geographic Information and Analysis, NCGIA.

Buttenfield, B. P., (1995) "Object-oriented map generalization: modelling and cartographic considerations", In: Müller, J.-C., Lagrange, J.-P., and Weibel, R., (eds.), *GIS and Generalization: Methodology and Practice, Gisdata* (1). London: Taylor & Francis, pp. 91–105.

Buttenfield, B. P., (1999) "Sharing Vector Geospatial Data on the Internet", In: Keller, C. P., (ed.), Proceedings, *19th International Cartographic Conference and 11th General Assembly of ICA, Touch the Past, Visualize the Future*, Ottawa, Canada, August 14–21, 1, 581–590.

Buttenfield, B. P., (2002) "Transmitting Vector Geospatial Data Across the Internet Progressively", *Proceedings, Second International GIScience Conference (GIScience 2002)*, Boulder, Colorado, pp. 51–64.

Buttenfield, B. P., and Delotto, J. S., (1989) *Multiple Representations, Scientific Report for the Specialist Meeting*. National Center for Geographic Information and Analysis, NCGIA, Technical Paper 89-3.

Buttenfield, B. P., and Mark, D. M., (1991) "Expert systems in cartographic design", In: Taylor, D. R. F., (ed.), *Geographic Information Systems: The Microcomputer and Modern Cartography*. Oxford: Pergamon Press, pp. 129–150.

Buttenfield, B. P., and McMaster, R. B., (eds.), (1991) *Map Generalization: Making Rules for Knowledge Representation*. London: Longman Group.

Cámara, M. A., and López, F. J., (2000) "Mathematical Morphology Applied to Raster Generalization of Urban City Block Maps", *Cartographica*, 37(1), 33–48.

Cardenas, A., (2004) *Utilisation de patrons géométriques comme support à la généralisation automatique à la volée*, M.Sc. Thesis. Dept. Geomatics Sciences, Laval University, Canada.

Carroll, J. M., Kellogg, W. A., and Rossen, M. B., (1991) "The Task Artifact Cycle", In: Carroll, J. M., (ed.), *Designing Interaction – Psychology at the Human Computer Interface*. Cambridge: Cambridge University Press, pp. 74–102.

Casakin, H., Barkowsky, T., Klippel, A., and Freksa, C., (2000) "Schematic Maps as Wayfinding Aids", In: Freksa, C., Brauer, W., Habel, C., and Wender, K.-F., (eds.), *Spatial Cognition II, Lecture Notes in Artificial Intelligence*, (1849). Berlin: Springer, pp. 54–71.

Castner, H., and Eastman, R., (1984) "Eye-Movement Parameters and Perceived Map Complexity – I", *The American Cartographer*, 11(2), 107–117.

Cecconi, A., (2003) *Integration of Cartographic Generalisation and Multi-Scale Databases for Enhanced Web Mapping*, Ph.D. Thesis. University of Zurich.

Cecconi, A., and Galanda, M., (2002) "Adaptive Zooming in Web Cartography", *Computer Graphics Forum*, 214, 787–799.

Cecconi, A., Weibel, R., and Barrault, M., (2002) "Improving Automated Generalisation for On-Demand Web Mapping by Multiscale Databases", In: Richardson, D. E., and van Oosterom, P., (eds.), *Advances in Spatial Data Handling, 10th International Symposium on Spatial Data Handling*, Ottawa, Canada, July 9–12. Berlin: Springer, pp. 515–531.

Chalmers, D., Sloman, M., and Dulay, N., (2001) "Map Adaptation for Users of Mobile Systems", *Proceedings 10th International Conference on World Wide Web*, Hong Kong, pp. 735–744.

CHANGE, (2004) Change – generalization of buildings and roads, Generalisation software, University of Hanover, Germany. Online: http://www.ikg.uni-hannover.de/forschung/change/index.en.html (16/06/06).

Chaudhry, O., and Mackaness, W. A., (2005) "Rural and urban road network generalization deriving 1:250,000 from OS MasterMap", *International Cartographic Conference 2005: Mapping Approaches into a Changing World*, A Coruña, Spain, July 10–16. CD-ROM: Theme 9: Cartographic Generalization and Multiple Representation, Session 4, 10 p.

Chen, P. P., (1976) "The Entity-Relationship Model – Towards a Unified View of Data", *ACM Transactions on Database Systems*, 1(1), 9–36.

Cheung, C. K., and Shi, W., (2004) "Estimation of the Positional Uncertainty in Line Simplification in GIS", *The Cartographic Journal*, 41(1), 37–45.

Cheverst, K., Davies, N., Mitchell, K., Friday, A., and Efstratiou, C., (2000) "Developing a Context-Aware Electronic Tourist Guide: Some Issues and Experiences", *Proceedings of the SIGCHI Conference on Human Factors in Computing Systems*, The Hague, The Netherlands, April 1–6, pp. 17–24.

Christ, F. A., (1978) "Program for the Fully Automated Displacement of Point and Line Features in Cartographic Generalization", *Nachrichten aus dem Karten- und Vermessungswesen*, II(35), 5–30.

Christensen, A., (1999) "Cartographic Line Generalization with Waterlines and Medial-Axes", *Cartography and Geographic Information Science*, 26(1), 19–32.

Claramunt, C., and Mainguenaud, M., (1995) "Spatial View: a dynamic and flexible vision of GIS database", *Proceedings of the DEXA International Conference and Workshop on Databases and Expert System Applications*, London, UK. London: Ominpress, pp. 483–493.

Claramunt, C., and Mainguenaud, M., (1997) "A Spatial Data Model for Navigation Knowledge", In: Kraak, M. J., Molenaar, M., and Fendel, E. M., (eds.), *Advances in GIS Research II, Proceedings of the 7th International Symposium on Spatial Data Handling*, Delft, The Netherlands, 1996. London: Taylor & Francis, pp. 767–784.

Clark, J. W., (1977) "Time-distance transformations of transportation networks", *Geographical Analysis*, 9, 195–205.

COGIT, (2005) COGIT ftp site. Online: ftp://ftp.ign.fr/ign/COGIT/ (28/06/06).

Cook, A. C., (1988) *Automatic Cartographic Name Placement Using Rule-Based Systems*, Dissertation. The Polytechnic of Wales, Pontypridd, Mid Glamorgan.

Coren, S., and Ward, L. M., (1989) *Sensation and Perception*. Orlando, FL: Harcourt Brace Jovanovich.

Crampton, J. W., (2002) "Interactivity Types for Geographic Visualization", *Cartography and Geographic Information Science*, 29(2), 85–98.

Cromley, R. G., and Campbell, G. M., (1992) "Noninferior Bandwidth Line Simplification: Algorithm and Structural Analysis", *Geographical Analysis*, 23(1), 25–38.

Cuthbert, A., (1998) "User Interaction with Geospatial Data", OpenGIS®Project Document 98-060.

CyberCity AG, (2005) CyberCity. Online: http://www.cybercity.tv/ (15/07/06).

Davis, E., (1986) *Representing and Acquiring Geographical Knowledge*. Los Altos, California: Morgan Kaufmann Pubs.

Deakin, A., (1996) "Landmarks as Navigational Aids on Street Maps", *Cartography and Geographic Information Systems*, 23(1), 21–36.

Deegree, (2004) Deegree – Building blocks for spatial data infrastructures. Online: http://deegree.sourceforge.net/ (28/06/06).

Delaney, D., and Leitner, H., (1996) "The Political Construction of Scale", *Political Geography*, 16, 93–97.

DeLucia, A., and Black, T. A., (1987) "Comprehensive approach to automatic feature generalization", *Proceedings of the 13th International Cartographic Conference*, Morelia, Mexico, October 12–21, pp. 168–191.

DeMarco, T., (1979) *Structured Analysis and System Specification.* New Jersey: Prentice-Hall.

Denis, M., Pazzaglia, F., Cornoldi, C., and Bertolo, L., (1999) "Spatial Discourse and Navigation: An Analysis of Route Directions in the City of Venice", *Applied Cognitive Psychology*, 13, 145–174.

Devogele, T., Parent, C., and Spaccapietra, S., (1998) "On spatial database integration", *International Journal of Geographical Information Science*, 12(4), 335–352.

Devogele, T., Trévisan, J., and Raynal, L., (1997) "Building a Multiscale Database with Scale-Transition Relationships", In: Kraak, M. J., Molenaar, M., and Fendel, E. M., (eds.), *Advances in GIS Research II, Proceedings of the 7th International Symposium on Spatial Data Handling*, Delft, The Netherlands, 1996. London: Taylor & Francis, pp. 337–351.

DiBiase, D., (1990) "Visualization in the earth sciences", *Earth and Mineral Sciences, Bulletin of the College of Earth and Mineral Sciences*, The Pennsylvania State University 59(2), 13–18.

Dey, A. K., (2001) "Understanding and Using Context", *Personal and Ubiquitous Computing*, 5(1), 4–7.

Diestel, R., (2000) *Graph Theory*. 2nd ed., New York: Springer.

Digest, (2005) Download DIGEST 2.1. Online: http://www.digest.org/DownloadDigest.htm (05/05/05).

Digital Chart of China, (2005) Fundamental GIS: Digital Chart of China, 1:1M, version 1. Online: http://sedac.ciesin.columbia.edu/china/geogmap/dcchina/dcchinadesc.html (19/06/06).

Dijkstra, E. W., (1959) "A note on two problems in connexion with graphs", *Numerische Mathematik*, 1, 269–271.

Dillemuth, J., (2005) "Evaluating maps for mobile display", In: Gartner, G., (ed.), Location-Based Services & Tele-Cartography, Proceedings of the Symposium 2005, *Geowissenschaftliche Mitteilungen*, Technische Universität Wien, November 28–30, (74), 143–147.

Ding, A., (2000) "An efficient fully-automated approach for extracting 3D building polygons from raw LIDAR data", *Proceedings of Urban and Regional Information Systems Association 2000*, August 19–23, Orlando, Florida. Online: http://www.urisa.org/store/urisa_conference_proceedings.htm (30/04/05).

Dodge, M., and Kitchin, R., (2001) *Mapping Cyberspace*. London: Routledge.

Doerschler, J. S., and Freeman, H., (1989) "An expert system for dense-map name placement", *Proceedings AUTO-CARTO 9, Ninth International Symposium on Computer-Assisted Cartography*, Baltimore, Maryland, April 2–7. ASPRS and ACSM, pp. 215–224.

Döllner, J., and Walther, M., (2003) "Real-Time Expressive Rendering of City Models", *Proceedings of IEEE International Conference on Information Visualization*, Seattle, Washington, October 19–21, London, pp. 245–250.

Dorling, D., and Fairbairn, D., (1997) "The History of Cartography", In: Dorling, D., and Fairbairn, D., *Mapping: Ways of representing the world.* Harlow: Longman, pp. 6–24.

Douglas, D. H., and Peucker, T. K., (1973) "Algorithms for the reduction of the number of points required to represent a line or its caricature", *The Canadian Cartographer*, 10(2), 112–123.

Downs, T. C., and Mackaness, W. A., (2002) "Automating the Generalisation of Geological Maps: The Need for an Integrated Approach", *The Cartographic Journal*, 39(2), 137–152.

Dransch, D., (2005) "Activity and Context – A Conceptual Framework for Mobile Geoservices", In: Meng, L., Zipf, A., and Reichenbacher, T., (eds.), *Map-based Mobile Services: Theories, Methods and Implementations.* Berlin: Springer, pp. 30–42.

Duchêne, C., (2003) "Automated map generalisation using communicating agents", *Proceedings of the 21st International Cartographic Conference (ICC), Cartographic Renaissance*, Durban, South Africa, August 10–16, pp. 160–169, CD-ROM.

Duchêne, C., (2004) *Généralisation cartographique par agents communicants: Le modèle CartACom*, Doctoral Thesis. LIP6, Université Paris 6 – Pierre et Marie Curie. Online: ftp://ftp.ign.fr/ign/COGIT/THESES (16/06/06).

Duchêne, C., Bard, S., Barillot, X., Ruas, A., Trévisan, J., and Holzapfel, F., (2003) "Quantitative and qualitative description of building orientation", *Fifth Workshop on Progress in Automated Map Generalisation*, Paris, April 28–30. Online: http://www.geo.unizh.ch/ICA/docs/paris2003/papers/duchene_et_al_v1.pdf (16/06/06).

Duckham, M., Kulik, L., and Worboys, M., (2003) "Imprecise navigation", *GeoInformatica*, 7(2), 79–94.

Duda, R. O., and Hart, P. E., (1972) "Use of the Hough Transformation to Detect Lines and Curves in Pictures", *Communications of the Association for Computing Machinery*, 15, 11–15.

Duda, R. O., Hart, P. E., and Stork, D. G., (2000) *Pattern Classification*. 2nd ed., New York: Wiley & Sons.

Dunkars, M., (2004) *Multiple representation databases for topographic information*, Doctoral Thesis. Royal Institute of Technology, Stockholm, Sweden.

Dutton, G., (1999) "Scale, Sinuosity and Point Selection in Digital Line Generalisation", *Cartography and Geographic Information Systems*, 26(1), 33–54.

Dykes, J., MacEachren, A. M., and Kraak, M.-J., (2005) *Exploring Geovisualization*. International Cartographic Association, London: Elsevier.

Eckert, M., (1908) "On the nature of maps and map logic", *Bulletin of the American Geographical Society*, 40(6), 344–351.

Edwardes, A., (2004) "Modelling space for the generalisation of point maps", *Proceedings of Geographical Information Systems Research UK 2004 Conference (GISRUK 2004)*, University of East Anglia, Norwich, UK, April 28–30.

Edwardes, A., and Mackaness, W. A., (2000) Intelligent generalisation of urban road networks. *Proceedings of Geographical Information Systems Research UK 2000 Conference (GISRUK 2000)*, University of York, April 5–7, pp. 81–85.

Edwardes, A., and Mackaness, W., (1999) "Modelling knowledge for automated generalisation of categorical maps – a constraint based approach", In: Atkinson, P., (ed.), *GIS and Geocomputation, Innovations in GIS*, (7), pp. 161–173.

Edwardes, A., Burghardt, D., and Weibel, R., (2003a) "WebPark – Location Based Services for Species Search in Recreation Area", *Proceedings of the 21st International Cartographic Conference (ICC), Cartographic Renaissance*, Durban, South Africa, August 10–16, pp. 1012–1021, CD-ROM.

Edwardes, A., Burghardt, D., and Weibel, R., (2005) "Portrayal and Generalisation of Point Maps for Mobile Information Services", In: Meng, L., Zipf, A., and Reichenbacher T., (eds.), *Map-based Mobile Services: Theories, Methods and Implementations*. Berlin: Springer, pp. 11–30.

Edwardes, A., Burghardt, D., Bobzien, M., Harrie, L., Lehto, L., Reichenbacher, T., Sester, M., and Weibel, W., (2003b) "Map Generalisation Technology: Addressing the Need for a Common Research Platform", *Proceedings of the 21st International Cartographic Conference (ICC), Cartographic Renaissance*, Durban, South Africa, August 10–16, pp. 170–180, CD-ROM.

Edwardes, A., Mackaness, W. A., and Urwin, T., (1998) "Self-Evaluating Generalisation Algorithms to Automatically Derive Multi Scale Boundary Sets", In: Poiker, T., and Chrisman, N., (eds.), *Proceedings of the 8th International Symposium on Spatial Data Handling*, Vancouver, Canada, July 12–15, pp. 361–372.

Egenhofer, M. J., and Frank, A. U., (1989) "Object-Oriented Modelling in GIS: Inheritance and Propagation", *AUTO-CARTO 9, Ninth International Symposium on Computer-Assisted Cartography*, Baltimore, Maryland, April 2–7. ASPRS and ACSM, pp. 588–598.

Egenhofer, M. J., Clementini, E., and Felice, P. D., (1994) "Evaluating Inconsistencies Among Multiple Representations", In: Waugh, T. C., and Healey, R. G., (eds.), *Advances in GIS Research, Proceedings of the 6th International Symposium on Spatial Data Handling, SDH'94*, Edinburgh, Scotland, September 5–9. IGU Commission on GIS and Association for Geographic Information, 2, 901–920.

Ehrliholzer, R., (1995) "Quality assessment in generalisation: integrating quantitative and qualitative methods", *Proceedings of the 17th International Cartographic Conference*, Barcelona, September 3–9, 2, 2241–2250.

El-Geresy, B. A., and Abdelmoty, A. I., (1998) "A Qualitative Approach to Integration in Spatial Databases", In: Quirchmayr, G., et al. (eds.), *Proceedings of the 9th International Conference on Database and Expert Systems Applications (DEXA'98)*, Vienna, August 24–28, *Lecture Notes in Computer Science*, (1460). Berlin: Springer, pp. 280–289.

Elias, B., (2002) "Automatic derivation of location maps", *Proceedings of the Joint International Symposium on "GeoSpatial Theory, Processing and Applications"*, ISPRS/Commission IV, Spatial Data Handling 2002, 95th Annual CIG Geomatics Conference, Ottawa, Canada, July 8–12, CD-ROM, WG IV/3, 6 p.

Elias, B., (2003a) "Determination of Landmarks and Reliability Criteria for Landmarks", *Papers of the Fifth Workshop on Progress in Automated Map Generalization*, Paris, France, April 28–30. Online: http://www.geo.unizh.ch/ICA/docs/paris2003/papers03.html (16/06/06).

Elias, B., (2003b) "Extracting Landmarks with Data Mining Methods", In: Kuhn, W., Worboys, M., and Timpf, S., (eds.), *Spatial Information Theory: Foundations of Geographic Information Science, Lecture Notes in Computer Science*, (2825). Berlin: Springer, pp. 398–412.

Elias, B., and Brenner, C., (2004) "Automatic Generation and Application of Landmarks in Navigation Data Sets", In: Fisher, P., (ed.), *Developments in Spatial Data Handling, 11th International Symposium on Spatial Data Handling*, Leicester, UK, August 23–25. Berlin: Springer, pp. 469–480.

Elias, B., Hampe, M., and Sester, M., (2005) "Adaptive Visualisation of Landmarks using an MRDB", In: Meng, L., Zipf, A., and Reichenbacher, T., (eds.), *Map-based Mobile Services: Theories, Methods and Implementations.* Berlin: Springer, pp. 73–86.

Erdös, P., and Rényi, A., (1960) "On the evolution of random graphs", *Publications of the Mathematical Institute of the Hungarian Academy of Sciences*, 5, 17–61.

ESRI, (2000) Map Generalization in GIS: Practical Solution with Workstation ArcInfo Software. Online: http://support.esri.com/index.cfm?fa=knowledgebase.whitePapers.gateway (28/06/06).

eSVG, (2004) Embedded Scalable Vector Graphics. Online: http://esvg.ultimodule.com/bin/esvg/templates/splash.asp?NC=1655X (28/06/06).

Falk City Guide, (2004). Online: https://www.falk.de/cityguide/ (18/06/06).

Faloutsos, M., Faloutsos, P., and Faloutsos, C., (1999) "On Power-Law Relationships of the Internet Topology", *Proceedings of the ACM SIGCOMM Conference on Network Architectures and Protocols*, Cambridge, MA, USA, August 30–September 3, pp. 251–262.

Farin, G., (2001) *Curves and Surfaces for CAGD: A Practical Guide.* 5th ed., San Francisco: Morgan Kaufmann.

FES, (2004) OpenGIS®Filter Encoding Implementation Specification. Online: http://www.opengis.org/techno/implementation.htm (05/05/05).

Fisher, P. F., (1998) "Is GIS Hidebound by the Legacy of Cartography?", *The Cartographic Journal*, 35, 5–9.

Fisher, P. F., and Mackaness, W. A., (1987) "Are cartographic expert systems possible?" In: Chrisman, N. R., (ed.), *Auto-Carto 8 Proceedings, Eighth International Symposium on Computer-Assisted Cartography*, Baltimore, Maryland, March 29–April 3, pp. 530–534.

Fitzke, J., Greve, K., Müller, M., and Poth, A., (2004) "Building SDIs with Free Software – the deegree project", *Proceedings 7th Conference Global Spatial Data Infrastructure*, Bangalore, India, February 2–6. Online: http://www.lat-lon.de/download/gsdi-7_full_paper_fitzke_greve_mueller_poth_2004-02-06.pdf (05/05/05).

Foldoc, (2005) Free On-line Dictionary of Computing. Online: http://foldoc.org/ (15/07/06).

Foley, J. D., van Dam, A., Feiner, S. K., and Hughes, J. F., (1996) *Computer Graphics: Principles and Practice.* 2nd ed., Reading, MA: Addison-Wesley.

Forberg, A., (2004a) "Generalization of 3D building data based on a scale-space approach", In: Altan, M. O., (ed.), *Proceedings of the XXth ISPRS Congress, The International Archives of the Photogrammetry, Remote Sensing and Spatial Information Sciences*, Istanbul, Turkey, July 12–23, XXXV(Part B), 194–199.

Forberg, A., (2004b) "Simplification of 3D building data", *ICA Workshop on Generalisation and Multiple Representation*, Leicester, UK, August 20–21. Online: http://ica.ign.fr/ (16/06/06).

Forberg, A., (2005) *Generalisierung dreidimensionaler Gebäudedaten auf der Basis von Maßstabsräumen*, Ph.D. Thesis. University of Bundeswehr, Munich.

Forberg, A., and Mayer, H., (2003) "Squaring and scale-space based generalization of 3D building data", *ICA Workshop on Progress in Automated Map Generalisation*, Paris, April 28–30. Online: http://www.geo.unizh.ch/ICA/docs/paris2003/papers03.html (16/06/06).

Forrest, D., (1993) "Expert systems and cartographic design", *The Cartographic Journal*, 30(2), 143–148.

Foucault, M., (1983) *This is not a pipe*. 3rd, rev. ed., Berkeley, Los Angeles: University of California Press.

Frank, A. U., (1990) "The National Center for Geographic Information and Analysis in the United States of America", *Proceedings of 19th Congress of International Federation of Surveyors*, Helsinki, Finland, June 10–19, pp. 50–64.

Frank, A. U., and Timpf, S., (1995) "Multiple Representations for Cartographic Objects in a Multi-Scale Tree – an Intelligent Graphical Zoom", *Computer and Graphics Special Issue: Modelling and Visualization of Spatial Data in Geographic Information Systems*, 18(6), 823–829.

Frank, A., and Egenhofer, M., (1988) *Object-Oriented Database Technology for GIS*, Seminar Workbook. San Antonio, Texas.

Frawley, W. J., Piatetsky-Shapiro, G., and Mathews, C. J., (1992) "Knowledge discovery in databases: An overview", *AI Magazine*, 13(3), 57–70.

Freeman, L. C., (1979) "Centrality in Social Networks: Conceptual Clarification", *Social Networks*, 1, 215–239.

Freitag, U., (1987) "Do We Need a New Cartography?", *Papers presented to the Thirteenth International Conference on Cartography*, International Cartographic Association (ICA), Morelia, Mexico, *Nachricten aus dem Karten- und Vermessungswesen.* Frankfurt a. M.: Verlag des Instituts für Angewandte Geodäsie, II(46), 51–59.

Freksa, C., (1999) "Spatial Aspects of Task-Specific Wayfinding Maps", In: Gero, J., and Tversky, B., (eds.), *Visual and Spatial Reasoning in Design*. Key Centre of Design Computing and Cognition, University of Sydney, pp. 15–32.

Friis-Christensen, A., (2003) *Issues in the Conceptual Modeling of Geographic Data*, Ph.D. Dissertation. Aalborg University, Denmark.

Friis-Christensen, A., Skogan, D., Jensen, C. S., Skagestein, G., and Tryfona, N., (2002) "Management of Multiply Represented Geographic Entities", In: Nascimento, M. A., Özsu, M. T., and Zaiane, O. R., (eds.), *Proceedings of the International Database Engineering & Applications Symposium (IDEAS'02)*, IEEE Computer Society, Edmonton, Canada, pp. 150–159.

Fritsch, E., (1997) *Représentations de la géométrie et des contraintes cartographiques pour la généralisation du linéaire routier*, Ph.D. Thesis. Université de Marne la Vallée, France. Online: ftp://ftp.ign.fr/ign/COGIT/THESES/FRITSCH/ (28/06/06).

Früh, C., (2002) *Automated 3D model generation for urban environments*, Ph.D. Thesis. University of Karlsruhe, Germany.

Fuhrmann, S., and Kuhn, W., (1999) "Interface Design Issues for Interactive Animated Maps", In: Keller, C. P. (ed.), *Proceedings, 19th International Cartographic Conference and 11th General Assembly of ICA, Touch the Past, Visualize the Future*, Ottawa, Canada, August 14–21, 1, 875–884.

Fuller, R., and Brown, N., (1996) "A CORINE map of Great Britain by automated means, Techniques for automatic generalization of the Land Cover Map of Great Britain", *International Journal of Geographical Information Systems*, 10(8), 937–953.

Gahegan, M., Wachowicz, M., Harrower, M., and Rhyne, T.-M., (2001) "The Integration of Geographic Visualisation with Knowledge Discovery in Databases and Geocomputation", *Cartography and Geographic Information Science*, 28(1), 29–44.

Galanda, M., (2003) *Automated Polygon Generalization in a Multi Agent System*, Doctoral Thesis. Department of Geography, University of Zurich.

Galanda, M., and Weibel, R., (2002) "An agent-based framework for polygonal subdivision generalization", In: Richardson, D. E., and van Oosterom, P., (eds.), *Advances in Spatial Data Handling, 10th International Symposium on Spatial Data Handling*, Ottawa, Canada, July 9–12. Berlin: Springer, pp. 121–135.

Galanda, M., and Weibel, R., (2003) "Using an Energy Minimization Technique for Polygon Generalization", *Cartography and Geographic Information Science*, 30(3), 259–275.

Gamma, E., Helm, R., Johnson, R., and Vlissides, J., (1995) *Design Patterns: Elements of reusable object-oriented software*. Reading, Mass.: Addison-Wesley.

Gardarin, G., (1999) *Bases de Données – Objet et Relationnel*. Paris: Editions Eyrolles.

Garland, K., (1994) *Mr Beck's Underground Map*. Harrow Weald, Middlesex: Capital Transport Publishing.

Gartner, G., (ed.), (2005) Location-Based Services & TeleCartography, Proceedings of the Symposium 2005, *Geowissenschaftliche Mitteilungen*, Technische Universität Wien, November 28–30 (74).

Gartner, G., and Uhlirz, S., (2005) "Cartographic Location Based Services", In: Meng, L., Zipf, A., and Reichenbacher, T., (eds.), *Map-based Mobile Services: Theories, Methods and Implementations*. Berlin: Springer, pp. 159–171.

Gartner, G., Frank, A., and Retscher, G., (2003) "Pedestrian Navigation System for Mixed Indoor/Outdoor Environments", In: Gartner, G., (ed.), Location Based Services & TeleCartography, Proceedings of the Symposium 2004, January 28–29, *Geowissenschaftliche Mitteilungen*, Technische Universität Wien, (66), 160–167.

Gbei E., Cosma, I., Moulin, B., and Jabeur, N., (2003a) "Modelling SVG and GML Data for the Cartographic Generalisation and the Multiple Representation on the Web", *Proceedings SVG Open 2003*, Vancouver, Canada.

Gbei, E., Moulin, B., Cosma, I., Jabeur, N., and Delval, N., (2003b) "Conception d'un prototype de service web géolocalisé appliqué à l'industrie récréo-touristique", *Revue Internationale de Géomatique, Special issue on Web GIS*, 13(3), 375–395.

GEOAPI, (2004), GeoApi Project. Online: http://geoapi.sourceforge.net/ (16/06/06).

GEOSERVER, (2004) The Geoserver project – the open internet gateway for geographic data. Online: http://geoserver.sourceforge.net/html/index.php (05/05/05).

GEOTOOLS, (2004) Geotools – The open source Java GIS toolkit. Online: http://geotools.codehaus.org/ (28/06/06).

Germanchis, T., and Cartwright, W., (2003) "The potential to use games engines and games software to develop interactive, three-dimensional visualisations of geography", *Proceedings of the 21st International Cartographic Conference (ICC), Cartographic Renaissance*, Durban, South Africa, August 10–16, pp. 352–357, CD-ROM.

Gill, G. A., and Trigg, A. D., (1988) "CANVAS: An intelligent system for colour selection on CRT displays", *International Journal of Remote Sensing*, 9(9), 1423–1432.
GiMoDig, (2001) Geospatial info-mobility service by real-time data-integration and generalisation, GiMoDig project, IST-2000-30090. Online: http://gimodig.fgi.fi/ (16/06/06).
GML, (2004) OpenGIS® Geography Markup Language (GML) Implementation Specification. Online: http://www.opengis.org/techno/implementation.htm (05/05/05).
GO-1, (2004) OpenGIS® Geographic Objects initiative. Online: http://ip.opengis.org/go1 (16/06/06).
Gold, C., and Dakowicz, M., (2003) "Digital Elevation Models from Contour Lines", *GIM International*, 17(2), 56–59.
Gold, C., and Snoeyink, J., (2001) "A One-Step Crust and Skeleton Extraction Algorithm", *Algorithmica*, 30, 144–163.
Gold, C., and Thibault, D., (2001) "Map Generalization by Skeleton Retraction", *Mapping the 21st Century: The 20th International Cartographic Conference*, Beijing, China, August 6–10, 3, 2072–2081.
Golledge, R., (1999) "Human Cognitive Maps and Wayfinding", In: Golledge, R. G., (ed.), *Wayfinding Behavior: Cognitive Mapping and Other Spatial Processes.* Baltimore: The John Hopkins University Press, pp. 5–46.
Goodchild, M. F., (2004) "Scales of Cybergeography", In: Sheppard, E., and McMaster, R. B., (eds.), *Scale and Geographic Inquiry: Nature, Society, and Method.* Oxford, UK: Blackwell Publishing, pp. 154–169.
Goodchild, M. F., and Proctor, J., (1997) "Scale in a digital geographic world", *Geographical and Environmental Modelling*, 1, 5–23.
Goodchild, M., and Jeansoulin, R., (1998) "Foreword", In: Goodchild, M., and Jeansoulin, R., (eds.), *Data quality in geographic information: from error to uncertainty*. Paris: Hermés, pp. 11–14.
Goodenough, D. G., Goldberg, M., Plunkett, G., and Zelek, J., (1987) "An Expert System for Remote Sensing", *IEEE Transactions on Geoscience and Remote Sensing*, G-25(3), 349–359.
Google Earth, (2005) A 3D interface to the planet. Online: http://earth.google.com/ (15/07/06).
Goudie, A., (ed.), (1985) *Encyclopedic Dictionary of Physical Geography*. Oxford: Blackwell.
Greenwood, J., and Mackaness, W. A., (2002) "Revealing Associative Relationships for Complex Object Creation", *Geoscience 2002 Abstracts: The Second International Conference on GIS*, Boulder, Colorado, pp. 48–55.
Gröger, G., Kolbe, T. H., Drees, R., Kohlhaas, A., Müller, H., Knospe, F., Gruber, U., and Krause, U., (2004) *Das interoperable 3D-Stadtmodell der SIG 3D der GDI NRW*, Version 2. Online: http://www.ikg.uni-bonn.de/sig3d/ (16/06/06).
Grünreich, D., (1985) "Computer-Assisted Generalization", In: *Papers CERCO-Cartography Course*. Institut für Angewandte Geodäsie, Frankfurt a. M.
Grünreich, D., (1992) "ATKIS – A topographic information system as a basis for GIS and digital cartography in Germany", In: Vinken, R., (ed.), *From Digital Map Series in Geosciences to Geo-Information Systems, Geologisches Jahrbuch*, Federal Institute of Geosciences and Resources, Hannover, A(122), 207–216.
Grünreich, D., (1995) "Development of Computer-Assisted Generalization", In: Müller, J.-C., Lagrange, J.-P., and Weibel, R., (eds.), *GIS and Generalization: Methodology and Practice.* London: Taylor & Francis, pp. 47–55.
Guptill, S. C., and Morrison, J. L., (eds.), (1995) *Elements of spatial data quality*. International Cartographic Association, Kidlington: Elsevier Science.
Haase, H., Dai, F., Strassner, J., and Göbel, M., (1997) "Immersive Investigation of Scientific Data", In: Nielson, G. M., Hagen, H., and Müller, H., (eds.), *Scientific Visualization: Overviews, Methodologies, Techniques*. IEEE Computer Society, pp. 35–58.
Hake, G., Grünreich, D., and Meng, L., (2002) *Kartographie – Visualisierung raum-zeitlicher Informationen*. 8th, rev. ed., Reading: de Gruyter.
Haken, H., and Portugali, J., (2003) "The face of the city is its information", *Journal of Environmental Psychology*, 23(4), 385–408.
Hampe, M., and Elias, B., (2003) "Integrating Topographic Information and Landmarks for Mobile Applications", In: Gartner, G., (ed.), Location Based Services & TeleCartography, Proceedings of the Symposium 2004, January 28–29, *Geowissenschaftliche Mitteilungen*, Technische Universität Wien, (66), 147–155.
Hampe, M., and Sester, M., (2002) "Real-time integration and generalization of spatial data for mobile applications", *Geowissenschaftliche Mitteilungen, Maps and the Internet*, Wien, Heft (60), pp. 167–175.
Hampe, M., Anders, K.-H., and Sester, M., (2003) "MRDB applications for data revision and real-time generalisation", *Proceedings of the 21st International Cartographic Conference (ICC), Cartographic Renaissance*, Durban, South Africa, August 10–16, pp. 192–202, CD-ROM.

Hampe, M., Sester, M., and Harrie, L., (2004) "Multiple Representation Databases to Support Visualization on Mobile Devices", In: Altan, M. O., (ed.), *Proceedings of the XXth ISPRS Congress, The International Archives of the Photogrammetry, Remote Sensing and Spatial Information Sciences*, Istanbul, Turkey, July 12–23, XXXV(B4:IV), 135–140.

Han, J., and Kamber, M., (2001) *Data Mining: Concepts and Techniques.* San Francisco: Morgan Kaufmann.

Hangouet, J.-F., (1998) *Approche et méthodes pour l'automatisation de la généralisation cartographique; application en bord de ville*, Ph.D. Thesis. University of Marne la Vallée / COGIT, France. Online: ftp://ftp.ign.fr/ign/COGIT/THESES (16/06/06).

Harary, F., (1969) *Graph Theory.* Reading, Mass.: Addison-Wesley.

Hardy, P. G., (1999) "Active Object Techniques for Production of Multiple Map and Geodata Products from a Spatial Database", In: Keller, C. P., (ed.), *Proceedings, 19th International Cartographic Conference and 11th General Assembly of ICA, Touch the Past, Visualize the Future*, Ottawa, Canada, August 14–21, 2, 1321–1340.

Hardy, P. G., and Woodsford, P. A., (1997) "Mapping with live features: object-oriented representation", *Proceedings 18th ICA/ACI International Cartographic Conference, ICC97*, Stockholm, Sweden, June 23–27, 3, 1682–1689.

Hardy, P., Hayles, M., and Revell, P., (2003) "Clarity – A New Environment for Generalisation, Using Agents, Java, XML and Topology", *Papers of the Fifth ICA Workshop on Progress in Automated Map Generalization*, Paris, April 28–30. Online: http://www.geo.unizh.ch/ICA/docs/paris2003/papers/hardy_et_al_v1.pdf (16/06/06).

Harrie, L., (1998) *Generalization Methods for Propagating Updates Between Cartographic Data Sets*, Licentiate Thesis. Department of Surveying, Lund Institute of Technology, Lund University, Lund, Sweden.

Harrie, L., (1999) "The Constraint Method for Solving Spatial Conflicts in Cartographic Generalization", *Cartography and Geographic Information Science*, 26(1), 55–69.

Harrie, L., (2001) *An Optimisation Approach to Cartographic Generalisation*, Doctoral Thesis. Department of Surveying, Lund Institute of Technology, Lund University, Lund, Sweden.

Harrie, L., (2003) "Weight-Setting and Quality Assessment in Simultaneous Graphic Generalisation", *The Cartographic Journal*, 40(3), 221–233.

Harrie, L., (2004) "Using Simultaneous Graphic Generalisation in a System for Real-Time Maps", *ICA Workshop on Generalisation and Multiple Representation*, Leicester, UK, August 20–21. Online: http://ica.ign.fr/ (16/06/06).

Harrie, L., (2005) "Integrating Graphic Generalisation Methods in a System for Real-Time Maps", In: Hauska, H., and Tveite, H., (eds.), *ScanGIS 2005, Proceedings of the 10th Scandinavian Research Conference on Geographical Information Science*, Stockholm, Sweden, June 13–15, pp. 63–74.

Harrie, L., and Hellström, A.-K., (1999) "A Prototype System for Propagating Updates between Cartographic Data Sets", *The Cartographic Journal*, 36(2), 133–140.

Harrie, L., and Johansson, M., (2003) "Real-time data generalisation and integration using Java", *Geoforum Perspektiv*, Feb. 2003, pp. 29–34.

Harrie, L., and Sarjakoski, T., (2002) "Simultaneous Graphic Generalization of Vector Data Sets", *GeoInformatica*, 6(3), 233–261.

Harrie, L., Sarjakoski, L. T., and Lehto, L., (2002a) "A Mapping Function for Variable-Scale Maps in Small-Display Cartography", *Journal of Geospatial Engineering*, 4(2), 111–123.

Harrie, L., Sarjakoski, L. T., and Lehto, L., (2002b) "A variable-scale map for small-display cartography", *Proceedings of the Joint International Symposium on "GeoSpatial Theory, Processing and Applications"*, ISPRS/Commission IV, Spatial Data Handling 2002, 95th Annual CIG Geomatics Conference, Ottawa, Canada, July 9–12, CD-ROM, WG IV/3, 6 p.

Harrie, L., Stigmar H., Koivula T., and Lehto, L., (2004) "An Algorithm for Icon Labelling on a Real-Time Map", In: Fisher, P., (ed.), *Developments in Spatial Data Handling, 11th International Symposium on Spatial Data Handling*, University of Leicester, August 23–25. Berlin: Springer, pp. 493–507.

Hartshorne, R., (1939) *The Nature of Geography: A critical survey of current thought in the light of the past.* Lancaster, PA: Association of American Geographers.

Hayes-Roth, F., Waterman, D., and Lenat, D., (1983) *Building Expert Systems.* Reading, MA: Addison-Wesley.

Head, G., (1984) "A map as a natural language – new insight into cartographic communication", *Cartographica*, 21(31), 1–32.

Heckbert, P., and Garland, M., (1997) "Survey of polygonal surface simplification algorithms", *Multiresolution Surface Modeling Course Notes #25, SIGGRAPH '97*, Los Angeles, August 3–8.

Hedley, N. R., (2003) "Empirical evidence of advanced geographic visualisation interface use", *Proceedings of the 21st International Cartographic Conference (ICC), Cartographic Renaissance*, Durban, South Africa, August 10–16, pp. 383–393, CD-ROM.

Heinzle, F., and Sester, M., (2004) "Derivation of Implicit Information from Spatial Data Sets with Data Mining", In: Altan, M. O., (ed.), *Proceedings of the XXth ISPRS International Congress, International Archives of Photogrammetry, Remote Sensing and Spatial Information Sciences*, Istanbul, Turkey, July 12–23, XXXV(B4/IV), 335–340.

Highway, (2004). Online: ftp://ftp.igsb.uiowa.edu/gis_library/IA_State/Infrastructure/Transportation/highway.zip (19/06/06).

Hjelm, J., (2002) *Creating Location Services for the Wireless Web: Professional Developer's Guide*. New York: Wiley.

Højholt, P., (1998) "Solving Local and Global Space Conflicts in Map Generalization Using a Finite Element Method Adapted from Structural Mechanics", *Proceedings of the 8th International Symposium on Spatial Data Handling*, Vancouver, Canada, July 12–15, pp. 679–689.

Højholt, P., (2000) "Solving Space Conflicts in Map Generalization: Using a Finite Element Method", *Cartography and Geographic Information Science*, 27(1), 65–73.

Horton, R. E., (1945) "Erosional development of streams and their drainage basins: hydrophysical approach to quantitative morphology", *Bulletin of the Geological Society of America*, 56, 275–370.

Hough, P. V. C., (1959) "Machine Analysis of Bubble Chamber Pictures", In: Kowarski, L., (ed.), *Proceedings of International Conference on High Energy Accelerators and Instrumentation*, CERN, Geneve, September 14–19, pp. 554–556.

Hubert, F., (2002) "Map Samples to Help GI Users Specify their Needs", In: Richardson, D. E., and van Oosterom, P., (eds.), *Advances in Spatial Data Handling, 10th International Symposium on Spatial Data Handling*, Ottawa, Canada, July 9–12. Berlin: Springer, pp. 533–545.

Hubert, F., (2003) *Modèle de traduction des besoins d'un utilisateur pour la dérivation de données géographiques et leur symbolisation par le web*, Ph.D Thesis. University of Caen / COGIT. Online: ftp://ftp.ign.fr/ign/COGIT/THESES (16/06/06).

Hui, W., Qingxin, X., and Yu, B., (2004) "A Knowledge-based restricted problem solving method in GIS applications", In: Brandt, S. A., (ed.), *Proceedings of the 12th International Conference on Geoinformatics – Geospatial Information Research: Bridging the Pacific and Atlantic*, University of Gävle, Sweden, June 7–9, pp. 179–186.

Hunt, E., and Waller, D., (1999) *Orientation and Wayfinding: A Review, ONR technical report, N00014-96-0380*. Arlington, VA: Office of Naval Research.

Hutchinson, G. E., (1953) "The Concept of Pattern in Ecology", *Proceedings of the Academy of Natural Sciences of Philadelphia*, (153), 1–12.

IBM, (2004) The Who, What, When, Where, Why and How of Becoming an On Demand Business. Online: http://www-306.ibm.com/e-business/ondemand/us/index.html (05/05/05).

ICA, (2005a) ICA Commission on Generalisation and Multiple Representation. Online: http://ica.ign.fr/ (17/07/06).

ICA, (2005b) ICA Commission for Maps and the Internet. Online: http://maps.unomaha.edu/ica/ (17/07/06).

ICA, (2005c) ICA Commission Visualization and Virtual Environments. Online: http://www.kartografie.nl/icavis/ (17/07/06).

ICA, International Cartographic Association, (1973) *Multilingual Dictionary of Technical Terms in Cartography*. Franz Steiner Verlag.

Illert, A., and Afflerbach, S., (2004) Global schema specification. GiMoDig-project, IST-2000-30090, Deliverable D5.3.1, Public EC report, 35 p. Online: http://gimodig.fgi.fi/deliverables.php (20/06/06).

Imhof, E., (1982) *Cartographic Relief Presentation*. Berlin: de Gruyter.

Inmon, W. H., (2002) *Building the Data Warehouse*. 3rd ed., New York: Wiley & Sons.

Inmon, W. H., Richard, D., and Hackathorn, D., (1996) *Using the Data Warehouse*. New York: Wiley & Sons.

Intergraph, (2003) DynaGEN Technical Overview. Online: http://imgs.intergraph.com/freebies/pdfopener.asp?item=wp&file=WP1009B.pdf (30/04/05).

ISO, (1996) ISO/TC 211/PT Terminology secretariat, (1996) Terms and definitions from CEN/TC 287 Geographic information, May 22, 1996. Online: http://www.don-imit.navy.mil/glossary/definition.asp (30/04/05).

ISO, (2005) General information ISO/TC 211 Geographic information/Geomatics. Online: http://www.isotc211.org/Outreach/Overview/Overview.htm (16/06/06).

IST, (2001) Information Society Technologies. Online: http://www.cordis.lu/ist/home.html (16/06/06).

Jäger, E., (1991) "Investigations on Automated Feature Displacement for Small Scale Maps in Raster Format", *Proceedings of the 15th Conference of the International Cartographic Association*, Bournemouth, September 22–October 1, 1, pp. 245–256.

Jahard, Y., Lemarie, C., and Lecordix, F., (2003) "The Implementation of New Technology to Automate Map Generalisation and Incremental Updating Processes", *Proceedings of the 21st International Cartographic Conference (ICC), Cartographic Renaissance*, Durban, South Africa, August 10–16, pp. 1449–1459, CD-ROM.

Jankowski, P., and Nyerges, T., (1989) "Design considerations for MaPKBS – a map projection knowledge-based system", *The American Cartographer*, 16(2), 85–95.

Jenks, G. F., (1989) "Geographic logic in line generalisation", *Cartographica*, 26(1), 27–42.

Jiang, B., and Claramunt, C., (2004a) "A Structural Approach to the Model Generalization of an Urban Street Network", *GeoInformatica*, 8(2), 157–171.

Jiang, B., and Claramunt, C., (2004b) "Topological analysis of urban street networks (small-world or not)", *Environment and Planning B: Planning and Design*, 31, 151–162.

João, E. M., (1995) "The importance of quantifying the effects of generalization", In: Müller, J.-C., Lagrange, J.-P., and Weibel, R., (eds.), *GIS and Generalization: Methodology and Practice.* London: Taylor & Francis, pp. 183–193.

João, E. M., (1998) *Causes and Consequences of Map Generalisation.* London: Taylor & Francis.

Jones, C. B., (1991) "Database Architecture for Multi-Scale GIS", *Auto-Carto 10: Technical Papers of the 1991 ACSM-ASPRS Annual Convention*, Baltimore, Maryland, 6, 1–14.

Jones, C. B., Abdelmoty, A. I., Lonergan, M. E., van der Poorten, P., and Zhou, S., (2000) "Multi-Scale Spatial Database Design for Online Generalisation", *Proceedings of the 9th Spatial Data Handling Symposium*, Beijing, pp. 7b.34–7b.44.

Jones, C. B., and Abraham, I. M., (1986) "Design Considerations for a Scale-Independent Cartographic Data Base", *Proceedings of 2nd International Symposium on Spatial Data Handling*, Seattle, Washington, pp. 384–398.

Jones, C. B., Bundy, G. L., and Ware, J. M., (1995) "Map Generalization with a Triangulated Data Structure", *Cartography and Geographic Information Systems*, 22(4), 317–331.

Jones, C. B., Kidner, D. B., Luo, L. Q., Bundy, G. L., and Ware, J. M., (1996) "Database Design for Multi-Scale Spatial Information System", *International Journal of Geographical Information Systems*, 10(8), 901–920.

Jones, C. B., (1997) *Geographical Information Systems and Computer Cartography.* Harlow: Longman.

Jones, C. B., Abdelmoty, A. I., and Fu, G., (2003) "Maintaining Ontologies for Geographical Information Retrieval on the Web", *Proceedings of OTM Confederated International Conferences CoopIS, DOA, and OOBASE*, Catania, Sicily, November 3–7, *Lecture Notes in Computer Science*, (2888), pp. 934–951.

JTS, (2005) JTS Topology Suite, Vivid Solutions. Online: http://www.vividsolutions.com/jts/jtshome.htm (16/06/06).

JUMP, (2005) The JUMP Project. Online: http://www.jump-project.org (16/06/06).

Kada, M., (2002) "Automatic generalization of 3D building models", *Proceedings ISPRS Commission IV Symposium 'Geospatial Theory, Processing and Applications', International Archives of the Photogrammetry, Remote Sensing and Spatial Information Sciences*, Ottawa, Canada, July 9–12, 34(4), 243–248.

Kaiser, M. K., (1993) "Introduction to Knowing", In: Ellis, S. R., Kaiser, M., and Grunwald, A. J., (eds.), *Pictorial Communication in Virtual and Real Environments.* London: Taylor and Francis, pp. 43–47.

Keahey, A., (1997) *Nonlinear magnification*, Ph.D. Dissertation. Indiana University, Computer Science.

Keates, J. S., (1989) *Cartographic Design and Production.* 2nd ed., Harlow, Essex: Longman.

Keates, J. S., (1996) *Understanding Maps.* 2nd ed., Harlow, Essex: Longman.

Kidner, D. B., and Jones, C. B., (1993) "Considerations for the Design of a Multiple Representation GIS", *Auto Carto 11 Proceedings, Technical papers of 11th International Symposium on Computer-Assisted Cartography*, October 30–November 1, Minneapolis, Minnesota, pp. 197–207.

Kidner, D. B., and Jones, C. B., (1994) "A Deductive Object-Oriented GIS for Handling Multiple Representations", In: Waugh, T. C., and Healey, R. G., (eds.), *Advances in GIS Research, Proceedings of the 6th International Symposium on Spatial Data Handling*, Edinburgh, Scotland, September 5–9, IGU Commission on GIS and Association for Geographic Information, 2, pp. 882–893.

Kilpeläinen, T., (1992) "Multiple Representations and Knowledge-Based Generalization of Topographical Data", *ISPRS XVII Congress, International Archives of Photogrammetry and Remote Sensing*, Commission III, Washington D.C., XXIX(B3), 954–964.

Kilpeläinen, T., (1994) "Updating Multiple Representation Geodata Bases by Incremental Generalisation", In: Ebner, H., Heipke, C. K., and Eder, K., (eds.), *ISPRS Commission III/IV Symposium, 'Spatial Information from*

Digital Photogrammetry and Computer Vision', International Archives of Photogrammetry and Remote Sensing, Munich, September 5–9, 30(3/1), 440–447.

Kilpeläinen, T., (1995) "Updating Multiple Representation Geodata Bases by Incremental Generalization", *Geo-Informations-Systeme*, 8(4), 13–18.

Kilpeläinen, T., (1997) "Multiple representation and generalization of geo-databases for topographic maps", Doctoral dissertation. *Publications of the Finnish Geodetic Institute*, Kirkkonummi, (124).

Kilpeläinen, T., (2000) "Knowledge Acquisition for Generalization Rules", *Cartography and Geographic Information Science*, 27(1), 41–50.

Kilpeläinen, T., (2001) "Maintenance of Multiple Representation Databases for Topographic Data", *The Cartographic Journal*, 37(2), 101–107.

Kilpeläinen, T., (ed.), (1999) *Map Generalisation in the Nordic Countries, Reports of the Finnish Geodetic Institute*, Kirkkonummi, (99:6), 58 p.

Kilpeläinen, T., and Sarjakoski, T., (1995) "Incremental Generalization for Multiple Representations of Geographical Objects", In: Müller, J.-C., Lagrange, J.-P., and Weibel, R., (eds.), *GIS and Generalization: Methodology and Practice*. London: Taylor & Francis, pp. 209–218.

Kimball, R., (2002) *The Data Warehouse Toolkit*. 2nd ed., New York: Wiley & Sons, 436 p.

Kimerling, A. J., (1989) "Cartography", In: Gaile, G. L., and Willmott, C. J., (eds.), *Geography in America*. Columbus Ohio: Merrill Publishing, pp. 686–717.

King, D. B., and Wertheimer, M., (2005) "Max Wertheimer and Gestalt Theory", London (U.K.): Transaction Publishers.

Klippel, A., (2003) *Wayfinding Choremes – Conceptualizing wayfinding and route direction elements*, Ph.D. Thesis. University of Bremen.

Klippel, A., Richter, K.-F., Barkowsky, T., and Freksa, C., (2005) "The Cognitive Reality of Schematic Maps", In: Meng, L., Zipf, A., and Reichenbacher, T., (eds.), *Map-based Mobile Services: Theories, Methods and Implementations*. Berlin: Springer, pp. 56–71.

Klippel, A., Tappe, H., and Habel, C., (2003) "Pictorial Representations of Routes: Chunking Route Segments During Comprehension", In: Freksa, C., Brauer, W., Habel, C., and Wender, K., (eds.), *Spatial Cognition III: Routes and Navigation, Human Memory and Learning, Spatial Representation and Spatial Learning, Lecture Notes in Computer Science*, (2685). Berlin: Springer, pp. 11–33.

Knöpfli, R., (1983) "Communication Theory and Generalization", In: Taylor, D. R. F., (ed.), *Graphic Communication and Design in Contemporary Cartography, Progress in Contemporary Cartography II*. New York: Wiley & Sons, pp. 177–218.

Köbben, B., (2003) "RIMapper – A test bed for online Risk Indicator Maps using data-driven SVG visualisation", In: Gartner, G., (ed.), Location Based Services & TeleCartography, Proceedings of the Symposium 2004, January 28–29, *Geowissenschaftliche Mitteilungen*, Technische Universität Wien, (66), 189–195.

Kolazny, A., (1969) "Cartographic Information – A Fundamental Concept and Term in Modern Cartography", *The Cartographic Journal*, 6, 47–49.

Kolbe, T. H., (2003) "Augmented Videos and Panoramas for Pedestrian Navigation", In: Gartner, G., (ed.), Location Based Services & TeleCartography, Proceedings of the Symposium 2004, January 28–29, *Geowissenschaftliche Mitteilungen*, Technische Universität Wien, (66), 45–52.

Kolbe, T. H., (2004) "Interoperable 3D-Visualisierung – 3D Web Map Server", *KS Band 9 – Der X-Faktor – Mehrwert für Geodaten und Karten*, Bonn, pp. 130–140.

Kolbe, T. H., and Gröger, G., (2003) "Towards unified 3D city models", *Proceedings of the ISPRS Comm. IV Joint Workshop on Challenges in Geospatial Analysis, Integration and Visualization II*, Stuttgart, CD-ROM.

Kopczynski, M., and Sester, M., (2004) "Representation of Sketch Data for Localisation in Large Data Sets", In: Altan, M. O., (ed.), *Proceedings of the XXth ISPRS Congress, The International Archives of the Photogrammetry, Remote Sensing and Spatial Information Sciences*, Istanbul, Turkey, July 12–23, XXXV(B3), 424–429.

Kopomaa, T., (2000) *The City in Your Pocket: Birth of the Mobile Information Society*. Helsinki: Gaudeamus.

Kraak, M.-J., and Brown, A., (2001) *Web Cartography: Developments and Prospects*. London: Taylor & Francis.

Kraak, M. J., and Ormeling, F., (2003) *Cartography: Visualization of Geospatial Data*. 2nd ed., Harlow: Prentice Hall.

Kreiter, N., (2002) *Multirepräsentationsdatenbank für die Kartographie*, Vertiefungsblockarbeit. Eidgenössische Technische Hochschule, Zürich.

Krippendorff, K., (1995) "On the essential contexts of artifacts or on the proposition that 'design is making sense (of things)' ", In: Margolin, V., and Buchanan, R., (eds.), *The Idea of Design*. Cambridge, Mass.: MIT Press, pp. 156–184.

Kulik, L., Duckham, M., and Egenhofer, M., (2005) "Ontology-driven map generalization", *Journal of Visual Languages and Computing*, 16, 245–267.

Label-EZ, (2005) MapText, Inc. Online: http://www.maptext.com/ (16/06/06).

Lagrange, J.-P., and Ruas, A., (1994) "Geographic Information Modelling: GIS and Generalization", In: Waugh, T. C., and Healey, R. G., (eds.), *Advances in GIS Research, Proceedings of the 6th International Symposium on Spatial Data Handling*, Edinburgh, Scotland, September 5–9, IGU Commission on GIS and Association for Geographic Information, 2, 1099–1117.

Lal, J., and Meng, L., (2003) "Aggregation on the basis of structure recognition", *5th Workshop on Progress in Automated Map Generalisation*, Paris, April 28–30. Online: http://www.geo.unizh.ch/ICA/docs/paris2003/papers03.html (16/06/06).

Lal, J., and Meng, L., (2004) "3D building recognition using artificial neural network", *ICA Workshop on Generalisation and Multiple Representation*, Leicester, UK, August 20–21. Online: http://ica.ign.fr/ (16/06/06).

Lamy, S., Ruas, A., Demazeau, Y., Jackson, M., Mackaness, W. A., and Weibel, R., (1999) "The Application of Agents in Automated Map Generalisation", In: Keller, C. P., (ed.), *Proceedings, 19th International Cartographic Conference and 11th General Assembly of ICA, Touch the Past, Visualize the Future*, Ottawa, Canada, August 14–21, 2, 1225–1234.

Lang, F., (1999) *Automatic geometric and semantic reconstruction of buildings from images by extraction of 3D-corners and their 3D-aggregation*, Ph.D. Thesis. Institute for Photogrammetry, Bonn University.

Lang, T., (1969) "Rules for the Robot Draughtsmen", *The Geographical Magazine*, 42(1), 50–51.

Laser-Scan, (2005). Online: http://www.laser-scan.com/ (29/06/06).

Laurini, R., and Thomson, D., (1992) *Fundamentals of Spatial Information Systems*. London: Academic Press.

Le Men, H., (1996) "Généralisation cartographique pour l'occupation du sol : Application au passage V1 à V2 pour la BD Carto®", *Revue Internationale de Géomatique*, 6(2–3), 227–248.

Lecordix, F., Plazanet, C., and Lagrange, J.-P., (1997) "A Platform for Research in Generalization: Application to Caricature", *GeoInformatica*, 1(2), 161–182.

Lecordix, F., Plazanet, C., Chirie, F., Lagrange, J. P., Banel, T., and Cras, Y., (1994) "Placement automatique des écritures d'une carte avec une qualité cartographique", In: Harts, J. J., et al., (eds.), *EGIS/MARI '94: Fifth European Conference and Exhibition on Geographical Information Systems EGIS*, Paris, France, March 29–April 1, 1, 22–32.

Lecordix, F., Regnauld, N., Meyer, M., and Fechir, A., (2005) "MAGNET Consortium", *The 8th ICA workshop on Generalisation and Multiple Representation*, A Coruña, Spain, June 7–8. Online: http://ica.ign.fr (26/09/05).

Ledgard, H., (1983) *ADA: An Introduction*. 2nd ed., New York: Springer-Verlag.

Lee, D., (1996) "Making Databases Support Map Generalization", *GIS/LIS Proceedings 96*, Denver, Colorado, November 19–21, pp. 467–480.

Lee, D., (1999) "Practical Solutions for Specific Generalization Tasks", *3rd Workshop on Progress in Automated Map Generalization*, Ottawa, Canada, August 12–14. Online: http://www.geo.unizh.ch/ICA/docs/ottawa1999/papers99.html (16/06/06).

Lee, D., (2000) Map Generalization in GIS: – practical solutions with Workstation ArcInfoTM Software. Online: http://downloads.esri.com/support/whitepapers/ao_/Map_Generalization.pdf. (16/06/06).

Lee, D., (2003) "Recent Generalization Development and Road Ahead", *5th Workshop on Progress in Automated Map Generalization*, Paris. Online: http://www.geo.unizh.ch/ICA/docs/paris2003/papers03.html (16/06/06).

Lehto, L., and Kilpeläinen, T., (2000) "Real-Time Generalization of Geodata in the Web", *ISPRS XIXth Congress, International Archives of Photogrammetry and Remote Sensing*, Amsterdam, July 16–22, XXXIII(B4/2), 559–566.

Lehto, L., and Kilpeläinen, T., (2001) "Generalizing XML-encoded Spatial Data on the Web", *Mapping the 21st Century: The 20th International Cartographic Conference*, Beijing, China, August 6–10, 4, 2390–2396.

Lehto, L., and Sarjakoski, L. T., (2005a) "Real-time generalization of XML-encoded spatial data for the Web and mobile devices", *International Journal of Geographical Information Science*, 19(8–9), 957–973.

Lehto, L., and Sarjakoski, T., (2004) "Schema Translations by XSLT for GML-Encoded Geospatial Data in Heterogeneous Web-Service Environment", In: Altan, M. O., (ed.), *Proceedings of the XXth ISPRS International Congress,*

International Archives of Photogrammetry, Remote Sensing and Spatial Information Sciences, Istanbul, Turkey, July 12–23, XXXV(B4/IV), 177–182.

Lehto, L., and Sarjakoski, T., (2005b) "XML in Service Architectures for Mobile Cartographic Applications", In: Meng, L., Zipf, A., and Reichenbacher, T., (eds.), *Map-based Mobile Services: Theories, Methods and Implementations*. Berlin: Springer, pp. 173–192.

Lehto, L., Kähkönen, J., and Kilpeläinen, T., (1997) "WWW-Technology as Means of Transfer and Visualization of Geographic Objects", *Proceedings of the 18th ICA/ACI International Cartographic Conference*, Stockholm, Sweden, June 23–27, 2, 681–689.

Leiner, U., Preim, B., and Ressel, S., (1997) "Entwicklung von 3D-Widgets – Überblicksvortrag", In: Deussen, O., and Lorenz, P., (eds.), *Simulation und Animation '97*, SCS-Society for Computer Simulation Int., SCS Europe, Delft, Belgium, pp. 171–188. Online: http://www.mevis.de/~bernhard/papers/maerz97.html/ (29/06/06).

Leitner, H., (2004) "The Politics of Scale and Networks of Spatial Connectivity: Transnational Interurban Networks and the Rescaling of Political Governance in Europe", In: Sheppard, E., and McMaster, B., (eds.), *Scale and Geographic Inquiry: Nature, Society, and Method*. Malden, MA: Blackwell Publishing, pp. 236–255.

Lemarie, C., and Badard, T., (2001) "Cartographic Database Updating", *Mapping the 21st Century: The 20th International Cartographic Conference*, Beijing, China, August 6–10, 2, 1376–1385.

Leontjew, A. N., (1978) *Activity, consciousness and personality*. Englewood Cliffs, NJ: Prentice Hall.

Levin, S. A., (1992) "The Problem of Pattern and Scale in Ecology", *Ecology*, 73, 1943–1965.

Levine, M., (1982) "You-Are-Here-Maps – Psychological Considerations", *Environment and Behavior*, 14(2), 221–237.

Li, Z., (1994) "Mathematical morphology in digital generalization of raster map data", *Cartographica*, 23(1), 1–10.

Li, Z., and Choi, Y. H., (2002) "Topographic map generalization: association of road elimination with thematic attributes", *The Cartographic Journal*, 39(2), 153–166.

Li, Z., Yan, H., Ai, T., and Chen, J., (2004) "Automated building generalization based on urban morphology and Gestalt theory", *International Journal of Geographical Information Science*, 18(5), 513–534.

Liao, P.-S., Chen, T.-S., and Chung, P.-C., (2001) "A Fast Algorithm for Multilevel Thresholding", *Journal of Information Science and Engineering*, 17, 713–727.

Lichtner, W., (1978) "Locational Characteristics and the Sequence of Computer Assisted Processes of Cartographic Generalization", *Nachrichten aus dem Karten- und Vermessungswesen*, Institut für Angewandte Geodäsie, Frankfurt a.M., II(35), 65–75.

Lichtner, W., (1979a) "Computer-assisted processes of cartographic generalization in topographic maps", *Geo-Processing*, 1, 183–199.

Lichtner, W., (1979b) "Kartennetztransformationen bei der Herstellung thematischer Karten", *Nachrichten aus dem Karten- und Vermessungswesen*, Institut für Angewandte Geodäsie, Frankfurt a.M, I(79), 109–119.

Linturi, R., and Simula, T., (2005) "Virtual Helsinki: Enabling the Citizen; Linking the Physical and Virtual", In: Besselaar, P., and Koizumi, S., (eds.), *Digital Cities III, Information Technologies for Social Capital: Cross-Cultural Perspectives*, Third International Digital Cities Workshop, Amsterdam, The Netherlands, September 18–19, *Lecture Notes in Computer Science*, (3081). Berlin: Springer, pp. 113–140.

Lisboa Filho, J., Iochpe, C., and Borges, K., (2002) "Analysis patterns for GIS data schema reuse on urban management applications", *CLEI Eletronic Journal*, 5(2).

Liu, Y., (2002) *Categorical Database Generalisation in GIS*, Ph.D. Thesis. Wageningen University, The Netherlands, *ITC Dissertation*, (88), The Netherlands.

Lonergan, M. E., and Jones, C. B., (2001) "An Iterative Displacement Method for Conflict Resolution in Map Generalization", *Algorithmica*, 30(2), 287–301.

Long, J., (1989) "Cognitive Ergonomics and Human-Computer Interaction", In: Long, J., and Whitefield, A., (eds.), *Cognitive Ergonomics and Human-Computer Interaction*. Cambridge, UK: Cambridge University Press.

Lovelace, K., Hegarty, M., and Montello, D., (1999) "Elements of Good Route Directions in Familiar and Unfamiliar Environments", In: Freksa, C., and Mark, M., (eds.), *Spatial Information Theory: Cognitive and Computational Foundations of Geographic Information Science*. Berlin: Springer, pp. 65–82.

LoVEUS, (2005) Location aware Visually Enhanced Ubiquitous Services, European Union, IST programme. Online: http://loveus.intranet.gr/ (16/06/06).

Lowe, D. G., (1985) *Perceptual Organisation and Visual Recognition*. Boston: Kluwer.

Lowe, D. G., (1989) "Organization of smooth image curves at multiple scales", *International Journal of Computer Vision*, 3(2), 119–130.

Luley, P. M., Almer, A., and Schnabel, T., (2003) "Geo-Data Presentation on Mobile Devices for Tourism Applications", In: Gartner, G., (ed.), Location Based Services & TeleCartography, Proceedings of the Symposium 2004, January 28–29, *Geowissenschaftliche Mitteilungen*, Technische Universität Wien, (66), 172–178.

Lynch, K., (1960) *The Image of the City.* Cambridge, Mass.: MIT Press.

Maaß, W., (1995) "How Spatial Information Connects Visual Perception and Natural Language Generation in Dynamic Environments: Towards a Computational Model", In: Frank, A., and Kuhn, W., (eds.), *Spatial Information Theory.* Berlin: Springer, pp. 223–240.

MacEachren, A., (1986) "A linear view of the world: Strip maps as a unique form of cartographic representation", *The American Cartographer*, 13(1), 7–25.

MacEachren, A., (1995) *How maps work: Representation, visualization and design.* New York: The Guildford Press.

MacEachren, A. M., and Ganter, J. H., (1990) "A Pattern Identification Model to Cartographic Visualization", *Cartographica*, 27, 64–81.

MacEachren, A. M., and Kraak, M. J., (1997) "Exploratory Cartographic Visualization: Advancing the Agenda", *Computers and Geosciences*, 23, 335–343.

MacEachren, A. M., Buttenfield, B., Campbell, J., DiBiase, D., and Monmonier, M. S., (1992) "Visualisation", In: Abler, R., Marcus, M., and Olson, J., (eds.), *Geographer's Inner Worlds: Pervasive Themes in Contemporary American Geography.* New Brunswick, NJ: Rutgers University Press, pp. 99–137.

MacEachren, A. M., Wachowicz, M., Edsall, R., and Haug, D., (1999) "Constructing Knowledge from Multivariate Spatiotemporal Data: Integrating Geographic Visualisation with Knowledge Discovery in Database Methods", *International Journal of Geographical Information Systems*, 13, 311–334.

Mackaness, W. A., (1991) "Integration and evaluation of map generalization", In: Buttenfield, B., and McMaster, R., (eds.), *Map Generalization: Making Rules for Knowledge Representation.* Harlow: Longman Group, pp. 217–226.

Mackaness, W. A., (1994) "An Algorithm for Conflict Identification and Feature Displacement in Automated Map Generalization", *Cartography and Geographic Information Systems*, 21(4), 219–232.

Mackaness, W. A., (1995a) "A Constraint Based Approach to Human Computer Interaction in Automated Cartography", *Proceedings of the 17th International Cartographic Conference*, Barcelona, Spain, September 3–9, 2, pp. 1423–1432.

Mackaness, W. A., (1995b) "Analysis of urban road networks to support cartographic generalization", *Cartography and Geographic Information Systems*, 22(4), 306–316.

Mackaness, W. A., and Beard, M. K., (1993) "Use of graph theory to support map generalization", *Cartography and Geographic Information Systems*, 20(4), 210–221.

Mackaness, W. A., and Edwards, G., (2002) "The Importance of Modelling Pattern and Structure in Automated Map Generalisation", *Joint ISPRS/ICA workshop on "Multi-Scale Representations of Spatial Data"*, Ottawa, Canada, July 7–8. Online: http://www.ikg.uni-hannover.de/isprs/Program_final.html (29/06/06).

Mackaness, W. A., and Fisher, P. F., (1987) "Automatic recognition and resolution of spatial conflicts in cartographic symbolisation", *AUTO-CARTO 8, Proceedings, Eighth International Symposium on Computer-Assisted Cartography*, Baltimore, Maryland, March 29 – April 3, pp. 709–718.

Mackaness, W. A., and Mackechnie, G. A., (1999) "Automating the detection and simplification of junctions in road networks", *GeoInformatica*, 3(2), 185–200.

Mackaness, W. A., and Rainsford, D., (2002) "Template Matching in Support of Generalisation of Rural Buildings", In: Richardson, D., and van Oosterom, P., (eds.), *Advances in Spatial Data Handling, 10th International Symposium on Spatial Data Handling*, Ottawa, Canada, July 9–12. Berlin: Springer, pp. 137–152.

Mackaness, W. A., Fisher, P. F., and Wilkinson, G. G., (1986) "Towards a cartographic expert system", *Proceedings AUTO-CARTO London, International Conference on the acquisition, management and presentation of spatial data*, London, September 14–19, 1, 578–587.

Mallot, H., (2000) *Computational Vision.* Cambridge, Mass.: MIT Press.

Mallot, H., Steck, S., and Loomis, J., (2002) "Mechanisms of spatial cognition: behavioural experiments in virtual environments", *KI* 4/02, 24–47.

Mantel, D., and Lipeck, U. W., (2004) "Matching Cartographic Objects in Spatial Databases", In: Altan, M. O., (ed.), *Proceedings of the XXth ISPRS Congress, The International Archives of the Photogrammetry, Remote Sensing and Spatial Information Sciences*, Istanbul, Turkey, July 12–23, XXXV(Part B4), 172–176.

Map24, (2005) Map 24. Online: http://www.map24.com (29/06/06).

MapWay, (2005) M-spatial products. Online: http://www.m-spatial.com/mapway.htm (01/02/05).

Mark, D. M., (1990) "Competition for Map Space as a Paradigm for Automated Map Design", *GIS/LIS Proceedings 90*, Anaheim, California, November 7–10, 1, 97–105.

Mark, D. M., (1991) "Object Modelling and Phenomenon-Based Generalization", In: Buttenfield, B. P., and McMaster R. B., (eds.), *Map Generalization: Making Rules for Knowledge Representation.* London: Longman Group, pp. 103–118.

Marlin Studios, (2005) Online: www.marlinstudios.com/products/bldgmodels (30/04/05).

Marston, S., (2000) "The Social Construction of Scale", *Progress in Human Geography*, 24, 219–242.

Martinez Casasnovas, J. A., and Molenaar, M., (1995) "Aggregation Hierarchies for Multiple Scale Representations of Hydrographic Networks in GIS", *Proceedings of 17th International Cartographic Conference*, Barcelona, September 3–9, 1, pp. 358–362.

McConalogue, D. J., (1970) "A quasi-intrinsic scheme for passing a smooth curve through a discrete set of points", *The Computer Journal*, 13(4), 392–396.

McCormick, B. H., DeFanti, T. A., and Brown, M. D., (eds.), (1987) "Visualization in Scientific Computing", *SIGGRAPH Computer Graphics Newsletter*, 21(6), October 14, appendices 61–70.

McGraw, K. L., and Harbison-Briggs, K., (1989) *Knowledge Acquisition: Principles and Guidelines.* 1st ed., Englewood Cliffs, NJ: Prentice-Hall.

McMaster, R. B., (1987) "Automated line generalization", *Cartographica*, 24(2), 74–111.

McMaster, R. B., (1989) "The integration of simplification and smoothing algorithms", *Cartographica*, 26(1), 101–121.

McMaster, R. B., (1991) "Conceptual Frameworks for Geographical Knowledge", In: Buttenfield, B. P., and McMaster R. B., (eds.), *Map Generalization: Making Rules for Knowledge Representation.* London: Longman Group, pp. 21–39.

McMaster, R. B., (1995) "Knowledge Acquisition for Cartographic Generalization: Experimental Methods", In: Müller, J.-C., Lagrange, J.-P., and Weibel, R., (eds.), *GIS and Generalization: Methodology and Practice*, *Gisdata*, (1). London: Taylor & Francis, pp. 161–179.

McMaster, R. B., and Barnett, L., (1993) "A Spatial-Object Level Organization of Transformations for Cartographic Generalization", *Auto-Carto 11 Proceedings, Technical Papers, 11th International Symposium on Computer-Assisted Cartography*, Minneapolis, Minnesota, October 30 – November 1, pp. 386–395.

McMaster, R. B., and Monmonier, M. S., (1989) "A conceptual framework for quantitative and qualitative raster-mode generization", *Proceedings of GIS/LIS '89*, Orlando, Florida, 2, pp. 390–403.

McMaster, R. B., and Shea, K. S., (1988) "Cartographic Generalization in a Digital Environment: A Framework for Implementation in a Geographic Information System", *GIS/LIS'88 Proceedings*, San Antonio, Texas, November 30 – December 2, pp. 240–249.

McMaster, R. B., and Shea, K. S., (1992) *Generalization in Digital Cartography.* Washington, DC: Association of American Geographers.

McMaster, R. B., and Sheppard, E., (2004) "Introduction: Scale and Geographic Inquiry", In: Sheppard, E., and McMaster, R. B., (eds.), *Scale and Geographic Inquiry: Nature, Society, and Method.* Malden, MA: Blackwell Publishing, pp. 1–22.

Meier, W., (2002) "eXist: An Open Source Native XML Database", *Revised papers from NODe 2002 Web and Database-Related Workshops on Web, Web-Services and Database Systems*, Erfurt, Germany, October 2002, *Lecture Notes in Computer Science*, (2593). Berlin: Springer, pp. 169–183.

Meng, L., (1993) *Erkennung der Kartenschrift mit einem Expertensystem*, Dissertation. *Wissenschaftliche Arbeiten der Fachrichtung Vermessungswesen der Universität Hannover*, (184).

Meng, L., (2002) "How can 3D geovisualization please users' eyes better?", *Geoinformatics*, 5(8), 34–35.

Meng, L., (2003) "Missing theories and methods in digital cartography", *Proceedings of the 21st International Cartographic Conference (ICC), Cartographic Renaissance*, Durban, South Africa, August 10–16, pp. 1887–1894, CD-ROM.

Meng, L., Zipf, A., and Reichenbacher, T., (eds.), (2005) *Map-Based Mobile Services: Theories, Methods and Implementations.* Berlin: Springer.

Meyer, U., (1986) "Software-Developments for Computer-Assisted Generalization", *Proceedings AUTO CARTO London, International Conference on the acquisition, management and presentation of spatial data*, September 14–19, London, UK, 2, 247–256.

Meyer, W. B., Gregory, D., Turner, B. L., and McDowell, P., (1992) "The Local-Global Continuum", In: Abler, R., Marcus, M., and Olson, J., (eds.), *Geography's Inner Worlds*. New Brunswick, NJ: Rutgers University Press, pp. 255–279.

Michon, P., and Denis, M., (2001) "When and Why Are Visual Landmarks Used in Giving Directions?", In: Montello, D., (ed.), *Spatial Information Theory, International Conference COSIT 2001, Proceedings*. Berlin: Springer, pp. 292–305.

Minsky, M., (1975) "A framework for representing knowledge", In: Winston, P., (ed.), *The psychology of computer vision*. New York: McGraw Hill, pp. 211–277.

Moellering, H., and Hogan, R., (eds.), (1997) *Spatial Database Transfer Standards 2: Characteristics for Assessing Standards and Full Descriptions of the National and International Standards in the World*. International Cartographic Association, London: Elsevier Science.

Moellering, H., Alders, H. J. and Crane, A., (eds.), (2005) *World Spatial Metadata Standards*, International Cartographic Association, Elsevier.

Molenaar, M., (1989) "Single valued vector maps – a concept in GIS", *Geo-Informationssysteme*, 2(1), 18–26.

Molenaar, M., (1993) "Object hierarchies and uncertainty in GIS or why standardisation is so difficult", *GeoInformations-Systeme*, 6(3), 22–28.

Molenaar, M., (1996a) "Multi-Scale Approaches for Geodata", *XVIII ISPRS Congress, International Archives of Photogrammetry and Remote Sensing*, Vienna, July 12–18, XXXI(B3), 542–554.

Molenaar, M., (1996b) "The Role of Topologic and Hierarchical Spatial Object Models in Database Generalization", In: Molenaar, M., (ed.), *Methods for the Generalization of Geo-Databases, Netherlands Geodetic Commission, Publications on Geodesy, New series*, Delft, (43), 13–35.

Molenaar, M., (1998a) *An introduction into the theory of topological and hierarchical object modelling for geo-information systems*. London: Taylor & Francis.

Molenaar, M., (1998b) "Composite Objects and Multiscale Approaches", In: Molenaar, M., (ed.), *An Introduction to the Theory of Spatial Object Modelling for GIS*. London: Taylor & Francis, pp. 161–191.

Monier, P., (1996) "Automated Generalization of Relief in the GIS Environment", *Proceedings of the 1st International Conference on Geographic Information Systems in Urban, Environmental and Regional Planning*, pp. 167–177.

Monmonier, M. S., (1982) *Computer-Assisted Cartography: Principles and Prospects*. Englewood Cliffs, NJ: Prentice-Hall.

Monmonier, M. S., (1983) "Raster Area Generalization for Land Use and Land Cover Maps", *Cartographica*, 20(4), 65–91.

Monmonier, M. S., (1984) "Geographic information and cartography", *Progress in Human Geography*, 8, 381–391.

Monmonier, M. S., (1991a) "Ethics and map design: Six strategies for confronting the traditional one-map solution", *Cartographic Perspectives*, 10, 3–8.

Monmonier, M. S., (1991b) *How to Lie with Maps*. Chicago: The University of Chicago Press.

Monmonier, M. S., (1999) "Epilogue", *Cartography and Geographic Information Science*, 26(3), 235–236.

Monmonier, M. S., and McMaster, R. B., (1991) "The Sequential Effects of Geometric Operators in Cartographic Line Generalization", *The International Yearbook of Cartography*, 30, 93–108.

Montello, D., (2005) "Navigation", In: Shah, P., and Miyake, A., (eds.), *The Cambridge Handbook of Visuospatial Thinking*. Cambridge: Cambridge University Press.

Moore, A., (2005) "Model-based cartographic generalisation with uncertainty", In: Whigham, P. A. (ed.), *SIRC 2005: A Spatio-temporal Workshop, Proceedings of the 17th Annual Colloquium of the Spatial Information Research Centre*, November 24–25, The University of Otago, Dunedin, New Zealand, pp. 131–136. Online: http://www.business.otago.ac.nz/SIRC05/conferences/index2005.html (29/06/06).

Morisset, B., and Ruas, A., (1997) "Simulation and agent modelling for road selection in generalization", *Proceedings of the 18th ICA/ACI International Cartographic Conference*, Stockholm, Sweden, June 23–27, 3, 1376–1380.

Morrison, J. L., (1974) "A theoretical framework for cartographic generalization with emphasis on the process of symbolization", *International Yearbook of Cartography*, 14, 115–127.

Morrison, J. L., (1997) "Topographic mapping for the twenty first century", In: Rhind, D., (ed.), *Framework for the World*. Cambridge, UK: GeoInformation International.

Mountain, D., and Raper, J., (2002) "Modelling human spatio-temporal behaviour: A challenge for location-based services", *Proceedings, GIS 2002: Annual Conference of the AGI*, London, UK, September 17–19, London.

Muehrcke, P. C., (1990) "Cartography and Geographic Information Systems", *Cartography and Geographic Information Systems*, 17(1), 7–15.

Muller, J. C., (1989) "Theoretical considerations for automated map generalisation", *ITC Journal*, 3–4, 200–204.

Muller, J. C., (1990) "Rule Based Generalization: Potentials and Impediments", In: *Proceedings of the 4th International Symposium on Spatial Data Handling*, July 23–27, Zurich, Switzerland, 1, 317–334.

Muller, J.-C., (1991) "Generalization of Spatial Databases", In: Maguire, D. J., Goodchild M. F., and Rhind, D., (eds.), *Geographical Information Systems 1: Principles and Applications*. Harlow, Essex: Longman, pp. 457–475.

Muller, J.-C., and Mouwes, P. J., (1990) "Knowledge Acquisition and Representation for Rule Based Map Generalization: An Example from the Netherlands", *GIS/LIS Proceedings 90*, Anaheim, California, 1, 58–67.

Müller, J.-C., and Wang, Z., (1992) "Area-patch generalization: a competitive approach", *The Cartographic Journal*, 29 (2), 137–144.

Müller, J.-C., and Zeshen, W., (1990) "A knowledge based system for cartographic symbol design", *The Cartographic Journal*, 27(1), 24–30.

Müller, J.-C., Lagrange, J.-P., and Weibel, R. (eds.), (1995a) *GIS and generalization: Methodology and Practice*, In: Masser, I., and Salgé, F., (series eds.), *Gisdata*, (1). London: Taylor & Francis.

Müller, J.-C., Weibel, R., Lagrange, J.-P., and Salgé, F., (1995b) "Generalization – state of the art and issues", In: Müller, J.-C., Lagrange, J.-P., and Weibel, R. (eds.), *GIS and Generalization: Methodology and Practice*. London: Taylor & Francis, pp. 3–17.

MurMur Consortium, (2002) Final Report of MurMur Project (Multi Representations – Multi Resolutions), European IST Project: 1999–1072, 2000–2002.

Mustière, S., (1998) "GALBE: Adaptive Generalisation, The Need for an Adaptive Process for Automated Generalisation, an Example on Roads", *GIS PLANET 1998, International Conference and Exhibition on Geographic Information*, Lisbon, Portugal, September 7–11, CD-ROM.

Mustière, S., (2001) *Apprentissage automatique pour la généralisation cartographique*, Ph.D. Thesis. Université Pierre et Marie Curie, Paris 6. Online: ftp://ftp.ign.fr/ign/COGIT/THESES/MUSTIERE/ (16/06/06).

Mustière, S., and Lecordix, F., (2002) "La généralisation du linéaire routier: des algorithmes et leur enchaînement", In: Ruas, A., (ed.), *Généralisation et représentation multiple*. Paris: Hermès Science Publications, pp. 241–255.

Navman GPS 3300 Terrain, (2005) Navman products. Online: http://www.navman-mobile.co.uk/ (29/06/06).

NCGIA, National Center for Geographic Information and Analysis, (1989) "The Research Plan of the National Center for Geographic Information and Analysis", *International Journal of Geographical Information Systems*, 3(2), 117–136.

Neun, M., Weibel, R., and Burghardt, D., (2004) "Data Enrichment for Adaptive Generalisation", *ICA Workshop on Generalisation and Multiple Representation*, Leicester, UK, August 20–21. Online: http://ica.ign.fr/. (16/06/06).

Newell, A., (1990) *Unified theories of cognition*. Cambridge, Massachusetts: Harvard University Press.

Nickerson, B. G., (1988) "Automated Cartographic Generalization for Linear Features", *Cartographica*, 25(3), 15–66.

Nickerson, B. G., (1991) "Knowledge Engineering for Generalization", In: Buttenfield, B. P., and McMaster, R. B., (eds.), *Map Generalization: Making Rules for Knowledge Representation*. London: Longman Group, pp. 40–55.

Nickerson, B. G., and Freeman, H., (1986) "Development of a Rule-Based System for Automatic Map Generalization", *Proceedings of the Second International Symposium on Spatial Data Handling*, Seattle, Washington, July 5–10, pp. 537–556.

Nivala, A.-M., and Sarjakoski, L. T., (2003) "Need for Context-Aware Topographic Maps in Mobile Devices", In: Virrantaus, K., and Tveite, H., (eds.), *ScanGIS'2003, Proceedings of the 9th Scandinavian Research Conference on Geographical Information Science*, June 4–6, Espoo, Finland, pp. 15–29.

Nivala, A.-M., and Sarjakoski, L. T., (2004) "Preventing Interruptions in Mobile Map Reading Process by Personalisation", *Papers of the 3rd International Workshop on 'HCI in Mobile Guides'*, in adjunction to: MobileHCI'04, 6th International Conference on Human Computer Interaction with Mobile Devices and Services, Glasgow, Scotland, September 13–16. Online: http://www.comp.lancs.ac.uk/computing/users/kc/mguides04/mGuidesHCI-papers.html (26/09/05).

Nivala, A.-M., and Sarjakoski, L. T., (2005) "Adapting Map Symbols for Mobile Users", *International Cartographic Conference 2005: Mapping Approaches into a Changing World*, A Coruña, Spain, July 10–16. CD-ROM: Theme 12: Internet Location-Based Services, Mobile Mapping and Navigation Systems, Session 5, 11 p.

Nivala, A.-M., Sarjakoski, L. T., Jakobsson, A., and Kaasinen, E., (2003) "Usability Evaluation of Topographic Maps in Mobile Devices", *Proceedings of the 21st International Cartographic Conference (ICC), Cartographic Renaissance*, Durban, South Africa, August 10–16, pp. 1903–1913, CD-ROM.

NOAA, National Geophysical Data Server, (1997). Online: http://web.ngdc.noaa.gov/seg/tools/gis/ondemand.shtml (05/05/05).

Nyerges, T. L., (1991) "Representing Geographical Meaning", In: Buttenfield, B. P., and McMaster, R. B., (eds.), *Map Generalization: Making Rules for Knowledge Representation.* Harlow, Essex: Longman Group, pp. 59–85.

O'Callaghan, J. F., and Mark, D. M., (1984) "The extraction of drainage networks from digital elevation data", *Computer Vision, Graphics and Image Processing*, 28(3), 323–344.

OGC, (1999a) OpenGIS® Simple Features Specification for SQL, Revision 1.1, Document 99-049, Release date: May 5, 1999

OGC, (1999b) Open GIS™ Abstract Specification, Topic 14: Semantics and Information Communities. Online: http://www.opengis.org (05/05/05).

OGC, (2002a) OpenGIS® Catalog Services Implementation Specification, rev. 1.1.1. OpenGIS®Consortium.

OGC, (2002b) OpenGIS® Web Feature Service Implementation Specification, v. 1.0.0.

OGC, (2003a) Open GIS™ Reference Model, Version 0.1.2. Abstract Specification OGC 03-040. Online: http://www.opengis.org (05/05/05).

OGC, (2003b) Open Geospatial Consortium, Inc., Geography Markup Language (GML) Implementation Specification, v. 3.0.0.

OGC, (2005) Open Geospatial Consortium. Online: http://www.opengeospatial.org/ (29/06/06).

Ogniewicz, R. L., and Ilg, M., (1992) "Voronoi skeletons: Theory and applications", *IEEE Conference on Computer Vision and Pattern Recognition*, Champaign, Illinois, June 1992, pp. 63–69.

OpenLS, (2005) Open Geospatial Specifications. Online: http://www.opengis.org/specs/?page=specs (12/07/05).

Oppermann, R., (ed.), (1994) *Adaptive User Support: Ergonomic Design of Manually and Automatically Adaptable Software, Computers, Cognition, and Work.* Hillsdale, New Jersey: Lawrence Erlbaum Associates.

Ormsby, D., and Mackaness, W., (1999) "The Development of Phenomenological Generalization Within an Object-Oriented Paradigm", *Cartography and Geographic Information Science*, 26(1), 70–80.

OSI, (2004) "Open Source Initiative OSI". Online: http://www.opensource.org (29/06/06).

Outdoor Navigator, (2005). Online: http://www.maptech.com/products/outdoornavigator/ (29/06/06).

Paiva, J. A., (1998) *Topological equivalence and similarity in multi-representation geographic databases*, Ph.D. Thesis in Spatial Information Science and Engineering. University of Maine.

Palmer, S., (1983) "The psychology of perceptual organisation: a transformational approach", In: Beck, J., (ed.), *Human and Machine Vision.* New York: Academic Press, pp. 269–339.

Papadimitriou, C. H., and Steiglitz, K., (1982) *Combinatorial optimization: algorithms and complexity.* Englewood Cliffs, NJ: Prentice Hall.

Parent, C., Spaccapietra, S., Zimanyi, E., Donini, P., Plazanet, C., Vangenot, C., (1998) "Modelling Spatial Sata in the MADS Conceptual Model", *Proceedings of the 8th International Symposium on Spatial Data Handling*, Vancouver, Canada, July 12–15, pp. 138–150.

Parsaye, K., and Chignell, M., (1988) *Expert Systems for Experts.* New York: Wiley & Sons.

Peng, W., (1997) *Automated Generalisation in GIS*, Ph.D. Thesis. Wageningen Agricultural University and ITC.

Peng, W., and Müller, J. C., (1996) "A dynamic decision tree structure supporting urban road network automated generalization", *The Cartographic Journal*, 33(1), 5–10.

Peng, W., Sijmons, K., and Brown, A., (1995) "Voronoi Diagram and Delaunay Triangulation Supporting Automated Generalization", *Proceedings of the 17th International Cartographic Conference*, Barcelona, September 3–9, 1, 301–310.

Persson, J., (2004) "Streaming of compressed multi-resolution geographic vector data", In: Brandt, S. A., (ed.), *Proceedings of the 12th International Conference on Geoinformatics – Geospatial Information Research: Bridging the Pacific and Atlantic*, University of Gävle, Sweden, June 7–9.

Peterson, M., (ed.), (2003) *Maps and the Internet.* International Cartographic Association, Cambridge: Elsevier.

Petzold, I., Gröger, G., and Plümer, L., (2003) "Fast Screen Map Labeling – Data-Structures and Algorithms", *Proceedings of the 21st International Cartographic Conference (ICC), Cartographic Renaissance*, Durban, South Africa, August 10–16, pp. 288–299, CD-ROM.

Peuquet, D. J., (1999), Time in GIS and geographical databases. In: Longley, P. A., Goodchild, M. F., Maguire, D. J., and Rhind, D. W., (eds.), *Geographical Information Systems, 1: Principles and Technical Issues.* 2nd ed., New York: Wiley & Sons, pp. 91–103.

Pfafstetter, O., (1989) *Classification of Hydrographic Basins: Coding Methodology.* Rio de Janeiro: DNOS.

Plazanet, C., (1995) "Measurement, Characterisation and Classification for Automated Linear Feature Generalisation", *Technical Papers for Auto Carto 12, ACSM/ASPRS Annual Convention & Exposition*, Charlotte, NC, February 27 – March 1, (4), pp. 59–68.

Plazanet, C., (1996) *Enrichissement des bases de données géographiques : analyse de la géométrie des objets linéaires pour la généralisation cartographique (application aux routes)*, Ph.D. Thesis. University of Marne-la-Vallée /COGIT, Paris. Online: ftp://ftp.ign.fr/ign/COGIT/THESES (16/06/06).

Plümer, K., (2002) "Level of Detail – Eine Übersicht", Eine Ausarbeitung im Rahmen des Hauptseminars 2002/03. Online: http://www.vermessung.uni-hannover.de/files/KonstantinPluemer.pdf (30/04/05).

Pomerol, J.-C., and Adam, F., (2004) "Practical Decision making – From the legacy of Herbert Simon to decision support systems", *Proceedings of Decision Support in an Uncertain and Complex world: the IFIP TC8/WG8.3 International Conference*, pp. 647–657.

Poole, J., Chang, D., Tolbert, D., and Mellor, D., (2002) *Common Warehouse Metamodel – An Introduction to the Standard for Data Warehouse Integration*. New York: Wiley & Sons.

Poslad, S., Laamanen, H., Malaka, R., Nick, A., Phil, B., and Zipf, A., (2001) "Crumpet: Creation of User-Friendly Mobile Services Personalised for Tourism", *Proceedings 3G 2001*, London, March 26–28.

Pospischil, G., Umlauft, M., and Michlmayr, E., (2002) "Designing LOL@, A Mobile Tourist Guide for UMTS", In: Paternò, F., (ed.), *Human Computer Interaction with Mobile Devices, Proceedings of 4th International Symposium, Mobile HCI 2002*, Pisa, Italy, September 18–20. Berlin: Springer, pp. 140–154.

Powitz, B., (1992) "Kartographische Generalisierung topographischer Daten in GIS", *Kartographische Nachrichten*, 43(6), 229–233.

Pratt, I., (1993) "Map Semantics", In: Frank, A. E., and Campari, I., (eds.), *Spatial Information Theory: A Theoretical Basis for GIS; Proceedings of COSIT '93*, Elba, Italy, September 1993. Berlin: Springer, pp. 77–91.

Proulx, M.-J., Larrivée, S., and Bédard, Y., (2002) "Représentation multiple et généralisation avec UML et l'outil Perceptory", In: Ruas, A., (ed.), *Généralisation et représentation multiple*. Paris: Hermès Sciences Publications, pp. 113–130.

Puppo, E., and Dettori, G., (1995) "Towards a formal model for multiresolution spatial maps", In: Egenhofer, M. J., and Herring J. R., (eds.), *Advances in Spatial Databases, SSD'95, Lecture Notes in Computer Science*, (951). Berlin: Springer, pp. 152–169.

Quinlan, J., (1986) "Induction of decision trees", *Machine Learning*, 1, 81–106.

Rahman, A.-A., and Khuan, C.-T., (2003) "3D analytical tools for terrain spatial objects", *Proceedings of the ISPRS Comm. IV Joint Workshop on Challenges in Geospatial Analysis, Integration and Visualization II*, Stuttgart, CD-ROM.

Rainsford, D., and Mackaness, W. A., (2002) "Template Matching in Support of Generalisation of Rural Buildings", In: Richardson, D. E., and van Oosterom, P., (eds.), *Advances in Spatial Data Handling, 10th International Symposium on Spatial Data Handling*. Berlin: Springer, pp. 137–152.

Raisz, E., (1962) *General Cartography*. New York: McGraw-Hall.

Ramos, F., Siret, D., and Musy, M., (2004) "A 3D GIS for managing building rehabilitation process", In: Brandt, S. A., (ed.), *Proceedings of the 12th International Conference on Geoinformatics – Geospatial Information Research: Bridging the Pacific and Atlantic*, University of Gävle, Sweden, June 7–9, pp. 518–524.

Raper, J., (1989) *Three dimensional applications in Geographical Information Systems*. London: Taylor & Francis.

Raper, J., Dykes, J., Wood, J., Mountain, D., Krause, A., and Rhind, D., (2002) "A framework for evaluating geographical information", *Journal of Information Science*, 28(2), 39–50.

Rappo, A., Cecconi, A., and Burghardt, D., (2004) *Fischaugenprojektionen für generalisierte Kartendarstellungen auf kleinen Bildschirmen, Kartographische Nachrichten*, Heft 2. Kirschbaum Verlag.

Ratajski, L., (1967) "Phénomènes des points de généralisation", *International Yearbook of Cartography*, 7, 143–151.

Raubal, M., and Winter, S., (2002) "Enriching Wayfinding Instructions with Local Landmarks", In: Egenhofer, M., and Mark, D., (eds.), *Geographic Information Science, Lecture Notes in Computer Science*, (2478). Berlin: Springer, pp. 243–259.

Regnauld, N., (1996) "Recognition of Building Clusters for Generalisation", *Proceedings of the 7th Spatial Data Handling Symposium*, Delft, The Netherlands, pp. 185–198.

Regnauld, N., (1998) *Généralisation du Bati: Structure Spatiale de Type Graphe et Représentation Cartographique*, Ph.D. Thesis. University of Provence, Aix Marseilles / COGIT. Online: ftp://ftp.ign.fr/ign/COGIT/THESES (16/06/06).

Regnauld, N., (2001a) "Contextual Building Typification in Automated Map Generalization", *Algorithmica*, 30(2), 312–333.

Regnauld, N., (2001b) "Constraint Based Mechanism to Achieve Automatic Generalisation Using Agent Modelling", *Proceedings of Geographical Information Systems Research – UK (GISRUK 2001)*, University of Glamorgan, UK.

Regnauld, N., (2003) "Algorithms for the Amalgamation of Topographic data", *Proceedings of the 21st International Cartographic Conference (ICC), Cartographic Renaissance*, Durban, South Africa, August 10–16, pp. 222–233, CD-ROM.

Regnauld, N., and Mackaness, W. A., (2006) "Creating a hydrographic network from its cartographic representation: a case study using Ordnance Survey MasterMap data", *International Journal of Geographical Information Science*, 20(6), 611–631.

Reichenbacher, T., (2001) "The World in Your Pocket – Towards a Mobile Cartography", *Mapping the 21st Century: The 20th International Cartographic Conference*, August 6–10, Beijing, China, 4, 2514–2521.

Reichenbacher, T., (2002) "SVG for adaptive visualisations in mobile situations", *Proceedings SVG Open*, Zürich, July 15–17, CD-ROM.

Reichenbacher, T., (2003), "Adaptive Methods for Mobile Cartography", *Proceedings of the 21st International Cartographic Conference (ICC), Cartographic Renaissance*, Durban, South Africa, August 10–16, pp. 1311–1322, CD-ROM.

Reichenbacher, T., (2004) *Mobile Cartography – Adaptive Visualisation of Geographic Information on Mobile Devices*, Dissertation. Department of Cartography, Technische Universität München. München: Verlag Dr. Hut.

Reichenbacher, T., (2005) "Adaptive egocentric maps for mobile users", In: Meng, L., Zipf, A., and Reichenbacher, T., (eds.), *Map-based Mobile Services: Theories, Methods and Implementations*. Berlin: Springer, pp. 141–158.

Reumann, K., and Witkam, A. P., (1974) "Optimizing Curve Segmentation in computer graphics", *International Computing Symposium*, pp. 467–472.

Reynes, J. L., (1997) *Selection du reseau routier en vue de la selection. DESS Report*, University Paris VI.

Ribelles, J., Heckbert, P. S., Garland, M., and Stahovich, T. F., (2001) "Finding and removing features from polyhedra", *Proceedings of DETC'01, ASME Design Engineering Technical Conferences*, Pittsburgh, Pennsylvania, USA. Online: http://www.me.cmu.edu/faculty1/stahovich/smarttools2/publications/publications.htm (30/04/05).

Richardson, D. E., (1993) *Automatic Spatial and Thematic Generalization Using a Context Transformation Model*, Doctoral Dissertation. Wageningen Agricultural University, The Netherlands. Ottawa: R&B Publications.

Richardson, D. E., (1994) "Generalization of spatial and thematic data using inheritance and classification and aggregation hierarchies", In: Waugh, T., and Healey, R., (eds.), *Advances in GIS Research, Proceedings of the 6th International Symposium on Spatial Data Handling*, Edinburgh, Scotland, September 5–9, IGU Commission on GIS and Association for Geographic Information, 2, pp. 957–972.

Richardson, D. E., and Müller, J.-C., (1991) "Rule selection for small-scale map generalization", In: Müller, J.-C., Lagrange, J.-P., and Weibel, R., (eds.), *GIS and generalization: Methodology and Practice, Gisdata, (1)*. London: Taylor & Francis, pp. 136–149.

Richardson, D. E., and Thomson, R. C., (1996) "Integrating thematic, geometric and topologic information in the generalization of road networks", *Cartographica*, 33(1), 75–83.

Richter, K., and Klippel, A., (2002) "Your-Are-Here-Maps: Wayfinding support as Location Based Services", In: Moltgen, J., and Wytzisk, A., (eds.), *GI-Technologien für Verkehr und Logistik, Beiträge zu den Münsteraner GI Tagen, IfGI Prints*, University of Muenster, (13), pp. 357–374.

Rieger, M. K., and Coulson, M. R., (1993) "Consensus for Confusion: Cartographers' Knowledge of Generalization", *Cartographica*, 30(2&3), 69–80.

Rivest, S., Bédard, Y., and Marchand, P., (2001) "Towards better support for spatial decision-making: Defining the characteristics of Spatial On-Line Analytical Processing (SOLAP)", *Geomatica*, 55(4), 539–555.

Robinson, A. H., Sale, R., and Morrison, J. L., (1978) *Elements of Cartography*. New York: Wiley & Sons.

Robinson, A. H., Sale, R. D., Morrison, J. L., and Muehrcke, P. C., (1984), *Elements of Cartography*. 5th ed., New York: Wiley & Sons.

Robinson, A. H., Morrison J. L., Muehrcke, P. C., Kimerling, A. J., and Guptill, S. C., (1995) *Elements of Cartography*. 6th ed., New York: Wiley & Sons.

Robinson, G. J., (1995) "A hierarchical top-down bottom-up approach to topographic map generalization", In: Müller, J. C., Lagrange, J. P., and Weibel, R. (eds.), *GIS and Generalisation: Methodology and Practice*, London: Taylor & Francis, pp. 235–245.

Robinson, G. J., and Zaltash, A., (1989) "Application of expert systems to topographic map generalisation", *Proceedings of the AGI 89 Conference*, Birmingham, A.3.1–A.3.6.

Rodriguez, M. A., Egenhofer, M. J., and Rugg, R. D., (1999) "Assessing Semantic Similarities Among Geospatial Feature Class Definitions", In: Vckovski, A., Brassel, K., and Schek, H.-J. (eds.), *The 2nd International Conference on Interoperating Geographic Information Systems, Interop '99*, March 10–22, Zürich, *Lecture Notes in Computer Science*, (1580). Berlin: Springer, pp. 189–202.

Roth, S., and Hefley, W., (1993) "Intelligent multimedia presentation systems: research and principles", In: Maybury, M. T., (ed.), *Intelligent multimedia interfaces*. Menlo Park, CA: AAAI Press, pp. 13–58.

Rottensteiner, F., (2001) *Semi-automatic extraction of buildings based on hybrid adjustment using 3D surface models and management of building data in a TIS*, Ph.D. Thesis. Technical University of Vienna.

Ruas, A., (1998a) "A method for building displacement in automated map generalisation", *International Journal of Geographical Information Science*, 12(8), 789–803.

Ruas, A., (1998b) "OO-Constraint Modelling to Automate Urban Generalization Process", *Proceedings of the Eighth International Symposium on Spatial Data Handling*, Vancouver, Canada, July 12–15, pp. 225–235.

Ruas, A., (1999) *Modèle de généralisation de données géographiques à base de constraints et d'autonomie*, Thèse de doctorat. L'université de Marne La Vallée, Paris. Online: ftp://ftp.ign.fr/ign/COGIT/THESES (16/06/06).

Ruas, A., (2000a) "Project AGENT: Overview and Results of a European R&D Project in Map Generalisation", *ICA Commission on Map Generalization, Seminar on On-demand Mapping*, Barcelona, Catalunya, Spain, September 21–23. Online: http://www.geo.unizh.ch/ICA/docs/barcelona2000/presentations00.html (29/06/06).

Ruas, A., (2000b) "The role of meso objects for generalization", *Proceedings of 9th International Symposium on Spatial Data Handling*, Beijing, China, August 10–12, Section 3b, pp. 50–63.

Ruas, A., (2001) *Automatic Generalisation Project: Learning Process from Interactive Generalisation*, OEEPE, European Organization for Experimental Photogrammetric Research, *Official Publication (39)*. Frankfurt a. M.: Bundesamt für Kartographie und Geodäsie.

Ruas, A., (2002a) *Généralisation et représentation multiple*. Paris: Hermès Science Publications.

Ruas, A., (2002b) "Les problématiques de l'automatisation de la généralisation", In: Ruas, A., (ed.), *Généralisation et représentation multiple*. Paris: Hermès Sciences Publications, pp. 75–90.

Ruas, A., and Holzapfel, F., (2003) "Automatic characterisation of building alignments by means of expert knowledge", *Proceedings of the 21st International Cartographic Conference (ICC), Cartographic Renaissance*, Durban, South Africa, August 10–16, pp. 1604–1615, CD-ROM.

Ruas, A., and Lagrange, J. P., (1995) "Data and knowledge modelling for generalization", In: Müller, J. C., Lagrange, J. P., and Weibel, R., (eds.), *GIS and Generalisation: Methodology and Practice, Gisdata*, (1). London: Taylor & Francis, pp. 73–90.

Ruas, A., and Mackaness, W. A., (1997) "Strategies for urban map generalization", *Proceedings of the 18th International Cartographic Conference*, Stockholm, Sweden, June 23–27, pp. 1387–1394.

Ruas, A., and Plazanet, C., (1997) "Strategies for Automated Generalization", In: Kraak, M. J., Molenaar, M., and Fendel, E. M., (eds.), *Advances in GIS Research II, Proceedings of the 7th International Symposium on Spatial Data Handling*, Delft, The Netherlands, 1996. London: Taylor & Francis, pp. 319–336.

Rumbaugh, J., Blaha, M. R., Premerlani, W., Eddy, F., and Lorensen, W., (1991) *Object-Oriented Modeling and Design*. Englewood Cliffs, NJ: Prentice-Hall.

Rusak Mazur, E., and Castner, H. W., (1990) "Horton's ordering scheme and the generalization of river networks", *The Cartographic Journal*, 27, 104–112.

Russell, S. J., and Norvig, P., (2003) *Artificial Intelligence: A Modern Approach*, (Prentice Hall Series in Artificial Intelligence). 2nd ed., Upper Saddle River, NJ: Prentice Hall.

Saafeld, A., (1999) "Topologically Consistent Line Simplification with Douglas-Peucker Algorithm", *Cartography and Geographic Information Science*, 26(1), 7–18.

Safe Software, (2004). Online: http://www.safe.com/ (29/06/06).

Salistschew, K. A., (1967) *Einführung in die Kartographie*. 2nd ed., Geographisch-Kartographische Anstalt Gotha, Leipzig: VEB Hermann Haack.

Salichtchev, K. A., (1976) "History and Contemporary Development of Cartographic Generalization", In: *International Yearbook of Cartography*. Bonn: Kirschbaum, pp. 158–172.

Salichtchev, K. A., (1983) "Cartographic communication: A theoretical survey", In: Taylor, D. R. F., (ed.), *Graphic communication and design in contemporary cartography*, (Progress in contemporary cartography). Chichester: Wiley, 2, 11–36.

Sarjakoski, L. T., and Nivala, A.-M., (2005) "Adaptation to Context – A Way to Improve the Usability of Mobile Maps", In: Meng, L., Zipf, A., and Reichenbacher, T., (eds.), *Map-based Mobile Services: Theories, Methods and Implementations*. Berlin: Springer, pp. 107–123.

Sarjakoski, L. T., Koivula, T., and Sarjakoski, T., (2005a) "A Knowledge-Based Map Specification Approach for Mobile Map Services", In: Gartner, G., (ed.), Location-Based Services & TeleCartography, Proceedings of the Symposium 2005, *Geowissenschaftliche Mitteilungen*, Technische Universität Wien, November 28–30, (74), 33–40.

Sarjakoski, L. T., Nivala, A.-M., and Hämäläinen, M., (2003) "Improving the Usability of Mobile Maps by Means of Adaption", In: Gartner, G., (ed.), Location Based Services & TeleCartography, Proceedings of the Symposium 2004, *Geowissenschaftliche Mitteilungen*, Technische Universität Wien, January 28–29, (66), 79–84.

Sarjakoski, T., and Kilpeläinen, T., (1999) "Holistic Cartographic Generalization by Least Squares Adjustment for Large Data Sets", In: Keller, C. P. (ed.), *Proceedings, 19th International Cartographic Conference and 11th General Assembly of ICA, Touch the Past, Visualize the Future*, Ottawa, Canada, August 14–21, 2, 1091–1098.

Sarjakoski, T., and Sarjakoski, L. T., (2005) The GiMoDig public final report. GiMoDig-project, IST-2000-30090, Deliverable D1.2.31, Public EC report. Online: http://gimodig.fgi.fi/deliverables.php (29/06/06).

Sarjakoski, T., Sester, M., Sarjakoski, L. T., Harrie, L., Hampe, M., Lehto, L., and Koivula, T., (2005b) "Web Generalisation Service in GiMoDig – towards a standardised service for real-time generalisation", In: Toppen, F., and Painho, M., (eds.), *Conference Proceedings of the 8th AGILE Conference on Geographic Information Science*, Estoril, Portugal, May 26–28, pp. 509–518.

Sarjakoski, T., Sarjakoski, L. T., Lehto, L., Sester, M., Illert, A., Nissen, F., Rystedt, R., and Ruotsalainen, R., (2002) "Geospatial Info-mobility Services – A Challenge for National Mapping Agencies", *Proceedings of the Joint International Symposium on "GeoSpatial Theory, Processing and Applications"*, ISPRS/Commission IV, Spatial Data Handling 2002, 95th Annual CIG Geomatics Conference, Ottawa, Canada, July 9–12, CD-ROM, WG IV/4, 6 p.

Saund, E., Mahoney, J., Fleet, D., and Larner, D., (2002) "Perceptual organization as a foundation for graphics recognition", In: Blostein, D., and Kwon, Y., (eds.), *Graphics Recognition: Algorithms and Applications, Lecture Notes in Computer Science*, (2390). New York: Springer, pp. 139–147.

Schilcher, M., Roschlaub, R., and Guo, Z., (1998) "Vom 2D-GIS zum 3D-Stadtmodell durch Kombination von GIS-, CAD- und Animationstechniken", *Proceedings ACS '98, Fachseminar Geoinformationssysteme*, November 12–14, Frankfurt a. M., CD-ROM.

Schilling, A., Coors, V., and Laakso, K., (2005) "Dynamic 3D Maps for Mobile Tourism Applications", In: Meng, L., Zipf, A., and Reichenbacher, T., (eds.), *Map-based Mobile Services: Theories, Methods and Implementations*. Berlin: Springer, pp. 227–239.

Schlender, D., Peters, O., and Wienhöfer, M., (2000) "The effects of maps and textual information on navigation in a desktop virtual environment", *Spatial Cognition and Computation*, 2(4), 421–433.

Schneider, D. C., (2001) "The Rise of the Concept of Scale in Ecology", *BioScience*, 51(7), 545–553.

Schylberg, L., (1992) "Cartographic amalgamation of area objects", *The XVIIth International Congress for Photogrammetry and Remote Sensing, International Archives of Photogrammetry and Remote Sensing*, Washington, D.C., Commission IV, XXIX(B4), 135–138.

Schylberg, L., (1993) "Computational Methods for Generalization of Cartographic Data in a Raster Environment", Doctoral Thesis. Department of Geodesy and Photogrammetry, Royal Institute of Technology, Stockholm, Sweden, *Photogrammetric Reports*, (60), *TRITA-FMI Report*, (1993:7).

Searby, S., (2003) "Personalisation – an overview of its use and potential", *BT Technology Journal*, 21(1), 13–19.

Searle, J., (1969) *Speech acts: An essay in the philosophy of language*. Cambridge: University Press.

Sedris, (2005) EDCS Reference Manual, Environmental Data Coding Specification (EDCS). Online: http://www.sedris.org/sdk_4.0/src/lib/edcs/docs/edcs.htm (29/06/06).

Sester, M., (1995) *Lernen struktureller Modelle für die Bildanalyse*, Ph.D. Thesis. Institute for Photogrammetry, University of Stuttgart, Deutsche Geodätische Kommission, München, C (441).

Sester, M., (2000) "Generalization Based on Least-Squares Adjustment", *The XIXth International Congress, Commission IV, International Archives of Photogrammetry and Remote Sensing*, Amsterdam, The Netherlands, XXXIII(B4), 931–938.

Sester, M., (2001) "Optimization Approaches for Generalization", In: Kidner, D. B., and Higgs, G., (eds.), *Proceedings of Geographical Information Systems Research – UK (GISRUK 2001)*, University of Glamorgan, Wales, April 18–20, pp. 32–35.

Sester, M., (2002) "Application Dependent Generalisation – the Case of Pedestrian Navigation", *Proceedings of the Joint International Symposium on "GeoSpatial Theory, Processing and Applications"*, ISPRS/Commission IV, Spatial Data Handling 2002, 95th Annual CIG Geomatics Conference, Ottawa, Canada, July 9–12, CD-ROM, WG IV/3, 6 p.

Sester, M., (2005a) "Optimization approaches for generalization and data abstraction", *International Journal of Geographic Information Science*, 19(8–9), 871–897.

Sester, M., (2005b) "Typification using Kohonen Feature Maps", In: *Self-Organising Maps: Applications for Geographic Information Science*, (to be published).

Sester, M., and Brenner, C., (2004) "Continuous Generalization for Visualization on Small Mobile Devices", In: Fisher, P., (ed.), *Developments in Spatial Data Handling – 11th International Symposium on Spatial Data Handling*. Berlin: Springer, pp. 355–368.

Sester, M., and Klein, A., (1999) "Rule Based Generalization of Buildings for 3D-Visualization", In: Keller, C. P. (ed.), *Proceedings, 19th International Cartographic Conference and 11th General Assembly of ICA, Touch the Past, Visualize the Future*, Ottawa, Canada, August 14–21, 2, pp. 214–224.

Sester, M., Sarjakoski, L. T., Harrie, L., Hampe, M., Koivula, T., Sarjakoski, T., Lehto, L., Elias, B., Nivala, A.-M., and Stigmar, H., (2004) *Real-time generalisation and multiple representation in the GiMoDig mobile service*, GiMoDig-project, IST-2000-30090, Deliverables D7.1.1, D7.2.1 and D7.3.1, Public EC report. Online: http://gimodig.fgi.fi/deliverables.php (29/06/06).

Shannon, C. E., and Weaver, W., (1949) *The Mathematical Theory of Communication*. Urbana: University of Illinois Press.

Shannon, C. E., and Weaver, W., (1998) *The Mathematical Theory of Communication*. Urbana: University of Illinois Press.

Shea, K. S., (1991) "Design Considerations for an Artificially Intelligent System", In: Buttenfield, B. P., and McMaster, R. B., (eds.), *Map Generalization: Making Rules for Knowledge Representation*. Harlow: Longman Group, pp. 3–20.

Shea, K. S., and McMaster, R. B., (1989) "Cartographic Generalization in a Digital Environment: When and How to Generalize", *AUTO-CARTO 9, Ninth International Symposium on Computer-Assisted Cartography*, Baltimore, Maryland, April 2–7, pp. 56–67.

Sheehan, R., (2001) "Generalisation at the Danish Mapping Agency: A Solution Based on the AGENT Research Project and Laser-Scan Object-Oriented Technology", *GIM International*, 15(7), 76–79.

Sheeren, D., (2003) "Spatial databases integration: Interpretation of multiple representations by using machine learning techniques", *Proceedings of the 21st International Cartographic Conference (ICC), Cartographic Renaissance*, Durban, South Africa, August 10–16, pp. 235–244, CD-ROM.

Sheeren, D., Mustière, S., and Zucker, J.-D., (2004) "Consistency Assessment between Multiple Representations of Geographical Databases: A Specification-Based Approach", In: Fisher, P., (ed.), *Developments in Spatial Data Handling, 11th International Symposium on Spatial Data Handling*, Leicester, UK, August 23–25. Berlin: Springer, pp. 617–628.

Sheppard, E., and McMaster, R. B., (2004) *Scale and Geographic Inquiry: Nature, Society, and Method*. Malden, MA: Blackwell Publishing.

Siegel, A., and White, S., (1975) "The development of spatial representations of large-scale environments", In: Reese, H., (ed.), *Advances in Child Development and Behaviour*, (10). New York: Academic Press, pp. 9–55.

Simon, H., (1977) *The New Science of Management Decision*. 3rd ed., Englewood Cliffs, NJ: Prentice-Hall.

Skogan, D., (2005) "Multi-Resolution Geographic Data and Consistency", Ph.D. Thesis. Faculty of Mathematics and Natural Sciences, University of Oslo, *Series of dissertations*, (421).

Skupin, A., and Fabrikant, S. I., (2003) "Spatialisation methods: A cartographic research agenda for non-geographic information visualisation", *Cartography and GIS*, 30, 99–119.

SLD, (2003) OpenGIS®Styled Layer Descriptor Implementation Specification. Online: http://www.opengis.org/techno/implementation.htm (30/12/04).

Slocum, T. A., McMaster, R. B., Kessler, F. C., and Howard, H. H., (2005) *Thematic Cartography and Geographic Visualization, Prentice Hall Series in Geographic Information Science*. 2nd ed., Upper Saddle River, NJ: Pearson/Prentice Hall.

Smith, J. M., and Smith, D. C. P., (1977) "Database Abstractions: Aggregation", *Communications of the ACM*, 20(6), 405–413.

Smith, J. M., and Smith, D. C. P., (1989) "Database Abstractions: Aggregation and Generalisation", In: Mylopoulos, J., and Brodie, M., (eds.), *Readings in Artificial Intelligence and Databases*. San Mateo, California: Morgan Kaufmann.

SOAP, (2003) SOAP Version 1.2 Part 1: Messaging Framework. Online: http://www.w3.org/TR/2003/REC-soap12-part1-20030624/ (29/06/06).

Sondheim, M., Gardels, K., and Buehler, K., (1999) "GIS interoperability", In: Longley, P. A., Goodchild, M. F., Maguire, D. J., and Rhind, D. W., (eds.), *Geographical Information Systems 1: Principles, Techniques, Management and Applications*. 2nd ed., New York: John Wiley & Sons, pp. 347–358.

Song, Z., Liu, Y., and Niu, W., (2004) "A new tetrahedral network (TEN) generation algorithm for 3-D GIS", In: Brandt, S. A., (ed.), *Proceedings of the 12th International Conference on Geoinformatics – Geospatial Information Research: Bridging the Pacific and Atlantic*, University of Gävle, Sweden, June 7–9, pp. 226–232.

Sorrows, M., and Hirtle, S., (1999) "The Nature of Landmarks for Real and Electronic Spaces", In: Freksa, C., and Mark, D., (eds.), *Spatial Information Theory: Cognitive and Computational Foundations of Geographic Information Science*. Berlin: Springer, pp. 37–50.

Spatial Corp., (2005) 3D Components. Online: http://www.spatial.com/ (15/07/06).

SSC, (2005) *Topographic Maps – Map Graphic and Generalization, Cartographic Publication Series*, (17), Swiss Society of Cartography, CD-ROM.

Star Informatic, (2005). Online: http://www.star.be/uk/solutions/MERCATOR.asp (05/05/05).

Staufenbiel, W., (1973), *Zur Automation der Generalisierung topographischer Karten mit besonderer Berücksichtigung großmaßstäbiger Gebäudedarstellungen*, Ph.D. Thesis. University of Hanover, Germany.

Steck, S., and Mallot, H., (2000) "The Role of Global and Local Landmarks in Virtual Environment Navigation", *Presence: Teleoperators and Virtual Environments*, MIT Press, 9(1), 69–83.

Stefanakis, K., and Tsoulos, L., (2005) "Structure and development of a knowledge base for cartographic composition", *International Cartographic Conference 2005: Mapping Approaches into a Changing World*, A Coruña, Spain, CD-ROM: Theme 3: Map Design and Production, Session 2, 10 p.

Stell, J. G., and Worboys, M. F., (1998) "Stratified map space: a formal basis for multi-resolution spatial databases", *Proceedings of the 8th Symposium on Spatial Data Handling, SDH'98*, Vancouver, Canada, July 12–15, pp. 180–189.

Stell, J. G., and Worboys, M. F., (1999) "Generalizing graphs using amalgamation and selection", *Advances in Spatial Database, 6th International Symposium, Proceedings SSD'99*, Hong Kong, China, July 20–23, *Lecture Notes in Computer Science*, (1651). Berlin: Springer, pp. 19–32.

Stonebracker, M., and Hellerstein, J. M., (1998) "Introduction to Distributed Database Systems", In: Stonebracker, M., and Hellerstein, J. M., (eds.), *Readings in Database Systems*, pp. 321–328.

Stonykova, A., (2003) "Design and Implementation of Federation of Spatio-Temporal Databases: Methods and Tools", Final Project Report of AMBER.

Strahler, A. N., (1952) "Dynamic basis of geomorphology", *Bulletin of the Geological Society of America*, 63, 923–938.

Streeter, L., Vitello, D., and Wonsiewicz, S., (1985) "How to Tell People Where to Go: Comparing Navigational Aids", *International Journal of Man-Machine Studies*, 22, 549–562.

Su, B., Li, Z., Lodwick, G., and Müller, J.-C., (1997) "Algebraic models for the aggregation of area features based upon morphological operators", *International Journal of Geographical Information Science*, 11(3), 233–246.

SVG, (2004) Scalable Vector Graphics (SVG), XML Graphics for the Web. Online: http://www.w3.org/Graphics/SVG/ (18/06/06).

Takagi, S., Kobayashi, A., and Tanaka, T., (2003) "Activities for realization of interoperability of location based services using SVG", *Proceedings SVG Open 2003*, Vancouver, Canada.

Taylor, D. R. F., (2005) *Cybercartography: Theory and Practice*. Amsterdam: Elsevier.

Taylor, P. J., (2004) "Is there a Europe of cities? World cities and the limitations of Geographical Scale Analyses", In: Sheppard, E., and McMaster, B., (eds.), *Scale and Geographic Inquiry: Nature, Society, and Method*. Malden, MA: Blackwell Publishing, pp. 213–235.

Thapa, K., (1989) "Data compression and critical points detection using normalized symmetric scattered matrix", *AUTO-CARTO 9, Ninth International Symposium on Computer-Assisted Cartography*, Baltimore, Maryland, April 2–7. ASPRS and ACSM, pp. 78–89.

Theobald, D. M., (2001) "Topology revisited: representing spatial relations", *International Journal of Geographical Information Science*, 15(8), 689–705.

Thiemann, F., (2004) "3D-Gebäude-Generalisierung", In: Koch, W.-G., (ed.), *Theorie 2003 – Kartographische Bausteine, Band 26*, Dresden, pp. 52–58.

Thiemann, F., and Sester, M., (2004), "Segmentation of buildings for 3D-generalisation", *ICA Workshop on Generalisation and Multiple Representation*, Leicester, UK, August 20–21. Online: http://ica.ign.fr/ (16/06/06).

Thomsen, E., Spofford, G., and Chase, D., (1999) *Microsoft OLAP Solutions*. New York: Wiley & Sons.

Thomson, R. C., and Brooks, R., (2000) "Efficient generalization and abstraction of network data using perceptual grouping", *Proceedings of the 5th International Conference on GeoComputation*, Greenwich, August 23–25. Online: http://www.geocomputation.org/2000/GC029/Gc029.htm. (29/06/06).

Thomson, R. C., and Brooks, R., (2002) "Exploiting perceptual grouping for map analysis, understanding and generalization: the case of road and river networks", In: Blostein, D., and Kwon, Y., (eds.), *Graphics Recognition: Algorithms and Applications, Lecture Notes in Computer Science*, (2390). New York: Springer, pp. 148–157.

Thomson, R. C., and Richardson, D. E., (1995) "A graph theory approach to road network generalization", *Proceedings of the 17th International Cartographic Conference*, Barcelona, September 3–9, 2, pp. 1871–1880.

Thomson, R. C., and Richardson, D. E., (1999) "The 'Good Continuation' Principle of Perceptual Organization Applied to the Generalization of Road Networks", In: Keller, C. P. (ed.), *Proceedings, 19th International Cartographic Conference and 11th General Assembly of ICA, Touch the Past, Visualize the Future*, Ottawa, Canada, August 14–21, 2, pp. 1215–1223.

Thorndyke, P., (1981) "Spatial Cognition and Reasoning", In: Harvey, J., (ed.), *Cognition, Social Behavior, and the Environment*. Hillsdale, New Jersey: Lawrence Erlbaum Associates, pp. 137–149.

Thulasiraman, K., and Swamy, M. N. S., (1992) *Graphs: Theory and Algorithms*. New York: Wiley & Sons.

Timpf, S., (1998) "Hierarchical Structures in Map Series", Doctoral Thesis. Department of Geoinformation, Technical University Vienna, Austria, *GeoInfo Series*, (13).

Timpf, S., and Frank, A. U., (1995) "A Multi-Scale DAG for Cartographic Objects", *Technical papers for ACSM/ASPRS, AutoCarto 12*, Charlotte, NC, February 27 – March 1, pp. 157–163.

Timpf, S., and Frank, A. U., (1997) "Exploring the Life of Screen Objects", *Technical Papers for ACSM/ASPRS, AutoCarto 13*, Seattle, Washington, 5, 194–203.

Timpf, S., Volta, G., Pollock, D., and Egenhofer, M., (1992) "A Conceptual Model of Wayfinding Using Multiple Levels of Abstraction", In: Frank, A., Campari, I., and Formentini, U., (eds.), *Theories and Methods of Spatio-Temporal Reasoning in Geographic Space, Lecture Notes in Computer Science*, (639). Berlin: Springer, pp. 348–367.

Tobler, W. R., (1970) "A computer movie simulating urban growth in the Detroit region", *Economic Geography*, 234–240.

Tom, A., and Denis, M., (2003) "Referring to Landmark or Street Information in Route Directions: What Difference Does It Make?" In: Kuhn, W., Worboys, M., and Timpf, S., (eds.), *Spatial information theory: Foundations of geographic information science, International Conference, COSIT 2003, Lecture Notes in Computer Science*, (2825). Berlin: Springer, pp. 362–374.

TomTom CityMaps, (2005) TomTom products. Online: http://www.tomtom.com (01/02/05).

Töpfer, F., and Pillewizer, W., (1966) "The principles of selection: a means of cartographic generalization", *The Cartographic Journal*, 3(1), 10–16.

Trévisan, J., (2004) "From DLM to Multi-Representation DCM – Modelling and Application on Buildings", *ICA Workshop on Generalisation and Multiple Representation*, Leicester, UK, August 20–21. Online: http://ica.ign.fr/ (16/06/06).

Tufte, E. R., (1983) *The visual display of quantitative information*. Cheshire, Connecticut: Graphics Press.

Tufte, E. R., (1990) *Envisioning Information*. Cheshire, CT: Graphics Press.

Turban, E., Aronson, J. E., and Liang, T. P., (2005) *Decision Support Systems and Intelligent Systems*. 7th ed., Englewood Cliffs, NJ: Prentice Hall.

Turnbull, D., (1989) *Maps are territories: Science is an atlas*. Geelong: Deakin University.

Tversky, B., (1990) "Where partonomies and taxonomies meet", In: Tsohatzidis, S. L., (ed.), *Meanings and Prototypes: Studies on Linguistic Categorisation*. London: Routledge, pp. 334–344.

Tversky, B., (1993) "Cognitive Maps, Cognitive Collages, and Spatial Mental Models", In: Frank, A., Campari, I., and Formentini, U., (eds.), *COSIT '93 Proceedings: Spatial Information Theory*, Elba, Italy. Berlin: Springer, pp. 14–24.

Tversky, B., and Lee, P., (1998) "How Space Structures Language", In: Freksa, C., Habel, C., and Wender, K. F., (eds.), *Spatial Cognition*. Berlin: Springer, pp. 157–176.

Tversky, B., and Lee, P., (1999) "Pictorial and Verbal Tools for Conveying Routes", In: Freksa, C., and Mark, D., (eds.), *Spatial Information Theory: Cognitive and Computational Foundations of Geographic Information Science, International Conference COSIT '99, Proceedings*. Berlin: Springer, pp. 51–64.

UDDI, (2004) Universal Description, Discovery, and Integration. Online: http://www.uddi.org/ (05/05/05).

Uitermark, H., (2001) *Ontology-Based Geographic Data Set Integration*, Ph.D. Thesis. Universiteit Twente, The Netherlands.

Urquhart, K., Miller, S., and Cartwright, W., (2003) "A user-centred research approach to designing useful geospatial representations for LBS", In: Gartner, G., (ed.), Location Based Services & TeleCartography, Proceedings of the Symposium 2004, January 28–29, *Geowissenschaftliche Mitteilungen*, Technische Universität Wien, (66), 69–78.

van der Steen, S. J. F. M., (2000) "Publishing on demand, technical possibilities and limitations", *ICA Commission on Map Generalization, Seminar on On-demand Mapping*, Barcelona, Catalunya, Spain, September 21–23. Online: http://www.geo.unizh.ch/ICA/docs/barcelona2000/presentations00.html (29/06/06).

van Kreveld, M. J., (2001) "Smooth generalisation for continuous zooming", *Mapping the 21st Century: The 20th International Cartographic Conference*, Beijing, China, August 6–10, 3, 2178–2185.

van Oosterom, P., (1990) *Reactive Data Structures for Geographic Information Systems*, Ph.D. Thesis. Department of Computer Science, Leiden University, The Netherlands.

van Oosterom, P., (1995) "The GAP-tree, an approach to 'on-the-fly' map generalization of an area partitioning", In: Müller, J.-C., Lagrange, J.-P., and Weibel, R., (eds.), *GIS and Generalization, Methodology and Practice, Gisdata*, (1). London: Taylor & Francis, pp. 120–147.

van Oosterom, P., (1999) "Spatial access methods", In: Longley, P. A., Goodchild, M. F., Maguire, D. J., and Rhind, D. W., (eds.), *Geographical Information Systems, 1: Principles and Technical Issues.* 2nd ed., New York: Wiley & Sons, pp. 385–400.

van Oosterom, P., and Schenkelaars, V., (1995) "The Development of an Interactive Multi-Scale GIS", *International Journal of Geographical Information Systems*, 9(5), 489–507.

van Oosterom, P., and Schenkelaars, V., (1996) "Applying Reactive Data Structures in an Interactive Multi-Scale GIS", In: Molenaar, M., (ed.), *Methods for the Generalization of Geo-Databases, Publications on Geodesy, New series*, Netherlands Geodetic Commission, Delft, (43), 37–56.

van Oosterom, P., Stoter, J., Quak, W., and Zlatanova, S., (2002) "The Balance between Geometry and Topology", In: Richardson, D. E., and van Oosterom, P., (eds.), *Advances in Spatial Data Handling, 10th International Symposium on Spatial Data Handling.* Berlin: Springer, pp. 209–224.

van Smaalen, J. W. N., (1996) "Spatial Abstraction Based on Hierarchical Re-classification", *Cartographica*, (*Monograph* 47), 33(1), 65–73.

van Smaalen, J. W. N., (2003) *Automated Aggregation of Geographic Objects: A New Approach to the Conceptual Generalisation of Geographic Databases*, Doctoral Dissertation. Wageningen University and Research Centre, The Netherlands.

Vangenot, C., (1998) "Représentation multi-résolution, Concepts pour la description de bases de données avec multi-représentation", *Revue Internationale de Géomatique*, 8(1–2), 121–147.

Vangenot, C., Parent, C., and Spaccapietra, S., (2002) "Modelling and Manipulating Multiple Representations of Spatial Data", In: Richardson, D. E., and van Oosterom, P., (eds.), *Advances in Spatial Data Handling, 10th International Symposium on Spatial Data Handling.* Berlin: Springer, pp. 81–93.

Vauglin, F., (2002) "A Practical Study on Precision and Resolution in Vector Geographic Databases", In: Shi, W., Fisher, P. E., and Goodchild, M. F., (eds.), *Spatial Data Quality.* London: Taylor & Francis, pp. 127–139.

Vckovski, A., (1998) *Interoperable and Distributed Processing in GIS*. London: Taylor & Francis.

Vickus, G., (1995) "Strategies for ATKIS-related cartographic products", In: Müller, J.-C., Lagrange, J.-P., and Weibel, R. (eds.), *GIS and generalization: Methodology and Practice, Gisdata* 1. London: Taylor & Francis, pp. 246–252.

Visvalingam, M., and Whyatt, J. D., (1993) "Line Generalization by Repeated Elimination of Points", *The Cartographic Journal*, 30(1), 46–51.

VIVID SOLUTIONS, (2004) JTS Topology Suite and JTS Conflation Suite. Online: http://www.vividsolutions.com/ (29/06/06).

W3C, (2001) Extensible Stylesheet Language (XSL), v.1.0. Online: http://www.w3c.org/TR/xsl (29/06/06).

W3C, (2003) Scalable Vector Graphics (SVG) 1.1 Specification. Online: http://www.w3c.org/TR/SVG11/ (29/06/06).

W3C, (2004) World Wide Web Consortium. Online: http://www.w3.org/ (29/06/06).

Walter, V., and Fritsch, D., (1999) "Matching Spatial Data Sets: A Statistical Approach", *International Journal of Geographical Information Science*, 13(5), 445–473.

Wang, Z., (1996) "Manual versus Automated Line Generalization", *Proceedings of GIS/LIS '96*, Denver, Colorado. Bethesda: ASPRS, pp. 94–106.

Wang, Z., and Lee, D., (2000) "Building Simplification Based on Pattern Recognition and Shape Analysis", *Proceedings 9th International Symposium on Spatial Data Handling*, Beijing, China, August 10–12, pp. 58–72.

Wang, Z., and Müller, J.-C., (1993) "Complex coastline generalization", *Cartography and Geographic Information Systems*, 20(2), 96–106.

Wang, Z., and Müller, J.-C., (1998) "Line generalisation based on an analysis of shape characteristics", *Cartography and Geographic Information Systems*, 25(1), 3–15.

Ware, J. M., and Jones, C. B., (1998) "Conflict Reduction in Map Generalization Using Iterative Improvement", *GeoInformatica*, 2(4), 383–407.

Ware, J. M., Jones, C. B., and Bundy, G. L., (1995) "A Triangulated Spatial Model for Cartographic Generalisation of Areal Objects", In: Frank, U., and Kuhn, W., (eds.), *Spatial Information Theory, A Theoretical Basis for GIS, COSIT '95, Lecture Notes in Computer Science*, (988), Semmering, Austria, September 21–23. Berlin: Springer, pp. 173–192.

Ware, J. M., Jones, C. B., and Thomas, N., (2003a) "Automated map generalization with multiple operators: a simulated annealing approach", *International Journal of Geographical Information Science*, 17(8), 743–769.

Ware, J. M., Wilson, I. D., and Ware, J. A., (2003b) "A Knowledge Based Genetic Algorithm Approach to Automating Cartographic Generalisation", *Knowledge-Based Systems*, 16, 295–303.

WEBPARK, (2004) WebPark – Location based services in natural areas. Online: http://www.webparkservices.info/ (29/06/06).

Weibel, R., (1991) "Amplified intelligence and rule-based systems", In: Buttenfield, B. P., and McMaster, R. B., (eds.), *Map Generalization: Making Rules for Knowledge Representation.* Harlow: Longman Group, pp. 172–186.

Weibel, R., (1992) "Models and Experiments for Adaptive Computer Assisted Terrain Generalization", *Cartography and Geographic Information Systems*, 19(3), 133–153.

Weibel, R., (1995) "Three essential building blocks for automated generalization", In: Müller, J.-C., Lagrange, J.-P., and Weibel, R., (eds.), *GIS and generalization: Methodology and Practice.* London: Taylor & Francis, pp. 56–69.

Weibel, R., (1997) "A Typology of Constraints to Line Simplification", In: Kraak, M. J., Molenaar, M., and Fendel, E. M., (eds.), *Advances in GIS Research II, 7th International Symposium on Spatial Data Handling*, Delft, The Netherlands, 1996. London: Taylor & Francis, pp. 533–546.

Weibel, R., and Dutton, G., (1998) "Constraint-based automated map generalization", *Proceedings of the 8th Spatial Data Handling Symposium*, Vancouver, Canada, July 12–15, pp. 214–224.

Weibel, R., and Dutton, G., (1999) "Generalising spatial data and dealing with multiple representations", In: Longley, P. A., Goodchild, M. F., Maguire, D. J., and Rhind, D. W., (eds.), *Geographical Information Systems, 1: Principles and Technical Issues.* 2nd ed., New York: Wiley & Sons, pp. 125–155.

Weibel, R., and Jones, C. B., (1998) "Computational Perspectives on Map Generalisation", *GeoInformatica*, 2(4): 301–314.

Weibel, R., Bernier, E., Bédard, Y., and Cecconi, A., (2002) "La généralisation à la volée", In: Ruas, A., (ed.), *Généralisation et représentation multiple.* Paris: Hermès Sciences Publications, pp. 319–335.

Weibel, R., Keller, S., and Reichenbacher, T., (1995) "Overcoming the Knowledge Acquisition Bottleneck in Map Generalization: The Role of Interactive Systems and Computational Intelligence", In: Frank, U., and Kuhn, W., (eds.), *Spatial Information Theory, A Theoretical Basis for GIS, COSIT '95*, Semmering, Austria, September 21–23, *Lecture Notes in Computer Science*, (988). Berlin: Springer, pp. 139–156.

Weiss, G., (ed.), (1999) *Multiagent Systems: A Modern Approach to Distributed Artificial Intelligence.* Cambridge, MA: MIT Press.

Wertheimer, M., (1938) "Laws of organization in perceptual forms", In: Ellis, W. D., (ed.), *A source book of Gestalt psychology.* New York: Harcourt & Brace, pp. 71–88.

WFS, (2002) Web Feature Service Implementation Specification. Online: http://www.opengis.org/docs/02-058.pdf (29/06/06).

WFS, (2004) OpenGIS®Web Feature Service Interfaces Implementation Specification. Online: http://www.opengis.org/techno/implementation.htm (05/05/05).

Wilbanks, T., (2002) "Geographic Scaling Issues in Integrated Assessments of Climate Change", *Integrated Assessment*, 3, (2–3), 100–114.

Wilson, I. D., Ware, J. M., and Ware, J. A., (2003) "A Genetic Algorithm Approach to Cartographic Map Generalisation", *Computers in Industry*, 52, 291–304.

Wilson, J. P., and Gallant, J., (2000) *Terrain Analysis*. New York: Wiley and Sons.

Winter, S., (2003) "Route Adaptive Selection of Salient Features", In: Kuhn, W., Worboys, M., and Timpf, S., (eds.), *Spatial Information Theory, Lecture Notes in Computer Science*, (2825). Berlin: Springer, pp. 320–334.

Winter, S., Raubal, M., and Nothegger, C., (2005) "Focalizing Measures of Salience for Wayfinding", In: Meng, L., Zipf, A., and Reichenbacher, T., (eds.), *Map-based Mobile Services: Theories, Methods and Implementations*. Berlin: Springer, pp. 125–139.

Witkin, A. P., and Tenenbaum, J. M., (1983) "On the role of structure in vision", In: Beck, J., Hope, B., and Rosenfeld, A., (eds.), *Human and Machine Vision*. New York: Academic Press.

WMS, (2004) OpenGIS®Web Map Service Interfaces Implementation Specification. Online: http://www.opengis.org/techno/implementation.htm (05/05/05).

Wood, J., (1996) "Scale-based characterisation of digital elevation models", In: Parker, D., (ed.), *Innovations in GIS 3*. London: Taylor and Francis, pp. 163–176.

Woodsford P. A., (2003) "MAGNET – Mapping Agencies Generalisation NETwork", *GeoInformatics*, 6(3), 20–23.

Worboys, M. F., (1999) "Relational databases and beyond", In: Longley, P. A., Goodchild, M. F., Maguire, D. J., and Rhind, D. W., (eds.), *Geographical Information Systems 1: Principles, Techniques, Management and Applications*. 2nd ed., New York: John Wiley & Sons, pp. 373–384.

Wright, J. K., (1942) "Map Makers Are Human", *Geographical Review*, 32, 527–544.

WSA, (2004) Web Services Architecture. Online: http://www.w3.org/TR/ws-arch/ (05/05/05).

WSDL, (2001) Web Services Description Language (WSDL) 1.1. Online: http://www.w3.org/TR/wsdl (05/05/05).

Yang, P., Lin, H., Mao, S. J., and Shen, D., (2004) "A study on algorithms of a 3D visualization dynamic modification system based on TIN", In: Brandt, S. A., (ed.), *Proceedings of the 12th International Conference on Geoinformatics – Geospatial Information Research: Bridging the Pacific and Atlantic*, University of Gävle, Sweden, June 7–9, pp. 347–354.

Zhang, Q., (2004) "Modeling Structure and Patterns in Road Network Generalization", *ICA Workshop on Generalisation and Multiple Representation*, Leicester, UK, August 20–21. Online: http://ica.ign.fr/ (29/06/06).

Zhizhuo, W., (1990) *Principles of Photogrammetry*. Beijing: Publishing House of Surveying and Mapping.

Zhou, S., and Jones, C. B., (2001) "Design and implementation of multi-scale spatial databases", *Proceedings of the 7th International Symposium on Spatial and Temporal Databases*, Redondo Beach, CA.

Zipf, A., (1998) "Deep Map – A prototype context sensitive tourism information system for the city of Heidelberg", *GIS Planet 98, International Conference and Exhibition on Geographic Information*, Lisbon, Portugal, September 7–11.

Zipf, A., and Richter, K. F., (2002) "Using Focus Maps to Ease Map Reading: Developing Smart Applications for Mobile Devices", *Künstliche Intelligenz (KI). Sonderheft Spatial Cognition*. 04/2002, pp. 35–37.

Zoraster, S., (1986) "Integer Programming Applied to the Map Label Placement Problem", *Cartographica*, 23(3), 16–27.

Zoraster, S., (1997) "Practical Results Using Simulated Annealing for Point Feature Label Placement", *Cartography and Geographic Information Systems*, 24(4), 228–238.

Author Index

Aalders, H. J., *see* Moellering, H. ix
Aasgaard, R. 34
Abdelmoty, A. I., *see* El-Geresy, B. A. 120
Abdelmoty, A. I., *see* Jones, C. B. 32, 202
Abowd, G. D. 139
Abraham, I. M., *see* Jones, C. B. 26
Adam, F., *see* Pomerol, J.-C. 272
Affholder, J.-G. 52
Afflerbach, S., *see* Illert, A. 167
Agrawala, M. 201, 206–208
Ai, T., *see* Li, Z. 47
Airault, S. 48
Akima, H. 52, 59, 62
Almer, A., *see* Luley, P. M. 140
Anders, K.-H. 251
Anders, K.-H., *see* Hampe, M. 31, 32, 168, 216, 288
Andersen, H., *see* Brodersen, L. 206
Appleyard, D. 202
Arikawa, M. 162
Arleth, M. 162
Armstrong, M. P. 19
Aronson, J. E., *see* Turban, E. 179, 184
Atkeson, C. G., *see* Abowd, G. D. 139
Austin, J. L. 280
Avelar, S. 162, 201, 202, 208
Averdung, C. 216
Ayasse, J. 219

Babaud, J. 50
Badard, T. 34, 120, 164, 172, 295, 321
Badard, T., *see* Lemarie, C. 296
Bader, M. 51, 52, 55, 56, 80, 81, 284, 290
Baeijs, C., *see* Barrault, M. 76, 78, 288
Balley, S. 26, 31, 32, 115, 120, 121
Banel, T., *see* Lecordix, F. 293
Bard, S. 100, 107–109, 271, 282
Bard, S., *see* Duchêne, C. 49
Barillot, X. 95
Barillot, X., *see* Duchêne, C. 49
Barkowsky, T. 162, 206
Barkowsky, T., *see* Casakin, H. 201
Barkowsky, T., *see* Klippel, A. 143
Barnett, L., *see* McMaster, R. B. 40
Barrault, M. 76, 78, 288, 293
Barrault, M., *see* Bader, M. 51, 55, 80
Barrault, M., *see* Cecconi, A. 32
Bartie, P. 209
Baudin, M., *see* Babaud, J. 50
Bauer, B. 10
Baus, J. 140, 141
Beard, K. 21, 40, 50, 70, 75, 162, 270, 275
Beard, M. K., *see* Mackaness, W. A. 53, 257
Bédard, Y. 114, 116, 118, 186
Bedard, Y., *see* Bernier, E. 31, 120, 184, 191
Bédard, Y., *see* Bernier, E. 191
Bédard, Y., *see* Proulx, M.-J. 186
Bédard, Y., *see* Rivest, S. 184
Bédard, Y., *see* Weibel, R. 178
Bernier, E. 31, 120, 184, 191
Bernier, E., *see* Bédard, Y. 186
Bernier, E., *see* Weibel, R. 178
Bertin, J. 3, 12, 14, 168
Bertolo, L., *see* Denis, M. 200–203, 206
Bertolotto, M. 141
Biederman, I. 219, 260
Bjerhammar, A. 86
Bjørke, J. T. 56, 92, 276
Black, T. A., *see* DeLucia, A. 259
Blaha, M. R., *see* Rumbaugh, J. 22, 26
Bobzien, M., *see* Edwardes, A. 162, 172, 320
Bodansky, E. 61
Boehler, W. 215
Boffet, A. 128, 271, 277, 278, 282
Bollobás, B. 235
Booch, G. 115
Booch, G. L. 26, 116, 275
Borges, K., *see* Lisboa Filho, J. 117
Brassel, K. E. 14–16, 38, 39, 67, 91, 270, 288
Braun, A. 31
Braun, A., *see* Badard, T. 164, 172, 321
Brazile, F. 107

Bregt, A. 133
Brenner, C. 216
Brenner, C., *see* Elias, B. 203, 205
Brenner, C., *see* Sester, M. 47, 231
Brewer, C. A. 5
Brodersen, L. 206
Brooks, R. 261, 263, 265
Brooks, R., *see* Thomson, R. C. 53–55, 262, 263, 266
Brophy, M. 50, 58
Brown, A., *see* Kraak, M.-J. 143
Brown, A., *see* Peng, W. 95
Brown, M. D., *see* McCormick, B. H. 6
Brown, N., *see* Fuller, R. 56
Bruegger, B. P. 25, 29, 30
Buehler, K., *see* Sondheim, M. 164
Bulens, J., *see* Bregt, A. 133
Bundy, G. L., *see* Jones, C. B. 22, 26, 47, 55
Bundy, G. L., *see* Ware, J. M. 42, 47
Burghardt, D. 46, 51, 80, 87, 141, 162, 164, 169, 171, 276
Burghardt, D., *see* Edwardes, A. 140, 141, 162, 170, 172, 320
Burghardt, D., *see* Neun, M. 168
Burghardt, D., *see* Rappo, A. 168
Burnett, G. 203, 209
Buttenfield, B. P. viii, 5, 13, 18–20, 28, 29, 31, 51, 69, 141
Buttenfield, B., *see* MacEachren, A. M. 6

Cámara, M. A. 47
Campbell, G. M., *see* Cromley, R. G. 78
Campbell, J., *see* MacEachren, A. M. 6
Cardenas, A. 197
Carroll, J. M. 318
Cartwright, W., *see* Germanchis, T. 211
Cartwright, W., *see* Urquhart, K. 141
Casakin, H. 201
Castner, H. 206
Castner, H. W., *see* Rusak Mazur, E. 256
Cecconi, A. 32, 165, 168, 177
Cecconi, A., *see* Burghardt, D. 46
Cecconi, A., *see* Rappo, A. 168
Cecconi, A., *see* Weibel, R. 178
Chalmers, D. 162
Chang, D., *see* Poole, J. 184
Chase, D., *see* Thomsen, E. 183
Chaudhry, O. 265
Chen, J., *see* Li, Z. 47
Chen, P. P. 115
Cheung, C. K. 93
Cheverst, K. 140
Chignell, M., *see* Parsaye, K. 20, 21
Chirie, F., *see* Lecordix, F. 293
Choi, Y. H., *see* Li, Z. 257
Christ, F. A. 13
Christensen, A. 50, 53
Claramunt, C. 120, 202
Claramunt, C., *see* Jiang, B. 251, 257
Clark, J. W. 259
Clementini, E., *see* Egenhofer, M. J. 31, 120
Cook, A. C. 19
Coors, V., *see* Schilling, A. 140, 141, 209
Coren, S. 260
Cornoldi, C., *see* Denis, M. 200–203, 206
Cosma, I., *see* Gbei, E. 169, 185, 186
Coulson, M. R., *see* Rieger, M. K. 20
Crampton, J. W. 162
Crane, A., *see* Moellering, H. ix
Cras, Y., *see* Lecordix, F. 293
Cromley, R. G. 78
Cuthbert, A. 167, 168

Dai, F., *see* Haase, H. 229
Dakowicz, M., *see* Gold, C. 56
Davies, N., *see* Cheverst, K. 140
Davis, E. 199
Deakin, A. 202
DeFanti, T. A., *see* McCormick, B. H. 6
Delaney, D. 3
Delotto, J. S., *see* Buttenfield, B. P. 28
DeLucia, A. 259
Delval, N., *see* Gbei, E. 185, 186
DeMarco, T. 33, 34
Demazeau, Y., *see* Barrault, M. 76, 78, 288
Demazeau, Y., *see* Lamy, S. 23, 62, 106, 266, 271, 288, 308
Denis, M. 200–203, 206
Denis, M., *see* Michon, P. 202, 203
Denis, M., *see* Tom, A. 202
Dettori, G., *see* Puppo, E. 257
Devogele, T. 25, 115, 125
Dey, A. K. 143
DiBiase, D. 6
DiBiase, D., *see* MacEachren, A. M. 6
Diestel, R. 235, 256, 257
Dijkstra, E. W. 247
Dillemuth, J. 141
Ding, A. 215
Dodge, M. 6
Doerschler, J. S. 19, 69
Dold, C., *see* Brenner, C. 216
Döllner, J. 229
Donini, P., *see* Parent, C. 114, 118, 119
Dorling, D. 3
Douglas, D. H. 13, 50, 58–60
Downs, T. C. 56, 97, 256

Dransch, D. 142
Drees, R., *see* Gröger, G. 213
Duchêne, C. 23, 49, 77, 78, 271, 279, 280
Duchêne, C., *see* Barrault, M. 76, 78, 288
Duckham, M. 205
Duckham, M., *see* Kulik, L. 24
Duda, R. O. 235, 250
Duda, R., *see* Babaud, J. 50
Dulay, N., *see* Chalmers, D. 162
Dunkars, M. 22, 26, 31, 34
Dutton, G. 5, 91
Dutton, G., *see* Weibel, R. 14, 16, 18, 25, 41, 57, 75, 106, 162, 168, 177, 270
Dykes, J. ix, 6
Dykes, J., *see* Raper, J. 162

Eastman, R., *see* Castner, H. 206
Eckert, M. 12, 37
Eddy, F., *see* Rumbaugh, J. 22, 26
Edsall, R., *see* MacEachren, A. M. 5, 6
Edwardes, A. 50, 53, 54, 93, 140, 141, 162, 170–172, 175, 265, 266, 320
Edwardes, A., *see* Burghardt, D. 141, 171
Edwards, G., *see* Mackaness, W. A. 235
Efstratiou, C., *see* Cheverst, K. 140
Egenhofer, M. J. 22, 31, 120
Egenhofer, M. J., *see* Rodriguez, M. A. 125, 133
Egenhofer, M., *see* Bertolotto, M. 141
Egenhofer, M., *see* Frank, A. 128
Egenhofer, M., *see* Kulik, L. 24
Egenhofer, M., *see* Timpf, S. 210
Ehrliholzer, R. 107
El-Geresy, B. A. 120
Elias, B. 162, 203, 205, 209
Elias, B., *see* Hampe, M. 210
Elias, B., *see* Sester, M. 32, 170
Erdös, P. 235

Fabrikant, S. I., *see* Skupin, A. 10
Fairbairn, D., *see* Dorling, D. 3
Faloutsos, C., *see* Faloutsos, M. 236
Faloutsos, M. 236
Faloutsos, P., *see* Faloutsos, M. 236
Farin, G. 61
Fechir, A., *see* Lecordix, F. 308
Feiner, S. K., *see* Foley, J. D. 167
Felice, P. D., *see* Egenhofer, M. J. 31, 120
Fisher, P. F. 6, 69, 321
Fisher, P. F., *see* Mackaness, W. A. 19, 20, 69, 270
Fitzke, J. 170
Fleet, D., *see* Saund, E. 260
Foley, J. D. 167
Forberg, A. 218, 223–225
Forrest, D. 20
Foucault, M. 99
Frank, A. 128
Frank, A. U. 25, 29, 30
Frank, A. U., *see* Bruegger, B. P. 25, 29, 30
Frank, A. U., *see* Egenhofer, M. J. 22
Frank, A. U., *see* Timpf, S. 30, 31
Frank, A., *see* Gartner, G. 140
Frawley, W. J. 5, 323
Freeman, H., *see* Doerschler, J. S. 19, 69
Freeman, H., *see* Nickerson, B. G. 14, 43, 69, 70
Freeman, L. C. 251
Freitag, U. 11
Freksa, C. 201, 205
Freksa, C., *see* Barkowsky, T. 206
Freksa, C., *see* Casakin, H. 201
Freksa, C., *see* Klippel, A. 143
Friday, A., *see* Cheverst, K. 140
Friis-Christensen, A. 30, 120, 126
Fritsch, D., *see* Walter, V. 125
Fritsch, E. 51
Früh, C. 215
Fu, G., *see* Jones, C. B. 202
Fuhrmann, S. 162
Fuller, R. 56

Gahegan, M. 6
Galanda, M. 24, 56, 78, 80, 84
Galanda, M., *see* Cecconi, A. 165
Gallant, J., *see* Wilson, J. P. 56
Gamma, E. 117, 166
Ganter, J. H., *see* MacEachren, A. M. 99
Gardarin, G. 114, 117
Gardels, K., *see* Sondheim, M. 164
Garland, K. 91, 208
Garland, M., *see* Heckbert, P. 227
Garland, M., *see* Ribelles, J. 220
Gartner, G. 140–142
Gbei, E. 169, 185, 186
Germanchis, T. 211
Gill, G. A. 69
Göbel, M., *see* Haase, H. 229
Gold, C. 52, 56
Goldberg, M., *see* Goodenough, D. G. 19
Golledge, R. 200
Goodchild, M. 90
Goodchild, M. F. 7
Goodenough, D. G. 19
Greenwood, J. 4
Gregory, D., *see* Meyer, W. B. 3
Greve, K., *see* Fitzke, J. 170
Gribov, A., *see* Bodansky, E. 61
Gröger, G. 213
Gröger, G., *see* Kolbe, T. H. 229

Gröger, G., *see* Petzold, I. 168
Gruber, U., *see* Gröger, G. 213
Grünreich, D. 11, 16, 17
Grünreich, D., *see* Hake, G. 217
Guo, Z., *see* Schilcher, M. 213, 215
Guptill, S. C. ix, 90
Guptill, S. C., *see* Robinson, A. H. 12

Haase, H. 229
Habel, C., *see* Klippel, A. 206, 209
Hackathorn, D., *see* Inmon, W. H. 179
Haire, K., *see* Barrault, M. 76, 78, 288
Hake, G. 217
Haken, H. 231
Hämäläinen, M., *see* Sarjakoski, L. T. 146
Hampe, M. 26, 31, 32, 120, 150, 168, 210, 216, 288
Hampe, M., *see* Elias, B. 203, 209
Hampe, M., *see* Sarjakoski, T. 87, 149, 166, 175
Hampe, M., *see* Sester, M. 32, 170
Han, J. 203
Hangouet, J.-F. 94, 96
Harary, F. 235
Harbison-Briggs, K., *see* McGraw, K. L. 19
Hardy, P. 34, 170, 175, 308
Hardy, P. G. 22, 26
Hardy, P., *see* Barrault, M. 76, 78, 288
Harrie, L. 22, 23, 26, 30, 34, 49, 55, 75, 80, 81, 86, 106, 142, 143, 155, 162, 164, 166, 168, 208, 276
Harrie, L., *see* Edwardes, A. 162, 172, 320
Harrie, L., *see* Hampe, M. 26, 31, 32, 150
Harrie, L., *see* Sarjakoski, T. 87, 149, 166, 175
Harrie, L., *see* Sester, M. 32, 170
Harrower, M., *see* Gahegan, M. 6
Hart, P. E., *see* Duda, R. O. 235, 250
Hartshorne, R. 3
Haug, D., *see* MacEachren, A. M. 5, 6
Hayes-Roth, F. 18, 69
Hayles, M., *see* Hardy, P. 34, 170, 175, 308
Head, G. 4
Heckbert, P. 227
Heckbert, P. S., *see* Ribelles, J. 220
Hedley, N. R. 218
Hefley, W., *see* Roth, S. 205
Hegarty, M., *see* Lovelace, K. 203
Heinzle, F. 251
Hellerstein, J. M., *see* Stonebracker, M. 126
Hellström, A.-K., *see* Harrie, L. 26, 34
Helm, R., *see* Gamma, E. 117, 166
Hilton, H. J., *see* Biederman, I. 260
Hirtle, S., *see* Sorrows, M. 202, 203
Hjelm, J. 164
Hogan, R., *see* Moellering, H. ix

Højholt, P. 22, 49, 55, 80, 128
Holzapfel, F., *see* Duchêne, C. 49
Holzapfel, F., *see* Ruas, A. 46, 123, 277
Hong, J., *see* Abowd, G. D. 139
Horton, R. E. 54, 256
Hough, P. V. C. 245
Howard, H. H., *see* Slocum, T. A. 13, 40
Hubert, F. 105, 106
Hughes, J. F., *see* Foley, J. D. 167
Hui, W. 21
Hummel, J. E., *see* Biederman, I. 260
Hunt, E. 200
Hutchinson, G. E. 2

Ilg, M., *see* Ogniewicz, R. L. 52
Illert, A. 167
Illert, A., *see* Sarjakoski, T. 140, 144
Imhof, E. 12
Inmon, W. H. 179, 184
Iochpe, C., *see* Lisboa Filho, J. 117

Jabeur, N., *see* Gbei, E. 169, 185, 186
Jackson, M., *see* Lamy, S. 23, 62, 106, 266, 271, 288, 308
Jäger, E. 51
Jahard, Y. 288
Jakobson, I., *see* Booch, G. L. 26, 116, 275
Jakobsson, A., *see* Nivala, A.-M. 145, 162
Jankowski, P. 69
Jeansoulin, R., *see* Goodchild, M. 90
Jenks, G. F. 50, 58
Jensen, C. S., *see* Friis-Christensen, A. 30
Jiang, B. 251, 257
João, E. M. viii, 24, 92, 107
Johansson, M., *see* Harrie, L. 164, 166
Johnson, R., *see* Gamma, E. 117, 166
Jones, C. B. 22, 26, 29, 32, 47, 55, 202
Jones, C. B., *see* Kidner, D. B. 25, 30, 31
Jones, C. B., *see* Lonergan, M. E. 48, 49
Jones, C. B., *see* Ware, J. M. 23, 42, 47, 49, 78, 79, 106, 162
Jones, C. B., *see* Weibel, R. 7
Jones, C. B., *see* Zhou, S. 31, 120
Jülge, K., *see* Brenner, C. 216

Kaasinen, E., *see* Nivala, A.-M. 145, 162
Kada, M. 228, 229
Kähkönen, J., *see* Lehto, L. 32
Kaiser, M. K. 2
Kambayashi, Y., *see* Arikawa, M. 162
Kamber, M., *see* Han, J. 203
Kawakita, H., *see* Arikawa, M. 162
Keahey, A. 208
Keates, J. S. 4, 12

Keller, S., *see* Weibel, R. 70
Kellogg, W. A., *see* Carroll, J. M. 318
Kessler, F. C., *see* Slocum, T. A. 13, 40
Khuan, C.-T., *see* Rahman, A.-A. 218
Kidner, D. B. 25, 30, 31
Kidner, D. B., *see* Jones, C. B. 22, 26
Kilpeläinen, T. 16–18, 20, 22, 24–34, 120, 123, 177, 288, 295
Kilpeläinen, T., *see* Lehto, L. 32, 144, 164, 165
Kilpeläinen, T., *see* Sarjakoski, T. 22, 49, 55, 86
Kimball, R. 179, 184
Kimerling, A. J. 5
Kimerling, A. J., *see* Robinson, A. H. 12
King, D. B. 260
Kitchin, R., *see* Dodge, M. 6
Klein, A., *see* Sester, M. 32, 228
Klippel, A. 143, 206, 209
Klippel, A., *see* Casakin, H. 201
Klippel, A., *see* Richter, K. 200
Knöpfli, R. 12
Knospe, F., *see* Gröger, G. 213
Kobayashi, A., *see* Takagi, S. 165
Köbben, B. 141
Kohlhaas, A., *see* Gröger, G. 213
Koivula, T., *see* Harrie, L. 155
Koivula, T., *see* Sarjakoski, L. T. 155
Koivula, T., *see* Sarjakoski, T. 87, 149, 166, 175
Koivula, T., *see* Sester, M. 32, 170
Kolazny, A. 11
Kolbe, T. H. 140, 213, 229
Kolbe, T. H., *see* Gröger, G. 213
Kooper, R., *see* Abowd, G. D. 139
Kopczynski, M. 201
Kopomaa, T. 142
Kraak, M. J. 3
Kraak, M. J., *see* MacEachren, A. M. 10
Kraak, M.-J. 143
Kraak, M.-J., *see* Dykes, J. ix, 6
Krause, A., *see* Raper, J. 162
Krause, U., *see* Gröger, G. 213
Kray, C., *see* Baus, J. 140
Kreiter, N. 24
Krippendorff, K. 8
Krüger, A., *see* Baus, J. 140, 141
Kuhn, W., *see* Fuhrmann, S. 162
Kulik, L. 24
Kulik, L., *see* Duckham, M. 205

Laakso, K., *see* Schilling, A. 140, 141, 209
Laamanen, H., *see* Poslad, S. 140
Lagrange, J. P., *see* Lecordix, F. 293
Lagrange, J. P., *see* Ruas, A. 131
Lagrange, J.-P. 30
Lagrange, J.-P., *see* Lecordix, F. 50
Lagrange, J.-P., *see* Müller, J.-C. viii, 43, 320
Lal, J. 222, 223
Lambert, M., *see* Bernier, E. 191
Lamy, S. 23, 62, 106, 266, 271, 288, 308
Lang, F. 218
Lang, T. 50, 58
Larner, D., *see* Saund, E. 260
Larrivée, S., *see* Bédard, Y. 118, 186
Larrivée, S., *see* Proulx, M.-J. 186
Latecki, L. J., *see* Barkowsky, T. 162
Laurini, R. 19, 21, 22, 30
Le Men, H. 56
Lecordix, F. 50, 293, 308
Lecordix, F., *see* Jahard, Y. 288
Lecordix, F., *see* Mustière, S. 255
Ledgard, H. 32
Lee, D. 20, 60, 72
Lee, D., *see* Wang, Z. 60
Lee, P., *see* Tversky, B. 200, 209
Lehto, L. 32, 144, 147, 164, 165, 167
Lehto, L., *see* Edwardes, A. 162, 172, 320
Lehto, L., *see* Harrie, L. 142, 143, 155, 168, 208
Lehto, L., *see* Sarjakoski, T. 87, 140, 144, 149, 166, 175
Lehto, L., *see* Sester, M. 32, 170
Leiner, U. 218
Leitner, H. 1
Leitner, H., *see* Delaney, D. 3
Lemarie, C. 296
Lemarié, C., *see* Badard, T. 120
Lemarie, C., *see* Jahard, Y. 288
Lenat, D., *see* Hayes-Roth, F. 18, 69
Leontjew, A. N. 142
Levin, S. A. 1, 2
Levine, M. 200
Li, Z. 47, 257
Li, Z., *see* Su, B. 47
Liang, T. P., *see* Turban, E. 179, 184
Lichtner, W. 14, 47, 208
Lin, H., *see* Yang, P. 218
Linturi, R. 32
Lipeck, U. W., *see* Mantel, D. 125
Lisboa Filho, J. 117
Liu, Y. 128
Liu, Y., *see* Song, Z. 220
Lodwick, G., *see* Su, B. 47
Lonergan, M. E. 48, 49
Lonergan, M. E., *see* Jones, C. B. 32
Long, J. 6
Long, S., *see* Abowd, G. D. 139
Loomis, J., *see* Mallot, H. 211
López, F. J., *see* Cámara, M. A. 47
Lorensen, W., *see* Rumbaugh, J. 22, 26

Lovelace, K. 203
Lowe, D. G. 50, 260
Luley, P. M. 140
Luo, L. Q., *see* Jones, C. B. 22, 26
Lynch, K. 201–203, 231

Maaß, W. 209
MacEachren, A. 3, 200, 206, 260
MacEachren, A. M. 5, 6, 10, 99
MacEachren, A. M., *see* Dykes, J. ix, 6
Mackaness, W. A. 19–21, 40, 48, 49, 53, 54, 69, 75, 92, 96, 235, 256, 257, 265, 270, 274
Mackaness, W. A., *see* Bartie, P. 209
Mackaness, W. A., *see* Chaudhry, O. 265
Mackaness, W. A., *see* Downs, T. C. 56, 97, 256
Mackaness, W. A., *see* Edwardes, A. 50, 53, 54, 93, 265, 266
Mackaness, W. A., *see* Fisher, P. F. 69
Mackaness, W. A., *see* Greenwood, J. 4
Mackaness, W. A., *see* Lamy, S. 23, 62, 106, 266, 271, 288, 308
Mackaness, W. A., *see* Rainsford, D. 175
Mackaness, W. A., *see* Regnauld, N. 52, 54
Mackaness, W. A., *see* Ruas, A. 256, 266
Mackaness, W., *see* Barrault, M. 76, 78, 288
Mackaness, W., *see* Edwardes, A. 175
Mackaness, W., *see* Ormsby, D. 8, 22, 66, 302
Mackechnie, G. A., *see* Mackaness, W. A. 54, 96, 256, 257, 265
Mahoney, J., *see* Saund, E. 260
Mainguenaud, M., *see* Claramunt, C. 120, 202
Malaka, R., *see* Poslad, S. 140
Mallot, H. 200, 201, 211
Mallot, H., *see* Steck, S. 203
Mannes, J., *see* Burghardt, D. 141
Mantel, D. 125
Mao, S. J., *see* Yang, P. 218
Marbs, A., *see* Boehler, W. 215
Marchand, P., *see* Rivest, S. 184
Mark, D. M. 19, 21, 27, 91
Mark, D. M., *see* Buttenfield, B. P. 69
Mark, D. M., *see* O'Callaghan, J. F. 56
Marston, S. 3
Martinez Casasnovas, J. A. 30
Mathews, C. J., *see* Frawley, W. J. 5, 323
Mathur, A., *see* Burghardt, D. 276
May, A., *see* Burnett, G. 203
Mayer, H., *see* Forberg, A. 223
McConalogue, D. J. 52, 60, 62
McCormick, B. H. 6
McDowell, P., *see* Meyer, W. B. 3
McGraw, K. L. 19
McMaster, R. B. viii, 1, 3, 4, 13–15, 20, 38–43, 50, 55, 92, 93, 270, 275
McMaster, R. B., *see* Brewer, C. A. 5
McMaster, R. B., *see* Buttenfield, B. P. viii, 13, 18–20, 28
McMaster, R. B., *see* Monmonier, M. S. 43
McMaster, R. B., *see* Shea, K. S. 13, 14
McMaster, R. B., *see* Sheppard, E. 1, 3
McMaster, R. B., *see* Slocum, T. A. 13, 40
Meier, S., *see* Burghardt, D. 51, 80, 162
Mellor, D., *see* Poole, J. 184
Meng, L. 20, 140, 162, 230
Meng, L., *see* Hake, G. 217
Meng, L., *see* Lal, J. 222, 223
Meyer, M., *see* Lecordix, F. 308
Meyer, U. 19
Meyer, W. B. 3
Michlmayr, E., *see* Pospischil, G. 140
Michon, P. 202, 203
Miller, S., *see* Urquhart, K. 141
Minsky, M. 8
Mitchell, K., *see* Cheverst, K. 140
Moellering, H. ix
Molenaar, M. 5, 30, 44, 127, 128, 133
Molenaar, M., *see* Martinez Casasnovas, J. A. 30
Monier, P. 56
Monmonier, M. S. 3, 6, 10, 21, 43, 92, 321
Monmonier, M. S., *see* MacEachren, A. M. 6
Monmonier, M. S., *see* McMaster, R. B. 40, 55
Montello, D. 200
Montello, D., *see* Lovelace, K. 203
Moore, A. 24
Morisset, B. 53, 259
Morrison, J. L. 6, 13
Morrison, J. L., *see* Guptill, S. C. ix, 90
Morrison, J. L., *see* Robinson, A. H. 4, 38, 39
Morrison J. L., *see* Robinson, A. H. 12
Moulin, B., *see* Gbei, E. 169, 185, 186
Mountain, D. 142
Mountain, D., *see* Raper, J. 162
Mouwes, P. J., *see* Muller, J.-C. 20
Muehrcke, P. C. 3
Muehrcke, P. C., *see* Robinson, A. H. 4, 12
Müller, H., *see* Ayasse, J. 219
Müller, H., *see* Gröger, G. 213
Muller, J. C. 8, 19
Müller, J. C., *see* Peng, W. 53, 266
Muller, J.-C. 18–20, 98
Müller, J.-C. viii, 19, 20, 43, 45, 320
Müller, J.-C., *see* Bruegger, B. P. 30
Müller, J.-C., *see* Richardson, D. E. 20
Müller, J.-C., *see* Su, B. 47
Müller, J.-C., *see* Wang, Z. 50, 53
Müller, M., *see* Avelar, S. 162
Müller, M., *see* Fitzke, J. 170

Mustière, S. 51, 163, 255
Mustière, S., *see* Sheeren, D. 34, 120
Musy, M., *see* Ramos, F. 214

Nadeau, M., *see* Bédard, Y. 118, 186
Neun, M. 168
Neun, M., *see* Burghardt, D. 87, 164, 169
Newell, A. 198
Nick, A., *see* Poslad, S. 140
Nickerson, B. G. 14, 20, 43, 51, 69, 70
Nilsen, S., *see* Bjørke, J. T. 56
Nissen, F., *see* Sarjakoski, T. 140, 144
Niu, W., *see* Song, Z. 220
Nivala, A.-M. 32, 143, 145, 162
Nivala, A.-M., *see* Sarjakoski, L. T. 140, 143, 145, 146
Nivala, A.-M., *see* Sester, M. 32, 170
Norvig, P., *see* Russell, S. J. 19, 20, 23, 86, 270, 273, 274
Nothegger, C., *see* Winter, S. 203
Nyerges, T. L. 20, 44
Nyerges, T., *see* Jankowski, P. 69

O'Callaghan, J. F. 56
Ogniewicz, R. L. 52
Oppermann, R. 141, 142
Ormeling, F., *see* Kraak, M. J. 3
Ormsby, D. 8, 22, 66, 302

Paiva, J. A. 120
Palmer, S. 260
Papadimitriou, C. H. 86
Parent, C. 114, 118, 119
Parent, C., *see* Balley, S. 26, 31, 32, 115, 120, 121
Parent, C., *see* Devogele, T. 25, 115, 125
Parent, C., *see* Vangenot, C. 31, 114, 120
Parsaye, K. 20, 21
Pazzaglia, F., *see* Denis, M. 200–203, 206
Peng, W. 53, 95, 128, 266, 321
Persson, J. 141
Peters, O., *see* Schlender, D. 209
Peterson, M. ix
Petzold, I. 168
Peucker, T. K., *see* Douglas, D. H. 13, 50, 58–60
Peuquet, D. J. 120, 125
Pfafstetter, O. 263
Phil, B., *see* Poslad, S. 140
Piatetsky-Shapiro, G., *see* Frawley, W. J. 5, 323
Pillewizer, W., *see* Töpfer, F. 13, 46, 91, 259
Pilouk, M., *see* Bodansky, E. 61
Pinkerton, M., *see* Abowd, G. D. 139
Plazanet, C. 51, 93
Plazanet, C., *see* Lecordix, F. 50, 293
Plazanet, C., *see* Parent, C. 114, 118, 119
Plazanet, C., *see* Ruas, A. 23, 24, 34, 74, 75, 163
Plümer, K. 228
Plümer, L., *see* Petzold, I. 168
Plunkett, G., *see* Goodenough, D. G. 19
Pollock, D., *see* Timpf, S. 210
Pomerol, J.-C. 272
Poole, J. 184
Portugali, J., *see* Haken, H. 231
Poslad, S. 140
Pospischil, G. 140
Poth, A., *see* Fitzke, J. 170
Powitz, B. 59
Pratt, I. 4
Preim, B., *see* Leiner, U. 218
Premerlani, W., *see* Rumbaugh, J. 22, 26
Proctor, J., *see* Goodchild, M. F. 7
Proulx, M. J., *see* Bédard, Y. 118, 186
Proulx, M.-J. 186
Proulx, M.-J., *see* Bédard, Y. 186
Puppo, E. 257
Purves, R., *see* Burghardt, D. 171

Qingxin, X., *see* Hui, W. 21
Quak, W., *see* van Oosterom, P. 305
Quinlan, J. 203

Rahman, A.-A. 218
Rainsford, D. 175
Rainsford, D., *see* Mackaness, W. A. 48
Raisz, E. 38
Ramos, F. 214
Raper, J. 56, 162
Raper, J., *see* Mountain, D. 142
Rappo, A. 168
Ratajski, L. 13
Raubal, M. 203
Raubal, M., *see* Winter, S. 203
Raynal, L., *see* Devogele, T. 25
Regnauld, N. 46, 47, 52, 54, 77, 95, 163
Regnauld, N., *see* Barrault, M. 76, 78, 288
Regnauld, N., *see* Lecordix, F. 308
Reichenbacher, T. 137, 140–143, 162, 165
Reichenbacher, T., *see* Edwardes, A. 162, 172, 320
Reichenbacher, T., *see* Meng, L. 140
Reichenbacher, T., *see* Weibel, R. 70
Rényi, A., *see* Erdös, P. 235
Ressel, S., *see* Leiner, U. 218
Retscher, G., *see* Gartner, G. 140
Reumann, K. 50, 58
Revell, P., *see* Hardy, P. 34, 170, 175, 308
Reynes, J. L. 259
Rhind, D., *see* Raper, J. 162
Rhyne, T.-M., *see* Gahegan, M. 6

Ribelles, J. 220
Richard, D., *see* Badard, T. 295
Richard, D., *see* Inmon, W. H. 179
Richardson, D. E. 13, 20, 29, 45, 133, 256, 259, 261, 263
Richardson, D. E., *see* Thomson, R. C. 53, 128, 238, 259–261
Richter, K. 200
Richter, K. F., *see* Zipf, A. 162
Richter, K.-F., *see* Barkowsky, T. 162
Richter, K.-F., *see* Klippel, A. 143
Rieger, M. K. 20
Riordan, O., *see* Bollobás, B. 235
Rivest, S. 184
Robinson, A. H. 4, 12, 38, 39
Robinson, G. J. 69, 131
Rodriguez, M. A. 125, 133
Roschlaub, R., *see* Schilcher, M. 213, 215
Rossen, M. B., *see* Carroll, J. M. 318
Roth, S. 205
Rottensteiner, F. 219
Ruas, A. viii, 22–24, 34, 43, 46, 48, 49, 53, 66, 73–75, 77, 93–96, 107, 123, 128, 131, 162, 163, 177, 256, 266, 270, 271, 274, 276, 277, 288, 308
Ruas, A., *see* Bard, S. 100
Ruas, A., *see* Barrault, M. 76, 78, 288
Ruas, A., *see* Duchêne, C. 49
Ruas, A., *see* Lamy, S. 23, 62, 106, 266, 271, 288, 308
Ruas, A., *see* Morisset, B. 53, 259
Ruas, *see* Lagrange, J.-P. 30
Rugg, R. D., *see* Rodriguez, M. A. 125, 133
Rumbaugh, J. 22, 26
Rumbaugh, J., *see* Booch, G. L. 26, 116, 275
Ruotsalainen, R., *see* Sarjakoski, T. 140, 144
Rusak Mazur, E. 256
Russell, S. J. 19, 20, 23, 86, 270, 273, 274
Rystedt, R., *see* Sarjakoski, T. 140, 144

Saafeld, A. 50
Sale, R. D., *see* Robinson, A. H. 4
Sale, R., *see* Robinson, A. H. 38, 39
Salgé, F., *see* Müller, J.-C. 43
Salichtchev, K. A. 12, 13
Salistschew, K. A. 13
Sarjakoski, L. T. 140, 143, 145, 146, 155
Sarjakoski, L. T., *see* Harrie, L. 142, 143, 168, 208
Sarjakoski, L. T., *see* Lehto, L. 144
Sarjakoski, L. T., *see* Nivala, A.-M. 32, 143, 145, 162
Sarjakoski, L. T., *see* Sarjakoski, T. 87, 139, 140, 144, 147, 149, 166, 175
Sarjakoski, L. T., *see* Sester, M. 32, 170
Sarjakoski, T. 22, 49, 55, 86, 87, 139, 140, 144, 147, 149, 166, 175
Sarjakoski, T., *see* Harrie, L. 80, 81, 106, 162, 276
Sarjakoski, T., *see* Kilpeläinen, T. 31–34
Sarjakoski, T., *see* Lehto, L. 147, 167
Sarjakoski, T., *see* Sarjakoski, L. T. 155
Sarjakoski, T., *see* Sester, M. 32, 170
Saund, E. 260
Schenkelaars, V., *see* van Oosterom, P. 30, 222
Schilcher, M. 213, 215
Schilling, A. 140, 141, 209
Schlender, D. 209
Schnabel, T., *see* Luley, P. M. 140
Schneider, D. C. 2
Schylberg, L. 19, 47, 69, 70
Searby, S. 142
Searle, J. 280
Sester, M. 23, 32, 46–49, 55, 60, 80, 162, 170, 208, 228, 231, 238
Sester, M., *see* Edwardes, A. 162, 172, 320
Sester, M., *see* Elias, B. 203, 209
Sester, M., *see* Hampe, M. 26, 31, 32, 120, 150, 168, 216, 288
Sester, M., *see* Heinzle, F. 251
Sester, M., *see* Kopczynski, M. 201
Sester, M., *see* Sarjakoski, T. 87, 140, 144, 149, 166, 175
Sester, M., *see* Thiemann, F. 218, 220, 221
Shannon, C. E. 12
Shea, K. S. 13, 14, 19
Shea, K. S., *see* McMaster, R. B. viii, 1, 4, 14, 15, 20, 38, 39, 41–43, 92, 93, 270, 275
Sheehan, R. 308
Sheeren, D. 34, 120
Shen, D., *see* Yang, P. 218
Sheppard, E. 1, 3
Sheppard, E., *see* McMaster, R. B. 3
Shi, W., *see* Cheung, C. K. 93
Siegel, A. 201
Sijmons, K., *see* Peng, W. 95
Simon, H. 272
Simula, T., *see* Linturi, R. 32
Siret, D., *see* Ramos, F. 214
Skagestein, G., *see* Friis-Christensen, A. 30
Skogan, D. 25, 31, 34
Skogan, D., *see* Friis-Christensen, A. 30
Skupin, A. 10
Slocum, T. A. 13, 40
Sloman, M., *see* Chalmers, D. 162
Smith, D. C. P., *see* Smith, J. M. 128, 129
Smith, D., *see* Burnett, G. 203
Smith, J. M. 128, 129
Snoeyink, J., *see* Gold, C. 52
Sondheim, M. 164

Song, Z. 220
Sorrows, M. 202, 203
Spaccapietra, S., *see* Balley, S. 26, 31, 32, 115, 120, 121
Spaccapietra, S., *see* Devogele, T. 25, 115, 125
Spaccapietra, S., *see* Parent, C. 114, 118, 119
Spaccapietra, S., *see* Vangenot, C. 31, 114, 120
Spofford, G., *see* Thomsen, E. 183
Stahovich, T. F., *see* Ribelles, J. 220
Staufenbiel, W. 59
Steck, S. 203
Steck, S., *see* Mallot, H. 211
Stefanakis, K. 21
Steiglitz, K., *see* Papadimitriou, C. H. 86
Stell, J. G. 25, 120, 257
Stigmar, H., *see* Harrie, L. 155
Stigmar, H., *see* Sester, M. 32, 170
Stonebracker, M. 126
Stonykova, A. 119
Stork, D. G., *see* Duda, R. O. 235
Stoter, J., *see* van Oosterom, P. 305
Strahler, A. N. 54, 256
Strassner, J., *see* Haase, H. 229
Streeter, L. 209
Su, B. 47
Swamy, M. N. S., *see* Thulasiraman, K. 235

Takagi, S. 165
Tanaka, T., *see* Takagi, S. 165
Tappe, H., *see* Klippel, A. 206, 209
Taylor, D. R. F. 6
Taylor, P. J. 1
Tenenbaum, J. M., *see* Witkin, A. P. 260
Thapa, K. 50
Theobald, D. M. 127
Thibault, D., *see* Gold, C. 56
Thiemann, F. 213, 216, 218, 220, 221
Thomas, N., *see* Ware, J. M. 79, 106
Thomsen, E. 183
Thomson, D., *see* Laurini, R. 19, 21, 22, 30
Thomson, R. C. 53–55, 128, 238, 259–263, 266
Thomson, R. C., *see* Richardson, D. E. 259, 261
Thorndyke, P. 201
Thulasiraman, K. 235
Timpf, S. 30, 31, 120, 210
Timpf, S., *see* Frank, A. U. 30
Tobler, W. R. 134
Tolbert, D., *see* Poole, J. 184
Tom, A. 202
Töpfer, F. 13, 46, 91, 259
Trévisan, J. 26
Trévisan, J., *see* Devogele, T. 25
Trévisan, J., *see* Duchêne, C. 49
Trigg, A. D., *see* Gill, G. A. 69
Tryfona, N., *see* Friis-Christensen, A. 30
Tsoulos, L., *see* Stefanakis, K. 21
Tufte, E. R. 3, 89
Turban, E. 179, 184
Turnbull, D. 3
Turner, B. L., *see* Meyer, W. B. 3
Tversky, B. 2, 200, 209

Uhlirz, S., *see* Gartner, G. 140, 141
Uitermark, H. 125
Umlauft, M., *see* Pospischil, G. 140
Urquhart, K. 141
Urwin, T., *see* Edwardes, A. 50, 93

van Dam, A., *see* Foley, J. D. 167
van der Poorten, P., *see* Jones, C. B. 32
van der Steen, S. J. F. M. 178
van Kreveld, M. J. 230
van Oosterom, P. 30, 114, 222, 305
van Smaalen, J. W. N. 22, 44, 95, 127, 128, 132, 134
Vangenot, C. 31, 114, 120, 177
Vangenot, C., *see* Parent, C. 114, 118, 119
Vauglin, F. 25
Vckovski, A. 164
Veblen, T., *see* Bauer, B. 10
Vickus, G. 16
Visvalingam, M. 50, 171
Vitello, D., *see* Streeter, L. 209
Vlissides, J., *see* Gamma, E. 117, 166
Volta, G., *see* Timpf, S. 210

Wachowicz, M., *see* Gahegan, M. 6
Wachowicz, M., *see* MacEachren, A. M. 5, 6
Wahlster, W., *see* Baus, J. 141
Waller, D., *see* Hunt, E. 200
Walter, V. 125
Walther, M., *see* Döllner, J. 229
Wang, Z. 50, 53, 60
Wang, Z., *see* Müller, J.-C. 55
Ward, L. M., *see* Coren, S. 260
Ware, J. A., *see* Ware, J. M. 79
Ware, J. A., *see* Wilson, I. D. 79
Ware, J. M. 23, 42, 47, 49, 78, 79, 106, 162
Ware, J. M., *see* Jones, C. B. 22, 26, 47, 55
Ware, J. M., *see* Wilson, I. D. 79
Waterman, D., *see* Hayes-Roth, F. 18, 69
Weaver, W., *see* Shannon, C. E. 12
Weibel, R. 7, 13, 14, 16, 18, 20, 21, 24, 25, 41, 56, 57, 70–72, 75, 106, 162, 168, 177, 178, 255, 270
Weibel, R., *see* Bader, M. 56, 80
Weibel, R., *see* Barrault, M. 76, 78, 288

Weibel, R., *see* Brassel, K. E. 14–16, 38, 39, 67, 91, 270, 288
Weibel, R., *see* Burghardt, D. 87, 164, 169
Weibel, R., *see* Cecconi, A. 32
Weibel, R., *see* Edwardes, A. 140, 141, 162, 170
Weibel, R., *see* Galanda, M. 24, 56, 80, 84
Weibel, R., *see* Lamy, S. 23, 62, 106, 266, 271, 288, 308
Weibel, R., *see* Müller, J.-C. viii, 43, 320
Weibel, R., *see* Neun, M. 168
Weibel, W., *see* Edwardes, A. 162, 172, 320
Weiss, G. 270
Wertheimer, M. 260
Wertheimer, M., *see* King, D. B. 260
White, S., *see* Siegel, A. 201
Whyatt, J. D., *see* Visvalingam, M. 50, 171
Wienhöfer, M., *see* Schlender, D. 209
Wilbanks, T. 2, 3
Wilkinson, G. G., *see* Mackaness, W. A. 20, 69, 270
Wilson, I. D. 79
Wilson, I. D., *see* Ware, J. M. 79
Wilson, J. P. 56
Winkler, J., *see* Bauer, B. 10
Winter, S. 203
Winter, S., *see* Raubal, M. 203
Witkam, A. P., *see* Reumann, K. 50, 58
Witkin, A. P. 260
Witkin, A., *see* Babaud, J. 50
Wonsiewicz, S., *see* Streeter, L. 209
Wood, J. 56
Wood, J., *see* Raper, J. 162
Woodsford, P. A., *see* Hardy, P. G. 22, 26
Woodsford P. A. 308
Worboys, M. F. 114, 116
Worboys, M. F., *see* Stell, J. G. 25, 120, 257
Worboys, M., *see* Duckham, M. 205
Wright, J. K. 37

Yan, H., *see* Li, Z. 47
Yang, P. 218
Yu, B., *see* Hui, W. 21

Zaltash, A., *see* Robinson, G. J. 69
Zelek, J., *see* Goodenough, D. G. 19
Zeshen, W., *see* Müller, J.-C. 19
Zhang, Q. 235
Zhizhuo, W. 86
Zhou, S. 31, 120
Zhou, S., *see* Jones, C. B. 32
Zimanyi, E., *see* Parent, C. 114, 118, 119
Zipf, A. 140, 162
Zipf, A., *see* Meng, L. 140
Zipf, A., *see* Poslad, S. 140
Zlatanova, S., *see* van Oosterom, P. 305
Zoraster, S. 78
Zucker, J.-D., *see* Sheeren, D. 34, 120

Subject Index

About Generalisation

Algorithms
  Aggregation 30, 42, 44, 45, 128, 132
  Amalgamation 42, 47, 53
  Caricature 50
  Classification 40, 44, 56, 310
  Clipping 49
  Collapse 42
  Displacement 13, 19, 43, 48, 51, 78, 171, 290
  Enhancement 43
  Enlargement 47
  Exaggeration 43
  Filtering 50
  Merging 42, 45, 129
  Refinement 42
  Rotation 49
  Schematisation 208
  Selection 46
  Simplification 13, 41, 47, 153, 223
  Skeletonisation 52
  Smoothing 42, 50, 153
  Squaring 48, 225
  Typification 46, 55, 171, 208
Cartographic Generalisation (map generalisation, graphic generalisation) 4, 12, 15, 44, 168, 217, 322
Conceptual cusp (abrupt change, abrupt transition) 98, 218, 230
Conceptual generalisation 14
Constraints (conflicts, cartographic conflicts, relational constraints) 7, 22, 23, 40, 46, 49, 53, 70, 73, 74, 76, 78, 79, 93, 124, 227, 265, 272, 274, 279, 289
  Typology of constraint 74
Contextual generalisation 277, 288
Continuous deformation 80, 106
Evaluation (type of, assessing, grading, controlling, editing, errors, describing evaluation criteria) 24, 78, 87, 89, 91, 92, 98, 99, 104–107, 124, 265, 297, 311, 319
  Real world 99, 100, 114, 120
  Ground truth 102
  Quality function 107
  Evaluation function 100
  Metadata 111
History of Generalisation 13, 37, 256, 270
Incremental generalisation (modular approach for generalisation) 32
Interpolation (generalisation by, interpolating generalisation) 229
Knowledge for generalisation (procedural knowledge, rules of generalisation, knowledge enrichment) 19, 20, 70, 217, 234, 262, 273–275, 317
Level of detail (scale, granularity, resolution, levels of abstraction) 1, 2, 7, 16, 25, 92, 101, 104, 119, 123, 184, 194, 202, 213, 277
  Meso (meso level, meso object) 84, 108, 131
  Pivot scale 181
Model generalisation 5, 15, 26, 43, 167, 217, 322
On demand generalisation (adaptative generalisation) 142, 282
On the fly generalisation (real time generalisation) 123, 137, 144, 170, 181, 227
Operators 14, 37, 153
Process (models of generalisation, conceptual model for generalisation, triggering algorithms, generalisation workflow) 7, 11, 67, 92, 156, 174, 309
  Process recognition 39, 91
  Process modelling 39
Radical law 13, 46, 259
Requirements (generalisation requirements) 86, 163, 216, 288, 308, 313
Research agenda (research challenges) 86, 282, 319
Setting (parameter setting, system setting, tuning, samples) 92, 97, 105
Statistical generalisation 15
Structural generalisation 14
User needs and interactions (user requirements, explicit/implicit needs, personalisation, user ori-

ented generalisation) 8, 92, 105, 100, 110, 142, 146, 150, 178, 192, 218, 319

Types of generalisation

3D generalisation 209, 211
Building generalisation 45, 211, 311
Line generalisation 49
Meso generalisation 276
Network generalisation (network simplification) 49, 53, 255, 256
Raster based generalisation (mathematical morphology operators) 19, 41, 47, 55
Relief generalisation (DTM, terrain generalisation) 38, 39, 56
River generalisation (river network simplification) 52, 54, 263
Road generalisation (road network simplification, street generalisation, road junction, interchanges, road simplification) 50, 53, 54, 96, 202, 257, 259, 262, 266, 278, 290, 291
Rural area generalisation 278
Rural feature generalisation (categorical map generalisation, lakes generalisation) 54, 55
Urban generalisation (urban block generalisation) 53, 278

Techniques for automation

Agent modelling (MAS, agent based technology, SGO – self generalising object, communicating agents) 23, 53, 56, 76, 93, 106, 197, 270, 278, 311
Amplified intelligence 21, 71
Batch processing (batch generalisation) 34, 68
Classification (IA Classification: neural network) 46, 222
Combinatorial generalisation (genetic algorithm, simulated annealing, simple gradient descent search) 49, 78, 79, 106
Computational efficiency (algorithm efficiency) 13, 57, 86
Condition-action modelling (knowledge based methods, expert system, rules bases system) 14, 18, 29, 68, 69, 73, 155
Constraint based modelling 22, 73, 84
Continuous deformation (rubber sheeting, flexibility graph, morphing techniques, fading techniques) 230, 290
Continuous optimisation (snakes, elastic beams, least-squares adjustment, finite element, mesh simplification) 22, 46, 49, 51, 56, 79, 80, 155, 290
Cost function 56
Human-interaction modelling 71
Interactive systems (interactive generalisation platform) 172
Knowledge acquisition techniques (machine learning, reverse engineering, knowledge revision) 20, 70, 85, 210, 284
Set theory 13

Spatial Analysis

Cartometric analysis (rose diagram) 90, 92, 110
Clustering (perceptual grouping, Gestalt, cluster analysis) 34, 42, 46, 49, 53, 54, 92, 98, 104, 222, 251, 257, 259, 260
Graph (graph theory, random graph, scale free graph, Delaunay triangulation, minimum spanning tree, proximity graph, skeleton) 23, 42, 47, 53, 54, 56, 82, 95, 233, 235, 251, 257
Measurements (measures, description of measures, template, typology, contextual measurement) 74, 76, 81, 92, 93, 103, 104, 257
  Compactness 249
  Congestion 92, 94, 95, 104
  Convexity 250
  Curvature 51
  Density 46, 53, 92, 104
  Granularity 104, 194
  Inflexion point 52
  Position 75, 82, 104
  Size 104
  Sinuosity 92, 291
  Shape 53, 76, 82, 92, 104
Pattern (types of, pattern recognition, extraction, spatial structures, structure recognition, Hough transformation, meta pattern) 1, 39, 46, 71, 76, 91, 95, 97, 104, 117, 217, 222, 125, 233, 245, 250, 261, 277
  Alignment 46, 104
  Building structure 222
  City centre 246, 250
  Grid pattern 240
  Road network pattern 237
  Ring shape 248
  Structure star pattern 246
Saliency (importance, significant feature) 51, 203, 220, 256, 260
Shortest path (Dijkstra algorithm, way finding algorithm) 140, 199, 247, 257
Stream order (main stream) 54, 256, 263
Strokes (good continuation) 53, 54, 238, 260
Structural knowledge 68
Template matching 48

Tesselation (Voronoï) 95, 96
Topology 75, 117, 127, 201
Visibility (measure of) 203

DB Modelling and matching

3D model 32, 218
Aggregation 30, 42, 44, 45, 128, 132
Association 44
Building model 215, 218
City model 211
Cartographic data base (DCM, DKM) 16, 27, 122, 288
Class composition (classification) 44
Class generalisation (class abstraction) 44, 129
Component object 134
Composite object 127, 128, 131, 134
Conceptual model 11, 38, 116, 306
CSG-tree (CSG) 219, 220
Data consistency (inconsistency, coherence, data maintenance) 29–31, 92, 104, 120, 181, 293, 302
Data schema (data model) 116, 303, 317
Datawarehouse (OLAP, drilling, federated data bases, datamarts) 126, 177, 179, 183, 184, 194
Ephemeral object 124
Geographic data base (DLM) 16, 122, 288
Integration (data integration, schema integration) 114, 119, 125, 181, 186, 295
Matching (data matching, schema matching, homologous object, equivalences) 30, 31, 34, 120, 125, 183, 187, 322
Modelling language (modelling system, VUEL, Perceptory, MADS, UML, Spatial_PVL) 26, 31, 116, 118, 186, 191
Multiple representation (MRDB, multiscale, multiresolution, multi-dimensional paradigm, scale space) 11, 18, 24, 87, 119, 170, 182, 184, 216, 223, 302, 320
Object hierarchy 22, 29, 128, 230
Object oriented 21, 26, 30, 115
Ontology 24, 182, 210
Partonomy 2, 133, 320
Stamps 120
Taxonomy 129, 320

About Cartography and representation

Cartographic design (Semiology, symbology, symbolisation) 3, 12, 40, 48, 110, 307
Coalescence (bend coalescence, symbol coalescence) 51, 92, 290, 291
Cognition 227
Communication (communication model, communication theory) 10, 11
Cybercartography 6
Exploration 10
Information theory 12, 92
Legibility 76, 82, 227, 279
Model of space 7
Name placement (text placement, label placement) 19, 78, 84, 168, 285, 293, 311
On demand mapping 177, 178
Perception (imperceptibility) 92, 227
Readability 47, 74
Real time visualisation 217
Representation (view, cartographic representation, representation rules) 27, 74, 99, 100, 110, 114
Small display (small devices) 145, 170, 185
Special maps (YAH, strip, sketch, schematic) 200, 201, 206
Symbol placement 15, 154
Telecartography 141

Web, LBS and Open system

Adaptative system 140, 142
Architecture (open architecture) 147, 164, 179, 191
Car navigation 140
Interoperability (format, protocol, XML, OGC, Standard, Formal data structure FDS) 30, 34, 87, 127, 148, 149, 164, 192, 320
Invasive simplification 227
LBS (Location based services) 137, 209
Mobile map service (mobile information system) 137, 170
Open system (library, open generalisation system) 161
Point of interest (PoIs, critical points, landmarks, important objects) 139–141, 147, 170, 201, 202, 209, 259
Research platform (Registry) 162, 168, 172, 175
Spatial chunking 209
Streaming (compression) 141
Web services (WMS, WFS, Geographic services, integration services, generalisation services, portrayal services, portal service, map services) 32, 126, 137, 147, 151, 167, 172, 192, 321

GIS System and modules for generalisation and matching

ACIS 221
Agent (agent prototype) 105, 270, 276

ArcGIS 9.0 (ESRI) 61
ArcInfo 9.0 (ESRI) 30, 60
CartACom 77, 279
CHANGE 59
Clarity 2.0 (Laser-Scan Ldt) 62, 276
DynaGen (Intergraph) 57, 58
FME 187
GALBE 51
GENEX 19
Lamps2 (Generalizer) (Laser-scan Ldt) 61, 71, 72, 286
Map Generalizer (Intergraph) 57, 71, 72
PlaGe 271
SIGERT 185
Stratège 271, 276
UMapIT 184, 191

Productions and projects

Cost and benefits 293, 298, 313
Generalisation requirements 163, 216, 288, 313
Evaluation and correction (interactive editing, visual assessment) 72, 104, 105, 297, 311
MAGNET group 308, 321
Map Series and DB (ATKIS, BDCarto©, BDTopo©, GeoDB, TOP10K, TOP100, TOP10vector) 16, 24, 34, 133, 287, 301, 302, 309
Production line (production workflow) 302
Projects (AGENT, Amber, Crumpet, CyberGuide, GEMURE, GiMoDig, LoVEUS, LOL@, Mur-Mur, Navio, WebPark) 23, 26, 31, 62, 76, 93, 118, 138, 140, 141, 144, 170, 181, 266, 270, 288
Updating (incremental updating) 24, 30, 31, 32, 120, 183, 285, 287, 295, 307

FSC
www.fsc.org
MIX
Papier | Fördert
gute Waldnutzung
FSC® C083411

Kolibri 360 GmbH
Ferdinand-Jühlke-Straße 7
99095 Erfurt, Deutschland
produktsicherheit@kolibri360.de